Topological Quantum Materials

Topological Quantum Materials
Concepts, Models, and Phenomena

Grigory Tkachov

Jenny Stanford
Publishing

Published by

Jenny Stanford Publishing Pte. Ltd.
101 Thomson Road
#06-01, United Square
Singapore 307591

Email: editorial@jennystanford.com
Web: www.jennystanford.com

British Library Cataloguing-in-Publication Data
A catalogue record for this book is available from the British Library.

Topological Quantum Materials: Concepts, Models, and Phenomena

ISBN 978-981-4968-12-6 (Hardcover)
ISBN 978-1-003-26641-9 (eBook)

To Vita and Stephan

Contents

Preface

The first edition of this book was published in 2015 under the title *Topological Insulators: The Physics of Spin Helicity in Quantum Transport*. It came into being along with large strides forward in the understanding of electronic properties of topological insulators—a class of materials that are insulators in the interior and conductors at the boundary. To characterize these materials, a twofold approach was adopted. On the one hand, building upon the seminal work of Thouless, Kohmoto, Nightingale, and den Nijs, topological invariants of a material's band structure were introduced. On the other hand, boundary states of two- and three-dimensional topological insulators were examined in detail. A special focus was on representative transport phenomena in topological insulators. The book gave a background on the quantum anomalous and spin Hall effects, which goes back to Haldane's pioneering work on the condensed-matter realization of the "parity anomaly". In particular, specific properties of topological edge and surface states were discussed, including nonlocal edge transport, backscattering of edge and surface states, weak antilocalization, induced triplet *p*-wave superconductivity, topological bound states, and Klein tunneling in Josephson junctions.

Much has happened since the publication of the first edition of the book. Our understanding of the quantum anomalous Hall effect has advanced, research on novel quantum materials (notably, on 2D transition metal dichalcogenides, Weyl semimetals and topological superconductors) has burgeoned, exotic theoretical concepts (e.g. emergent axion electrodynamics) have been tested, and, of course, a Nobel prize in Physics has been awarded "for theoretical discoveries of topological phase transitions and topological phases of matter."

To take account of all these changes is hardly possible. Nevertheless, I have made many additions and amendments. First, I have added new sections on quantum anomalous and spin Hall insulators. The goal is to explore different models of 2D topological insulators (e.g. Dirac, tight-binding square-lattice and Bernevig–Hughes–Zhang ones) and to establish the connection to Thouless–Kohmoto–Nightingale–den Nijs' and Haldane's theories. Second, this edition highlights topological aspects of 2D transition metal dichalcogenides and Weyl semimetals. The former hold promise for the room-temperature quantum spin Hall effect, while the latter display a non-trivial topology with emergent monopole/antimonopole pairs in 3D. Third, I have added new sections on the surface quantum Hall effect, emergent axion electrodynamics and topological magnetoelectric effect, and a whole new chapter on topological superconductors and Majorana modes. These are all subjects of current experimental and theoretical effort. Finally, I have expanded the introductory chapter and added a number of new figures and end-of-chapter problems.

As the scope of the book now extends beyond topological insulators, I have adopted a broader notion of topological quantum materials. It reflects the synergy between two burgeoning fields—topological phases of matter and quantum materials. This broader notion provides a common thread linking diverse topics covered in the book, thereby helping to present a wealth of the underlying concepts, models, and phenomena.

A bulk of this edition of the book is based on a course of lectures that I have given at Wuerzburg and Augsburg Universities. I have enjoyed many illuminating discussions and wish to thank everybody involved. I am also forever grateful and indebted to my patient and supportive family.

Grigory Tkachov
Wuerzburg
January 2022

Chapter 1

Introducing Topological Materials: Mind the Time Reversal

This chapter acquaints the reader with the notion of topological insulators and related quantum materials. Topological insulators are two- or three-dimensional materials with an insulating interior and a conducting edge or surface supporting mobile carriers. As we shall see, the emergence of topological insulators is a vivid example of the importance of fundamental symmetries in physics. In this case, it is the symmetry with respect to the reversal of the arrow of time. We, therefore, begin with a discussion of general ideas about time-reversal invariance and its violation, which will lead us step by step to the notion of topological insulators. Their conducting states possess robust quantum properties such as helicity and chirality. These will be the central concepts in our discussion of the two-dimensional topological insulators—quantum spin Hall and anomalous Hall insulators—and their three-dimensional analogues. Similar types of conducting states are encountered in other quantum materials, notably in Weyl semimetals and topological superconductors. These will be introduced in this chapter on an equal footing with topological insulators. The general goal is to focus

Topological Quantum Materials: Concepts, Models, and Phenomena
Grigory Tkachov

ISBN 978-981-4968-12-6 (Hardcover), 978-1-003-26641-9 (eBook)
www.jennystanford.com

on basic ideas and concepts rather than on theoretical details and comprehensive literature review.

1.1 Time-Reversal Symmetry, Antiunitary $\mathbb{T}$ Operation and Kramers' Theorem

Time-reversal symmetry (TRS) means the invariance of physical properties of a system under a change in time direction,

$$t \mapsto -t.$$

For a particle with linear momentum $\boldsymbol{p}$ and spin $\boldsymbol{s}$ located at position $\boldsymbol{r}$, time reversal (TR) implies flipping both vectors $\boldsymbol{p}$ and $\boldsymbol{s}$ at the same position:

$$\boldsymbol{r} \mapsto \boldsymbol{r}, \qquad \boldsymbol{p} \mapsto -\boldsymbol{p}, \qquad \boldsymbol{s} \mapsto -\boldsymbol{s}.$$

The system is said to be TR invariant if its Hamiltonian, $\hat{h}$, does not change under above transformations. This is obviously true if the Hamiltonian is an even function of the momentum and independent of the spin, e.g. the Hamiltonian of a free non-relativistic particle. A more interesting example of the TRS is a Hamiltonian that is odd in both momentum and spin, e.g.

$$\hat{h} = v\boldsymbol{\sigma} \cdot \boldsymbol{p} = \frac{2v}{\hbar}\boldsymbol{s} \cdot \boldsymbol{p}, \qquad \boldsymbol{s} = \frac{\hbar}{2}\boldsymbol{\sigma}, \tag{1.1}$$

where $\hbar = h/(2\pi)$ is Planck's constant, and vector $\boldsymbol{\sigma} = (\sigma_x, \sigma_y, \sigma_z)$ consists of 2×2 Pauli matrices

$$\sigma_x = \begin{bmatrix} 0 & 1 \\ 1 & 0 \end{bmatrix}, \qquad \sigma_y = \begin{bmatrix} 0 & -i \\ i & 0 \end{bmatrix}, \qquad \sigma_z = \begin{bmatrix} 1 & 0 \\ 0 & -1 \end{bmatrix}.$$

Hamiltonian (1.1) describes a massles spin-1/2 particle moving with velocity v. More generally, TRS holds for any intrinsically nonmagnetic system in the absence of an external magnetic field.

Irrespective of the details of the system, TRS can formally be defined as the invariance of the Hamiltonian under a certain operation, $\mathbb{T}$:

$$\mathbb{T}\hat{h}\mathbb{T}^{-1} = \hat{h} \qquad \Longleftrightarrow \qquad \mathbb{T}\hat{h} = \hat{h}\mathbb{T},$$

under conditions

$$\mathbb{T}\boldsymbol{r}\mathbb{T}^{-1} = \boldsymbol{r}, \qquad \mathbb{T}\boldsymbol{p}\mathbb{T}^{-1} = -\boldsymbol{p}, \qquad \mathbb{T}\boldsymbol{s}\mathbb{T}^{-1} = -\boldsymbol{s}, \tag{1.2}$$

where $\mathbb{T}^{-1}$ is the inverse operation. In order to find a mathematical representation for $\mathbb{T}$, we notice that with $\boldsymbol{p} = -i\hbar\nabla$ and $\boldsymbol{s} = \frac{\hbar}{2}\boldsymbol{\sigma}$ Eqs. (1.2) reduce to

$$\mathbb{T}i\mathbb{T}^{-1} = -i, \tag{1.3}$$

$$\mathbb{T}\sigma_x\mathbb{T}^{-1} = -\sigma_x, \qquad \mathbb{T}\sigma_y\mathbb{T}^{-1} = -\sigma_y, \qquad \mathbb{T}\sigma_z\mathbb{T}^{-1} = -\sigma_z. \tag{1.4}$$

Since the operation $\mathbb{T}$ flips the sign of the imaginary unit i, it must involve complex conjugation, K, so we can write

$$\mathbb{T} = U\,K.$$

The remaining operation, U, must be unitary to preserve the norm of the spinor, i.e. $U^{-1} = U^{\dagger}$. Its properties follow from Eqs. (1.4):

$$U\,\sigma_x U^{-1} = -\sigma_x, \qquad U\,\sigma_y U^{-1} = \sigma_y, \qquad U\,\sigma_z U^{-1} = -\sigma_z.$$

These are the properties of a spin rotation by 180° about the y-axis, e.g.

$$U = \exp\left(i\pi\frac{s_y}{\hbar}\right) = \exp\left(i\pi\frac{\sigma_y}{2}\right) = i\sigma_y.$$

We can therefore choose the TR operation to be

$$\mathbb{T} = i\sigma_y K. \tag{1.5}$$

This is an example of *antiunitary* operators introduced to physics by Wigner [1]. Generally, an operator $\mathbb{A}$ is called antiunitary if it maps the Hilbert space on itself, preserving the norm, and the scalar product of any two states $\mathbb{A}|a\rangle$ and $\mathbb{A}|b\rangle$ is given by

$$\langle a|\mathbb{A}^{\dagger}\mathbb{A}|b\rangle = \langle a|b\rangle^* = \langle b|a\rangle. \tag{1.6}$$

This differs from the unitary case by an extra complex conjugation (denoted above by the asterisk, whereas † means Hermitian conjugation). In addition to Eq. (1.6), the $\mathbb{T}$ operation has the property

$$\mathbb{T}^2 = -1 \qquad \Longleftrightarrow \qquad \mathbb{T}^{-1} = -\mathbb{T}. \tag{1.7}$$

We are now prepared to prove a theorem that has important implications for TR-invariant quantum mechanical systems. It is called Kramers' theorem. It states that each energy level of a TR-invariant system with a half-integer spin is at least *double degenerate.*

Consider two states $|\pm\rangle$ related to each other by the TR operation:

$$|+\rangle = \mathbb{T}|-\rangle. \tag{1.8}$$

If $\mathbb{T}$ is a symmetry of the Hamiltonian, the states $|\pm\rangle$ correspond to the same energy. To prove this, we take an equation for one Kramers partner, e.g.

$$\hat{h}|-\rangle = E|-\rangle$$

and apply the $\mathbb{T}$ operation commuting with $\hat{h}$:

$$\mathbb{T}\hat{h}|-\rangle = \hat{h}\mathbb{T}|-\rangle = \hat{h}|+\rangle = E|+\rangle.$$

That is, the $\mathbb{T}$ operation returns the same equation for the other Kramers partner.

We still need to prove that $|\pm\rangle$ are two independent states, which is the case if they are orthogonal to each other, i.e. have a vanishing scalar product

$$\langle +|-\rangle = 0.$$

Using Eqs. (1.6), (1.7) and (1.8), we transform the scalar product as

$$\langle +|-\rangle = \langle -|\mathbb{T}^{\dagger}\mathbb{T}^{-1}|+\rangle = -\langle -|\mathbb{T}^{\dagger}\mathbb{T}|+\rangle = -\langle -|+\rangle^{*} = -\langle +|-\rangle.$$

This necessarily leads to the conclusion that product $\langle +|-\rangle$ vanishes identically, thus proving the theorem. As we shall see, many properties of topological quantum materials can be understood as a consequence of Kramers' theorem.

1.2 Broken Time-Reversal Symmetry, Chiral Edge States and Quantum Hall Effect

Breaking the TRS can have dramatic consequences. An important (and very relevant to our subject) example of the TRS breaking is the quantum Hall effect [2, 3]. It occurs in a two-dimensional (2D) electron system subject to a perpendicular external magnetic field $\boldsymbol{B}$, as illustrated in Fig. 1.1. In the classical picture, the magnetic field forces a charged particle to move along curved trajectories that can be divided into two groups. One of them includes closed orbits

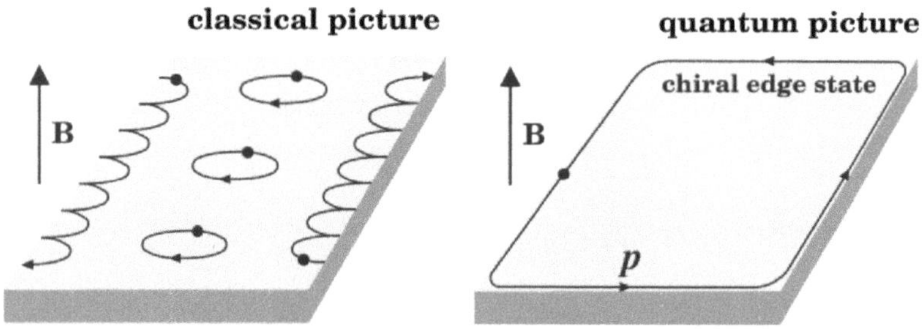

Figure 1.1 Two-dimensional electron system in a perpendicular external magnetic field ***B***. The formation of a chiral extended edge state manifests broken time-reversal symmetry.

appearing in the interior (bulk) of the sample away from its edge. These are quantized into discrete *Landau levels*. The other group consists of open electron trajectories that repeatedly hit the sample edge. The open orbits accommodate specific electronic states that can skip all the way around the sample perimeter. These skipping trajectories indicate the formation of propagating *edge states* in a magnetic field and can be viewed as bended Landau levels [4, 5], shown in Fig. 1.2.

In a strong magnetic field, when the periodic carrier motion is quantized, a skipping trajectory transforms into a quasi-one-dimensional edge channel encircling the interior of the system. The

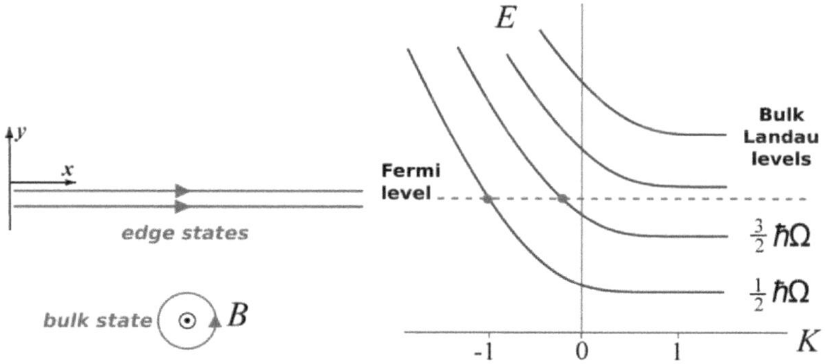

Figure 1.2 Edge channels in a quantizing magnetic field. Edge-state energies evolve from bulk Landau levels $\frac{1}{2}\hbar\Omega$, $\frac{3}{2}\hbar\Omega$, ..., bending upwards and crossing the Fermi level as function of the wave number along x.

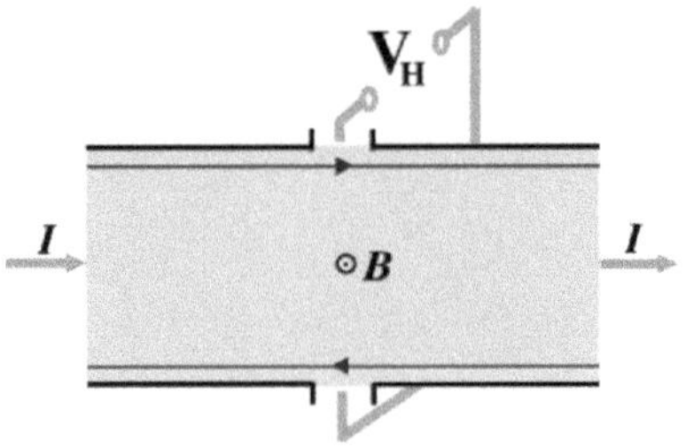

Figure 1.3 Schematic of a Hall bar in a strong magnetic field.

direction of the edge-state momentum $\boldsymbol{p}$ is tied to the orientation of the magnetic field. Such directiveness of the edge state protects it against disorder which is inevitably present, to a certain degree, in any realistic system. At low temperatures the main source of disorder is random fluctuations of the electrostatic background that tend to induce elastic momentum scattering with $\boldsymbol{p} \mapsto -\boldsymbol{p}$. For an edge state, however, such *backscattering* processes are rigorously forbidden because of the absence of a suitable counter-propagating state. Although we generally have more than one propagating channel at the edge, they are all one-way movers. Such robust directiveness of edge states is called *chirality*. It is a rather nontrivial manifestation of the broken TRS that explains the dissipationless character of the well-known macroscopic quantum phenomenon—the quantum Hall effect.

In the Hall effect a voltage difference, V_H, in a direction transverse to the electric current, I, is produced upon applying an external magnetic field (see also Fig. 1.3). In strong magnetic fields the Hall conductivity $\sigma_H = I/V_H$ exhibits *quantization* in integer units of quantum $2e^2/h$ [3],

$$\sigma_H = \frac{2e^2}{h} n, \tag{1.9}$$

where the factor of 2 is the spin degeneracy, e is the electron charge and $n = \pm 1, \pm 2, \ldots$ is integer, while the conductivity in the direction of the current (longitudinal conductivity) is vanishingly small. In this quantum regime the electric current produces no resistive heating since the directions of the current flow and electric field are perpendicular to each other. Such *dissipationless* quantum

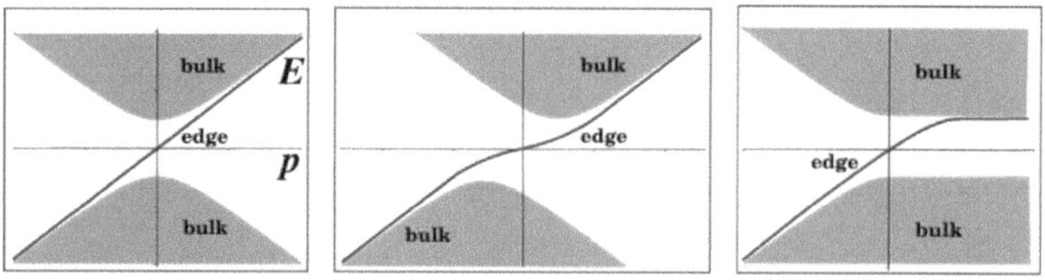

Figure 1.4 Examples of the band structure of an insulator with a chiral in-gap edge mode. All three examples correspond to the same quantum Hall state as they differ by continuous deformations.

Hall transport is naturally attributed to the chiral edge states that are capable of carrying electric current without scattering and, consequently, without generating resistance along the edge. The integer $|n|$ coincides with the number of the chiral edge channels. The necessary condition is the absence of the mobile bulk carriers at the Fermi level. In Fig. 1.2, for example, the edge modes of a quantum Hall system occur in the energy gap between the bulk Landau levels.

More generally, we can think of a bulk band insulator with in-gap edge states, as sketched in Fig. 1.4. In this case, the quantum Hall effect is expected as well, albeit in the absence of an external magnetic field and Landau levels, a possibility first suggested by Haldane [6]. Moreover, as long as the bulk band gap remains intact, we expect material's physical properties, such as the in-gap electron transport etc., to be robust against continuous deformations of the band structure. These are the hallmarks of 2D topological insulators (TIs) discussed next.

1.3 From Chiral to Helical: Quantum Spin Hall Insulators

Kane and Mele [7, 8] have identified 2DTIs as bulk band insulators with a *Kramers pair* of gapless states at the edge. Unlike the quantum Hall systems discussed above, such 2DTIs host edge states in a zero magnetic field. This becomes possible due to intrinsic spin–orbit coupling effects in TI materials. In the simplest picture the spin–orbit coupling can be viewed as an intrinsic effective magnetic field

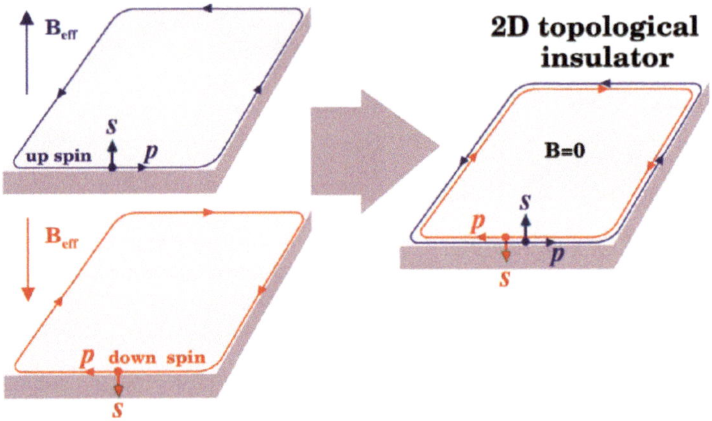

Figure 1.5 Schematic of helical edge states in a two-dimensional topological insulator.

$\boldsymbol{B}_{\text{eff}}$ that points in the opposite directions for the up- and down-spin species, as sketched in Fig. 1.5. The two spin subsystems can be viewed as two copies of Haldane's quantum Hall insulator with a gapless edge mode each, resulting in a pair of edge states in total in a zero magnetic field. The defining feature of the edge states in 2DTIs is the locking between the spin and momentum directions, also referred to as *helicity*. The edge states themselves are often called helical. The two edge modes form a Kramers doublet with crossing energy-momentum dispersions, as plotted in Fig. 1.6.

Although in 2DTIs the edge states are available in both directions, the scattering between them is forbidden for any perturbation potential preserving the TRS. To gain a deeper understanding, we notice that the helical edge states transform into each other upon a simultaneous reversal of the momentum and spin directions. Since this implies the reversal of time's arrow, the two edge states are related by the TR operation: $|+\rangle = \mathbb{T}|-\rangle$, where $|\pm\rangle$ refer to the right- and left-moving states, respectively (see also Fig. 1.7). The absence of scattering between the edge states means the vanishing of the matrix element

$$\langle +|\hat{V}|-\rangle = 0,$$

where $\hat{V}$ is a perturbation (scattering) potential. We can prove this by analogy with Kramers' theorem. Consider an operator $\hat{V}$

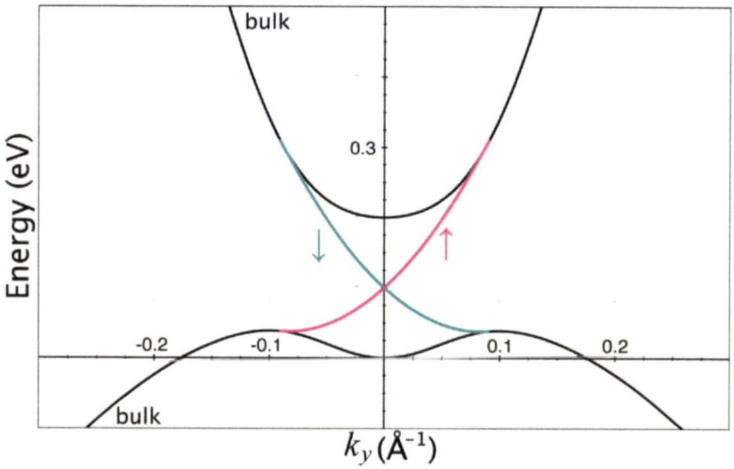

Figure 1.6 Typical band structure of a 2DTI with helical edge states in the gap between the bulk conduction and valence bands. For details, see Bernevig–Hughes–Zhang (BHZ) model in Chapter 2.

commuting with $\mathbb{T}$:

$$\mathbb{T}\hat{V}\mathbb{T}^{-1} = \hat{V} \quad \Longleftrightarrow \quad \mathbb{T}\hat{V} = \hat{V}\mathbb{T}. \tag{1.10}$$

Using Eqs. (1.6), (1.7), (1.8), and (1.10), we can recast the scattering matrix element as follows:

$$\begin{aligned}\langle +|\hat{V}|-\rangle &= \langle -|\mathbb{T}^{\dagger}\hat{V}\mathbb{T}^{-1}|+\rangle = -\langle -|\mathbb{T}^{\dagger}\hat{V}\mathbb{T}|+\rangle = -\langle -|\mathbb{T}^{\dagger}\mathbb{T}\hat{V}|+\rangle \\ &= -\langle -|\hat{V}|+\rangle^{*} = -\langle +|\hat{V}^{\dagger}|-\rangle.\end{aligned}$$

Clearly, for a Hermitian perturbation ($\hat{V} = \hat{V}^{\dagger}$) matrix element $\langle +|\hat{V}|-\rangle$ is identically zero.

As a consequence of that fact, irregularities of the potential landscape of the sample cannot flip the propagation directions of the helical edge states [see Fig. 1.7(left)]. Even though the shape of their trajectories may change near certain obstacles [see Fig. 1.7(right)], they still remain the "time-reversed partners" in the sense of Kramers' theorem, and their global conduction capability is just the same as in the ideal system. This falls into the general notion of topological objects as such whose global characteristics are invariant under continuous local deformations. Further aspects of 2DTIs, such as the *skyrmion* band topology, *band inversion* and topological *invariants*, will be discussed in the next chapter.

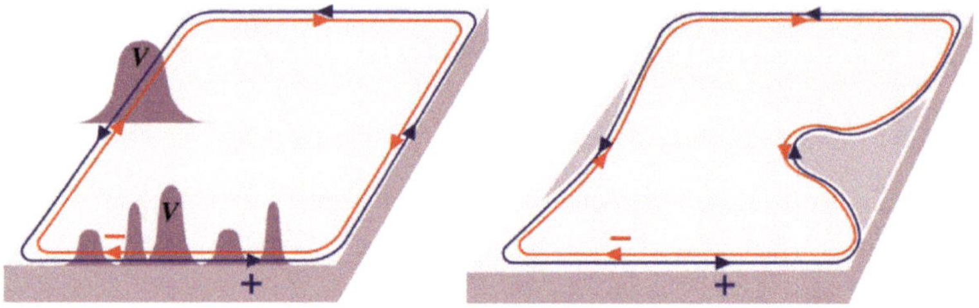

Figure 1.7 Illustration of robust helical edge states in an irregular potential landscape.

In order to appreciate the topological robustness of the helical edges, it is informative to compare them with one-dimensional ordinary metals in the presence of potential disorder, e.g. random impurities. In ordinary metallic conductors, all propagating states become spatially localized due to backscattering off the random potential, a phenomenon known as *Anderson localization* [9]. It occurs for arbitrary weak disorder provided that the system is large enough. The vanishing of $\langle +|\hat{V}|-\rangle$ implies no Anderson localization in TIs, which distinguishes them qualitatively from ordinary metallic conductors. By the way, this conclusion holds not only for 2DTIs, but also for their three-dimensional counterparts discussed later in this chapter.

2DTIs are frequently called quantum spin Hall insulators (QSHIs), as they are expected to have a quantized spin Hall conductivity if spin component s_z is conserved. We can readily compute the spin Hall conductivity for linear edge modes $E_{\uparrow,\downarrow} = \pm\hbar vk$, where v is the mode velocity. Consider the carrier densities $\Delta N_{\uparrow,\downarrow}$ induced at the edge upon an adiabatic change of the chemical potential from μ to $\mu + eV_H$, where V_H is the Hall (transverse) voltage. Each spin channel is akin to a non-degenerate one-dimensional metal, so we can write $\Delta N_{\uparrow,\downarrow} = \Delta k_{\uparrow,\downarrow}/(2\pi)$, where $\Delta k_{\uparrow,\downarrow}$ are the changes in the Fermi wave numbers of the two spin channels. On the other hand, the energy variation at the Fermi level is $\pm\hbar v\Delta k_{\uparrow,\downarrow} = eV_H$. Solving for $\Delta k_{\uparrow,\downarrow}$, we find the induced carrier densities as

$$\Delta N_{\uparrow,\downarrow} = \pm\frac{eV_H}{hv}. \tag{1.11}$$

Since $\Delta N_\uparrow + \Delta N_\downarrow = 0$, there is no charge edge current, only the spin one $I_s = \frac{\hbar}{2} v(\Delta N_\uparrow - \Delta N_\downarrow) = \sigma_{sH} V_H$, with the quantized spin Hall conductivity

$$\sigma_{sH} = \frac{e}{2\pi}. \tag{1.12}$$

It is a manifestation of the quantum spin Hall effect in which a quantized spin of $\hbar$, rather than the charge, is accumulated at the edge.

1.4 Graphene, Related 2D Materials, Inverted Semiconductors and 1T′ Transition Metal Dichalcogenides

The first theory of the quantum spin Hall effect [7] was done for graphene shortly after the experimental discovery of this 2D material [10]. The model used a tight-binding approach including spin–orbit coupling, albeit very weak in the pristine material. Subsequently, the effect was theoretically identified in semiconductor materials with an inverted band ordering [11–13], notably in HgTe/CdTe quantum wells. Following this, König *et al* reported experimental evidence for a QSHI state in HgTe/CdTe heterostructures [14]. Other inverted semiconductors, such as InAs/GaSb heterostructures, were predicted to be QSHIs as well [15]. A transport experiment of Knez *et al* reported evidence for the helical edge states in InAs/GaSb quantum wells, despite the fact that the parallel bulk contribution had to be subtracted from the total conductance [16].

In passing, we mention a few 2D materials other than graphene that have been proposed as candidate QSHIs, e.g. monolayers of Si, Ge, Sn and Bi. Considerable experimental effort has been devoted to such single-element monolayer materials (e.g. silicene, germanene, stanine), which are prepared mostly by molecular beam epitaxy (MBE) and characterized *in situ* using scanning tunneling microscopy (STM) or angle resolved photoemission spectroscopy (ARPES).

An alternative approach to obtaining QSHIs is to exfoliate layered compounds with strong spin–orbit interactions, such as transition metal dichalcogenides (TMDs) of the MX_2 family (e.g. M = W or Mo, while X = Te, Se or S). According to theory [17], monolayer TMDs are topological in the distorted octahedral (so called 1T′) crystallographic phase. Guided by this prediction, experimental work has been done on 1T′ - TMDs, most notably on the bulk and edge states in tungsten ditelluride (WTe_2) and tungsten diselenide (WSe_2). For example, an ARPES experiment and first-principles calculations of Tang *et al* [18] found that 1T′ - WTe_2 had a bulk energy gap of 55 meV, much larger than those in HgTe/CdTe and InAs/GaSb quantum wells. An even larger bulk band gap (120 meV) has been reported in the ARPES and STM experiments [19] on 1T′ - WSe_2. A large bulk band gap of 1T′ - TMDs can be an asset in the ongoing quest for room-temperature *topological electronics*. Currently, the edge conductance in these materials is observable up to the temperature of 100 K [20]. Besides, the overall shape of the energy spectrum of 1T′ - TMDs can be quite different from the typical prediction of the Bernevig–Hughes–Zhang (BHZ) model (cf. Figs. 1.6 and 1.8).

1.5 Quantum Anomalous Hall Insulators

As mentioned above, it is helpful to view a QSHI as two species of Haldane's quantum Hall insulator subject to effective magnetic fields $\boldsymbol{B}_{\rm eff}$ and $-\boldsymbol{B}_{\rm eff}$, respectively. The signs here are dictated by TRS. If magnetic interactions are introduced into a QSHI, the effective magnetic fields are generally no longer equal in magnitude: $\boldsymbol{B}_{\rm eff}+\boldsymbol{B}_{\rm m}$ and $-\boldsymbol{B}_{\rm eff}+\boldsymbol{B}_{\rm m}$, where an extra term $\boldsymbol{B}_{\rm m}$ accounts for the intrinsic magnetism. If $\boldsymbol{B}_{\rm m}$ is weak, the topological state of the system does not change. However, for $\boldsymbol{B}_{\rm m} \rightarrow \boldsymbol{B}_{\rm eff}$, one of the QSHI subsystems becomes free of the magnetic field and loses its edge state, while the edge state in the other subsystem survives. This is in fact a *topological phase transition* from a QSHI to a quantum anomalous Hall insulator (QAHI) defined as a 2D bulk insulator with a chiral mode at the edge. The adjective "anomalous" is an allusion to the

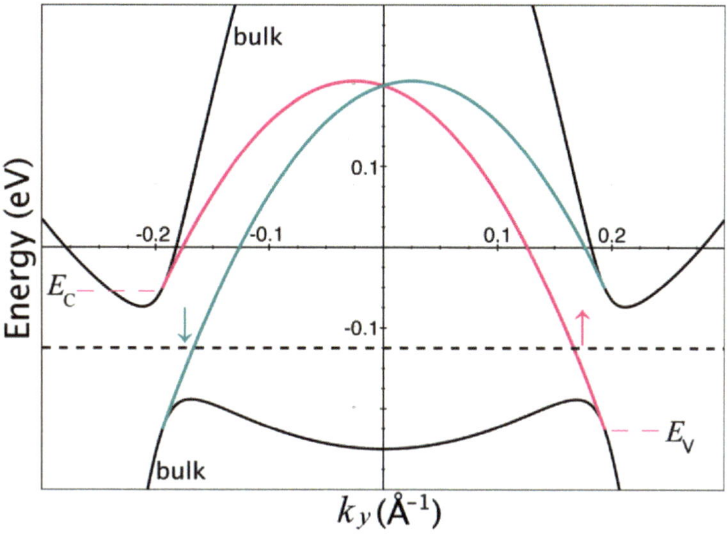

Figure 1.8 Example of the energy spectrum of a 1T′ - TMD. The edge modes disperse between the bulk conduction and valence bands, forming a Kramers pair in the band gap. E_c and E_v are the energies at the termination points ($\approx \pm 0.2\,\text{Å}^{-1}$) of the edge spectrum (adapted from [21]). For details, see Chapter 2.

Hall physics without external magnetic fields, as exemplified by Haldane's model [6] and recent studies [22, 23].

We can compute the ensuing Hall conductivity from the induced carrier density $\Delta N_\uparrow$ in Eq. (1.11). The charge edge current is $I = ev\Delta N_\uparrow = \sigma_H V_H$, with the quantized Hall conductivity

$$\sigma_H = \frac{e^2}{h}. \tag{1.13}$$

A more accurate and general derivation based on the Thouless–Kohmoto–Nightingale–den Nijs formula [3] will be given in Chapter 3. By the same token, the spin edge current is $I_s = \frac{\hbar}{2} v\Delta N_\uparrow = \sigma_{sH} V_H$, where the quantized spin Hall conductivity is half that in Eq. (1.12):

$$\sigma_{sH} = \frac{e}{4\pi}. \tag{1.14}$$

These results just mean that a chiral edge mode in a QAHI carries the charge of e and spin of $\hbar/2$. An example of a quasi-2D magnetic insulator that shows the quantum anomalous Hall effect is Cr-doped $(Bi,Se)_2Te_3$ thin films [23]. Achieving a quantized σ_H requires tuning the Fermi level into an energy gap generated by the exchange interaction with the magnetic dopants.

1.6 Three-Dimensional Topological Insulators

The higher dimensional analogues of QSHIs are three-dimensional topological insulators (3DTIs) [24–26]. A 3DTI can be viewed as a stack of 2D QSHIs in which a continuum of Kramers' pairs resides on the surface of a bulk material, while its inner part remains an insulator (see Fig. 1.9). A surface carrier has a freedom to move in two dimensions, but, similar to the edge states, its spin $\boldsymbol{s} = \frac{\hbar}{2}\boldsymbol{\sigma}$ is tied to the direction of its momentum $\boldsymbol{p}$. The simplest Hamiltonian describing the surface states with spin-momentum locking is $\hat{h} = v\boldsymbol{\sigma} \cdot \boldsymbol{p}$, as introduced earlier in Eq. (1.1), where the particle momentum $\boldsymbol{p}$ should be treated as a two-dimensional vector on the surface, e.g.

$$\boldsymbol{p} = (p_x, p_y, 0).$$

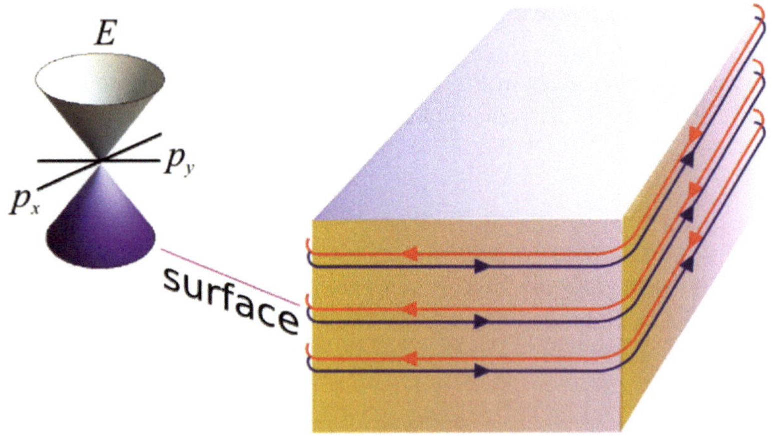

Figure 1.9 3D topological insulator as a stack of 2D quantum spin Hall insulators with a continuum of Kramers' pairs on the surface of the material.

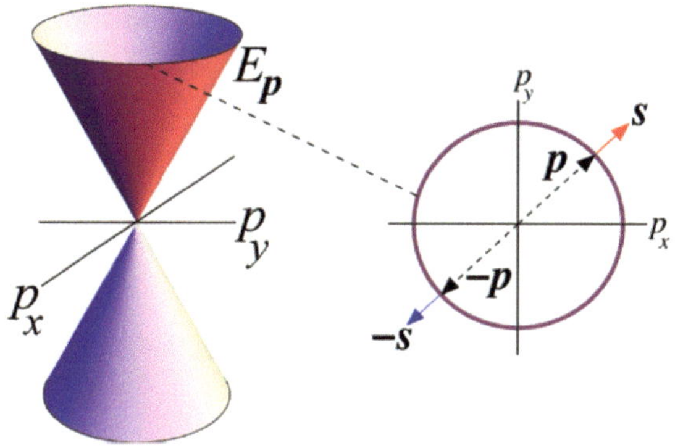

Figure 1.10 Three-dimensional plot of surface state spectrum (1.15) (left) and an example of a pair of protected states on a constant-energy line (right). Arrows show the locking of momentum $\boldsymbol{p}$ and spin $\boldsymbol{s}$ vectors.

The surface Hamiltonian can then be written as a 2×2 matrix:

$$\hat{h} = v(p_x\sigma_x + p_y\sigma_y) = v\begin{bmatrix} 0 & p_x - ip_y \\ p_x + ip_y & 0 \end{bmatrix}.$$

The energy, $E_{\boldsymbol{p}}$, of a surface state with momentum $\boldsymbol{p}$ is obtained by solving the eigenvalue equation

$$v\boldsymbol{\sigma} \cdot \boldsymbol{p}\psi = E_{\boldsymbol{p}}\psi, \qquad \psi = \begin{bmatrix} \psi_\uparrow \\ \psi_\downarrow \end{bmatrix},$$

where wave function ψ is a spinor with two components $\psi_\uparrow$ and $\psi_\downarrow$. Solving for $E_{\boldsymbol{p}}$ yields

$$E_{\boldsymbol{p}} = \pm v|\boldsymbol{p}| \equiv \pm v\sqrt{p_x^2 + p_y^2}. \tag{1.15}$$

The dispersion relation $E_{\boldsymbol{p}}$ has the form of a cone, with signs $\pm$ corresponding to positive- and negative-energy branches (see also Fig. 1.10). These play the role of the conduction and valence bands for the surface carriers. The conical energy spectrum in Eq. (1.15) resembles that of an ultrarelativistic Dirac electron. For this reason, surface states of 3DTIs are frequently called *Dirac fermions*.

The family of materials and heterostructures that can host surface states is pretty large [27, 28]. The surface states were first predicted at interfaces between normal and inverted bulk semiconductors (e.g. Refs. [29, 30]). More recently, a number of other candidate materials to host a 3DTI state have been identified,

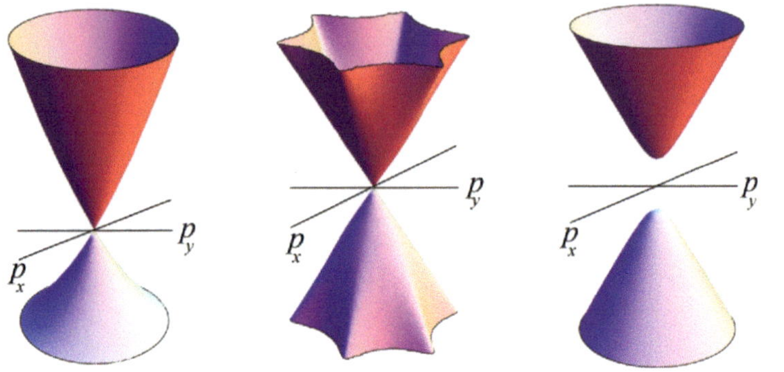

Figure 1.11 Illustration of topological states with broken particle-hole symmetry (left), hexagonal warping (middle) and energy gap (right).

e.g. the semiconducting alloy $Bi_{1-x}Sb_x$ [25], strained layers of α-Sn and HgTe [25], tetradymite compounds, e.g. Bi_2Se_3, Bi_2Te_3 and Sb_2Te_3 [31], thallium-based ternary chalcogenides $TlBiTe_2$ and $TlBiSe_2$ [32–34] as well as Pb-based layered chalcogenides [35, 36]. Experimentally, topological surface states have been observed by ARPES in $Bi_{1-x}Sb_x$ [37, 38], Bi_2Se_3 [39], Bi_2Te_3 [40], $TlBiSe_2$ [41–43], $TlBiTe_2$ [43], $Pb(Bi_{1-x}Sb_x)_2Te_4$ [44] and $PbBi_2Te_4$ [45], just to name a few.

It is important to keep in mind that the band structure of actual TI materials is more complex than the Dirac cone in Eq. (1.15). Away from the Dirac point $\boldsymbol{p} = 0$ the linear surface dispersion is modified by higher-order terms $\propto \boldsymbol{p}^2$. These break the *particle-hole symmetry*, i.e. the symmetry between the positive- and negative-energy branches of Eq. (1.15), as illustrated in Fig. 1.11(left). In the tetradymite compounds Bi_2Se_3, Bi_2Te_3 and Sb_2Te_3 the surface bands become also anisotropic in momentum space due to *hexagonal warping* effects [46–49] [see also Fig. 1.11(middle)]. Furthermore, in thin films of TI materials the band topology undergoes a qualitative change: instead of a gapless Dirac cone, a semiconductor-type spectrum with an *energy gap* at $\boldsymbol{p} = 0$ is formed [see, e.g. Ref. [50, 51] and Fig. 1.11(right)]. This can be understood as lifting the band degeneracy at $\boldsymbol{p} = 0$ by a perturbation of the form $m\sigma_z$:

$$\hat{h} = v(p_x\sigma_x + p_y\sigma_y) + m\sigma_z.$$

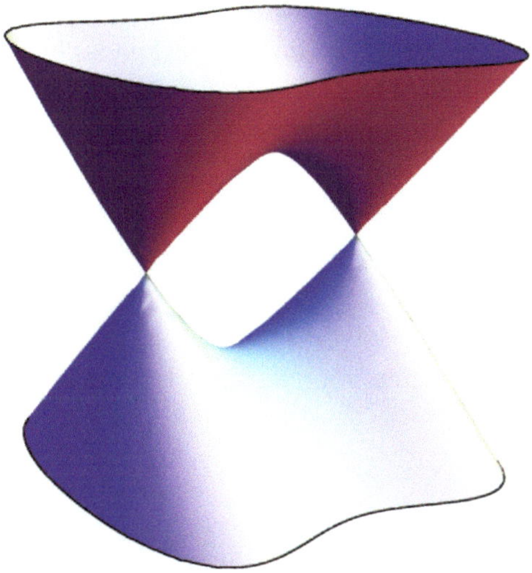

Figure 1.12 Example of the band structure of a Weyl semimetal near a pair of band degeneracy points (Weyl nodes).

Now, energy bands $\pm\sqrt{v^2(p_x^2 + p_y^2) + m^2}$ are separated by a gap of $2|m|$. In Chapter 5, we shall discuss how various modifications of the material's band structure affect its transport properties.

1.7 Weyl Semimetals

A nontrivial generalization of the surface Dirac cone is a degeneracy point between two bands in 3D momentum space. It can still be described by a 2×2 model, e.g.

$$\hat{h} = v(p_x\sigma_x + p_y\sigma_y + p_z\sigma_z),$$

but has a new distinct property: It cannot be gapped by a perturbation of the form $m\sigma_z$ (or a combination of the Pauli matrices), unlike in the 2D case. The perturbed model

$$\begin{aligned}\hat{h} &= v(p_x\sigma_x + p_y\sigma_y + p_z\sigma_z) + m\sigma_z \\ &= v(p_x\sigma_x + p_y\sigma_y + [p_z + m/v]\sigma_z)\end{aligned}$$

still has a node, albeit shifted to $[0, 0, -m/v]$. A 3D degeneracy point is protected, which is an analogue of the monopole [52], a point-like topological defect with a quantized charge. Such protected degeneracy points are referred to as *Weyl nodes* (see also Fig. 1.12); the material itself is called a Weyl semimetal (WS). A typical example discussed in literature is tantalum arsenide (TaAs) and related materials (see e.g. [53–56]).

A more general form of the WS model [57] can be deduced from the Hamiltonian

$$\hat{h} = d^x_{\boldsymbol{p}}\sigma_x + d^y_{\boldsymbol{p}}\sigma_y + d^z_{\boldsymbol{p}}\sigma_z$$

by linearizing coefficients $d^x_{\boldsymbol{p}}$, $d^y_{\boldsymbol{p}}$ and $d^z_{\boldsymbol{p}}$ near a degeneracy point $\boldsymbol{p} = \boldsymbol{p}_0$:

$$\hat{h} = \sum_{i=x,y,z} d^i_{\boldsymbol{p}}\sigma_i \approx \sum_{i,j=x,y,z} v_{ij}(p_j - p_{0j})\sigma_i,$$

where v_{ij} are the elements of a 3×3 velocity matrix. Compared to the Dirac cone in Eq. (1.15), the energy spectrum of WSs is generally anisotropic. The determinant of the velocity matrix, $\det(v_{ij})$, yields the *chirality* or *topological charge* of the node, $\mathrm{sgn}[\det(v_{ij})]$. For example, taking a diagonal matrix $v_{ij} = v_i\delta_{ij}$, we have

$$\mathrm{sgn}(v_x v_y v_z) = \pm 1,$$

e.g. the nodes with $v_x = v_y = v_z = v$ or $v_x = -v_y = -v_z = v$ have the positive chirality, while for the nodes with $v_x = v_y = v_z = -v$ or $v_x = v_y = -v_z = v$ the chirality is negative (assuming $v > 0$).

For the Brillouin zone to remain "topologically neutral", Weyl nodes must come in *monopole/antimonopole pairs* in momentum space. This is easy to see if the crystal has a center of inversion. If near a Weyl node $\boldsymbol{p}_0$ we have $+v\boldsymbol{\sigma} \cdot (\boldsymbol{p} - \boldsymbol{p}_0)$, then inversion $\boldsymbol{p} \to -\boldsymbol{p}$ produces the opposite-chirality partner $-v\boldsymbol{\sigma} \cdot (\boldsymbol{p} + \boldsymbol{p}_0)$ at $-\boldsymbol{p}_0$. A monopole/antimonolope pair is the minimal number of the Weyl nodes provided that TR is not a symmetry. Taken at the same point ($\boldsymbol{p}_0 = 0$), a pair of opposite-chirality Weyl species makes up a usual Dirac fermion. A similar reasoning shows that if TR is a symmetry, while inversion is not, the number of Weyl points must be a multiple of four: for every pair of Weyl nodes that are TR-partners with positive chirality, another TR-pair with negative chirality must exist so that the total chirality is zero.

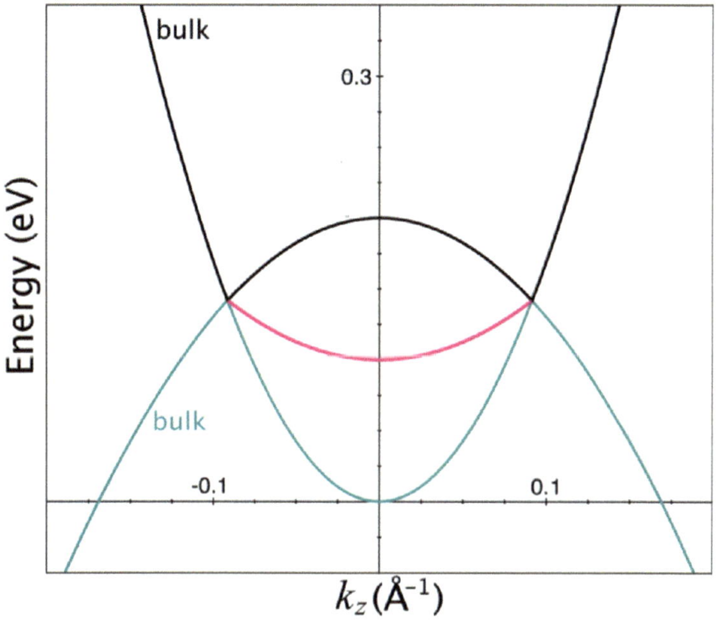

Figure 1.13 A section of 3D energy spectrum of a Weyl semimetal through the band degeneracy points. Two Weyl nodes on k_z-axis are connected by a surface Fermi arc. For details of theoretical model, see Chapter 4.

While a parallel to Dirac's theory is apparent, it should not be taken too literally. Unlike the electronic bands in WSs, the relativistic energy-momentum relation has no energy bounds. This manifests itself, in particular, in the chiral (or Adler–Bell–Jackiw) anomaly whereby Weyl fermions of one type are converted to the opposite-chirality species. In solids, Weyl fermions are only emergent quasiparticles. Nevertheless, an analogue of the chiral anomaly in the presence of a magnetic field has been explored in WSs as well [57].

Finally, the nontrivial bulk topology in WSs comes along with topological boundary states. Their Fermi surface is not a closed loop, but rather an arc connecting a pair of opposite-chirality Weyl nodes. An example is shown in Fig. 1.13. Such unusual segments of the Fermi surface are called *Fermi arcs*. In Chapter 4, we shall explore a theoretical model for these interesting properties of WSs.

1.8 Topological Superconductors

In the preceding sections, we have dealt with normal topological materials, i.e. materials that exhibit no superconductivity. This phenomenon has its origin in attractive pairwise interactions among electrons which bring about radical changes in material's properties [58]. Notably, below a certain *critical* temperature T_c a normal metallic state turns into a condensate of electron pairs, called *Cooper pairs*, capable of conducting electricity without resistance.

Similar to band insulators, superconductors can be divided into ordinary and topological [59]. We can think of a topological superconductor (TS) as an electronic material with an energy gap in the bulk and a gapless mode at the boundary, as illustrated in Fig. 1.14. Now, the bulk energy gap and the boundary mode both come from the pairing interaction between the electrons in the material. In particular, the bulk gap characterizes the energy cost for breaking a Cooper pair into two single-particle excitations, a particle above the Fermi level and a hole below it. Besides, the pairing interaction brings about an exact symmetry between the particle and hole excitations, which has an interesting implication for the gapless boundary mode: It behaves as both a particle and a hole at the same time. An instructive analogy is the *Majorana fermion*. In relativistic field theory, it embodies both a particle and

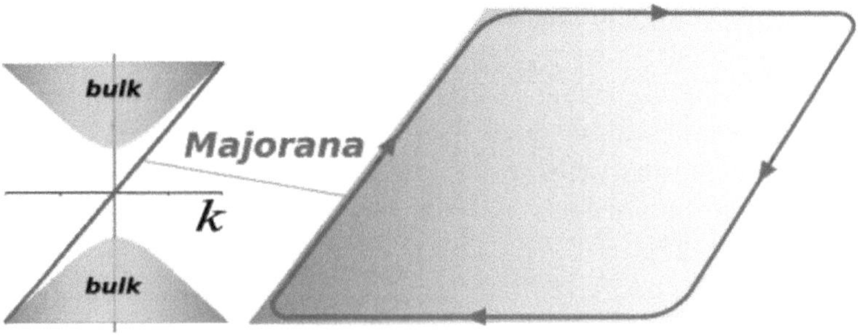

Figure 1.14 Schematic of energy bands in a TS. Pairing interaction among electrons opens an energy gap in the bulk, while a gapless Majorana mode appears at the boundary (adapted from [60]).

an antiparticle [61]. The possibility of Majorana-like modes in TSs has caused a surge of interest in their properties and potential for topological quantum computation [62].

To be precise, we are focusing here on a chiral *p*-wave TS. It is a 2D state with broken parity and TRS, which, akin to a QAHI, requires a magnetic order [63]. Perhaps the first material for which such a possibility has been considered is the layered perovskite ruthenate Sr_2RuO_4. It is a quasi-2D material in which ruthenium oxides tend to become ferromagnetic, which favors unconventional superconductivity [64]. The expectation that Sr_2RuO_4 harbors chiral superconductivity has been strengthened by the muon spin rotation experiments [65] which detected internal magnetic fields below T_c. While the scenario of the chiral *p*-wave superconductivity in Sr_2RuO_4 may be debatable, the tunneling experiment [66] has reported an enhanced zero-bias conductance consistent with the existence of the boundary modes. A recent review on the point-contact spectroscopy as a means of detecting topological superconductivity is given in Ref. [67].

Apart from Sr_2RuO_4, a number of other candidate materials to host topological superconductivity have been identified in the past decade, most notably $Cu_xBi_2Se_3$, $Sn_{1-x}In_xTe$ as well as some noncentrosymmetric superconductors in which the *p*-wave gap is larger than the *s*-wave one (for specific details, see review article [59]). As opposed to the chiral superconductivity, these materials are expected to host TR symmetric (helical) topological phases.

In the above-mentioned materials, superconductivity is an intrinsic property emerging below a certain critical temperature, e.g. $T_c \sim 1.5$K for Sr_2RuO_4. Alternatively, superconductivity can be induced in a normal conductor brought into electric contact with an intrinsic superconductor, e.g. as pictured in Fig. 1.15. Although the normal conductor has no pairing interaction of its own, it acquires the properties of a superconductor through the phenomenon known as the *proximity effect.* Its microscopic origin lies in *Andreev reflection* [68]. This is a scattering process whereby a particle incident at the superconductor is reflected as a hole back into the normal region, while a Cooper pair passes across the interface into the superconductor (see also Fig. 1.15). For

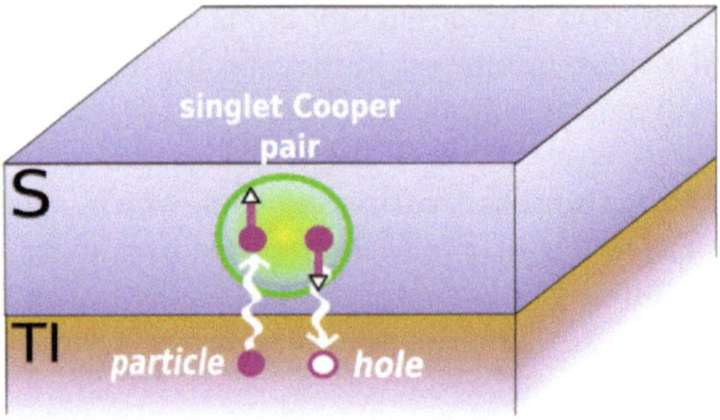

Figure 1.15 Schematic of a hybrid structure created by placing a conventional superconductor (S) on top of a topological insulator (TI) material. Andreev reflection at the S/TI boundary gives rise to the superconducting proximity effect in the TI (adapted from [60]).

such hybrid structures, one typically chooses conventional (singlet) superconductors described by Bardeen–Cooper–Schrieffer (BCS) theory [58]. These points will be addressed in Chapter 6.

The superconductor proximity effect offers an attractive alternative to intrinsic superconductivity, as topological phases can be "engineered", using broken symmetries of the normal system. Research into the induced superconductivity has intensified with the discovery of TIs and related topological materials, with a flurry of theoretical as well as experimental activity that followed. A prominent example is the theoretical proposal of 2008 [69] for a chiral TS and Majorana fermions at the surface of a 3DTI proximitized by a conventional superconductor. Since then, TI materials with induced superconducting and magnetic orders have been a fertile ground for modeling the chiral TS and related phenomena. In particular, in the series of papers [70, 71] and [72], a chiral p-wave TS with a Majorana edge mode has been discovered theoretically in QAHI/superconductor structures. The Majorana edge channel was predicted to have the half-integer longitudinal conductance

$$\sigma_{xx} = 0.5\frac{e^2}{h}, \tag{1.16}$$

which could be used as an experimental test. More recently, progress has been achieved in fabricating and characterizing hybrid structures of superconductors and Cr-doped $(Bi,Se)_2Te_3$ thin films [73, 74]. The latter are magnetic topological insulators which, in the absence of the superconducting pairing, exhibit the quantum anomalous Hall effect [23]. Experiments [73, 74] have reported the observation of the Majorana edge modes in such magnetic thin films with superconducting Nb contacts. We shall review these findings in Chapter 7.

Chapter 2

Two-Dimensional Topological Insulators

In the preceding chapter, we used simple intuitive arguments to introduce the two-dimensional topological insulators as 2D systems with an insulating interior and a Kramers pair of conducting edge states. Such systems are also known as quantum spin Hall insulators. We now intend to construct theoretical models that would elucidate the origin of quantum spin Hall insulators. We have already mentioned the original theoretical works on the quantum spin Hall effect in the context of graphene [7, 8] and semiconductor materials [11–13]. These models are closely related to the notion of Chern insulators. It goes back to Haldane's paper [6] on a condensed matter realization of the quantum electrodynamical "parity anomaly" in $2 + 1$ dimensions. Chern insulators have a nontrivial band structure characterized by a topological invariant, the Chern number, hence the name. These systems exhibit broken time-reversal symmetry, chiral edge states and a quantum Hall effect in the absence of any external magnetic field and Landau levels. Albeit unusual, this was envisioned by the theory of Thouless, Kohmoto, Nightingale and den Nijs [3]. It establishes a general relation between an observable—the Hall conductivity—and the topological Chern number. We shall explore different models of Chern insulators and use them as building blocks for quantum spin

Topological Quantum Materials: Concepts, Models, and Phenomena
Grigory Tkachov

ISBN 978-981-4968-12-6 (Hardcover), 978-1-003-26641-9 (eBook)
www.jennystanford.com

Hall insulators. In particular, a nontrivial band topology will be explained in terms of an intrinsic band inversion in reciprocal space [12, 13]. This key attribute of topological insulators reflects their striking quantum mechanics combining the elements of both Dirac and Schroedinger equations.

2.1 Chern Insulators I

2.1.1 Edge Mode and Band Inversion

The simplest Chern insulator is a 2D Dirac system described by the Hamiltonian

$$\hat{h}_{\boldsymbol{k}} = \boldsymbol{d}_{\boldsymbol{k}} \cdot \boldsymbol{\sigma}, \qquad \boldsymbol{d}_{\boldsymbol{k}} = (\hbar v k_x, \hbar v k_y, m), \tag{2.1}$$

where $\boldsymbol{\sigma}$ is the Pauli matrix vector, v is the particle velocity (chosen positive throughout) and m characterizes the particle mass. The eigenstates and eigenvalues of the Hamiltonian can be found from the equation

$$E\psi_{\boldsymbol{k}} = \boldsymbol{d}_{\boldsymbol{k}} \cdot \boldsymbol{\sigma}\, \psi_{\boldsymbol{k}} = \sqrt{A^2\boldsymbol{k}^2 + m^2}\; (\boldsymbol{u}_{\boldsymbol{k}} \cdot \boldsymbol{\sigma})\psi_{\boldsymbol{k}}, \tag{2.2}$$

where we define a unit vector $\boldsymbol{u}_{\boldsymbol{k}} = \boldsymbol{d}_{\boldsymbol{k}}/|\boldsymbol{d}_{\boldsymbol{k}}|$. It is convenient to use the polar coordinate system in which $k_x = k\cos\phi$, $k_y = k\sin\phi$ and $\boldsymbol{u}_{\boldsymbol{k}} = (u_{\parallel}\cos\phi, u_{\parallel}\sin\phi, u_z)$ with

$$u_{\parallel} = \sqrt{1-u_z^2}, \qquad u_z = \frac{m}{\sqrt{(\hbar v\boldsymbol{k})^2 + m^2}}. \tag{2.3}$$

We can choose the solutions to be the eigenstates of the spin matrix $\boldsymbol{u}_{\boldsymbol{k}} \cdot \boldsymbol{\sigma}$. It is the projection of the spin on the direction of unit vector $\boldsymbol{u}_{\boldsymbol{k}}$. This matrix is called helicity. Its eigenstates satisfy the equation

$$\boldsymbol{u}_{\boldsymbol{k}} \cdot \boldsymbol{\sigma}\, \psi_{\boldsymbol{k}s} = s\psi_{\boldsymbol{k}s}, \tag{2.4}$$

with real eigenvalues $s = \pm 1$. The solutions of Eq. (2.4), normalized to unity, are

$$\psi_{\boldsymbol{k}s} = \frac{\sigma_0 + s\,\boldsymbol{u}_{\boldsymbol{k}} \cdot \boldsymbol{\sigma}}{\sqrt{2(1+su_z)}} \begin{bmatrix} 1 \\ 0 \end{bmatrix} = \frac{1}{\sqrt{2}} \begin{bmatrix} \sqrt{1+su_z} \\ se^{i\phi}\sqrt{1-su_z} \end{bmatrix}. \tag{2.5}$$

The corresponding eigenvalues follow from Eqs. (2.2) and (2.4),

$$E_{\boldsymbol{k}s} = s\sqrt{A^2\boldsymbol{k}^2 + m^2}. \tag{2.6}$$

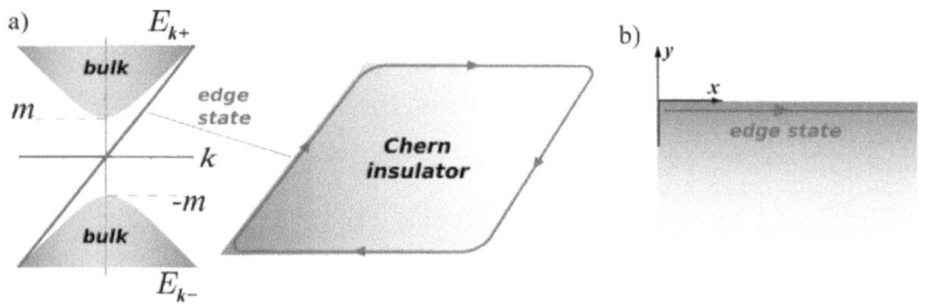

Figure 2.1 (a) Schematic of bulk and edge states of a Chern insulator [see Eqs. (2.6) and (2.14)] and (b) system geometry. For concreteness, we consider a positive mass term m (hence, edge chirality $\mathrm{sgn}(m) = +1$).

In Eq. (2.6), the positive ($s = +1$) and negative ($s = -1$) energies correspond to the bulk conduction and valence bands, respectively. The mass term opens an energy gap of $2m$ at the Dirac point $\boldsymbol{k} = 0$ between the bands (see also Fig. 2.1a). That is, the bulk behaves as a 2D band insulator. It is a nontrivial insulator, though, as it appears to have a gapless mode localized at the edge of the system. Similar localized modes were first found in relativistic theory of 1D fermions (the so-called fermion zero modes) [75]. To see how this comes about in our system, we consider the real-space Dirac equation

$$E\,\psi(x, y) = [-i\hbar v(\sigma_x \partial_x + \sigma_y \partial_y) + m\sigma_z]\psi(x, y) \tag{2.7}$$

in a 2D semi-space defined by $-\infty < y \leq 0$ and $-\infty < x < \infty$ [see also Fig. 2.1b]. We seek the solution in the form of a plane wave propagating along the edge with a wave vector k and decaying exponentially in the bulk on a length λ:

$$\psi(x, y) = \Phi\, e^{ikx + y/\lambda}, \tag{2.8}$$

with Φ being a position-independent spinor. Inserting Eq. (2.8) into Eq. (2.7) we have

$$E\Phi = \hbar v k \sigma_x \Phi + \left(-i\frac{\hbar v}{\lambda}\sigma_y + m\sigma_z\right)\Phi. \tag{2.9}$$

This equation can be split into two equations

$$E\Phi = \hbar v k \sigma_x \Phi, \tag{2.10}$$

and

$$\left(-\frac{\hbar v}{\lambda} i\sigma_y + m\sigma_z\right)\Phi = 0 \quad \Longleftrightarrow \quad \sigma_z\left(-\frac{\hbar v}{\lambda}\sigma_x + m\right)\Phi = 0. \tag{2.11}$$

The second equation in (2.11) is obtained with the help of the identity $i\sigma_y = \sigma_z\sigma_x$. This equation serves to find Φ and λ. We choose Φ to be an eigenstate of σ_x, satisfying the equation

$$\sigma_x \Phi = \mathrm{sgn}(m)\, \Phi. \tag{2.12}$$

Then, Eq. (2.11) reduces to $-\mathrm{sgn}(m)\hbar v/\lambda + m = 0$, yielding the decay length

$$\lambda = \frac{\hbar v}{|m|} > 0. \tag{2.13}$$

We note that the inclusion of the sign factor sgn(m) in Eq. (2.12) guaranties a positive λ, which is a requirement for the existence of the edge state. Combining now Eqs. (2.10) and (2.12), we find the edge-state spectrum

$$E(k) = \mathrm{sgn}(m)\, v\hbar k. \tag{2.14}$$

The obtained solution has several notable properties:

(i) It is gapless, spanning the energy range within the gap, $-|m| < E < |m|$, which is forbidden for the bulk states (see Fig. 2.1a).

(ii) The direction of its velocity is determined by the sign of the mass term, sgn(m). It is the mathematical expression for the edge chirality and a rather nontrivial consequence of the broken TRS.

(iii) The occurrence of a gapless edge mode can be explained by the inversion of the band structure at the edge. Indeed, any state $\psi(x, y)$ in the negative semispace $y < 0$ can be extended to the positive semispace $y > 0$ by the parity transformation

$$\psi'(x, y) = \sigma_x \psi(x, -y). \tag{2.15}$$

Such a mirror reflection changes the sign of the mass term in the Dirac equation (2.7):

$$E\, \psi'(x, y) = [-i\hbar v(\sigma_x \partial_x + \sigma_y \partial_y) - m\sigma_z]\psi'(x, y). \tag{2.16}$$

The mirror image has the same conduction- and valence-band spectra as in Eq. (2.6), but the ordering of the conduction and valence bands is opposite to the original (see Fig. 2.2). In other words, there is a band inversion at the

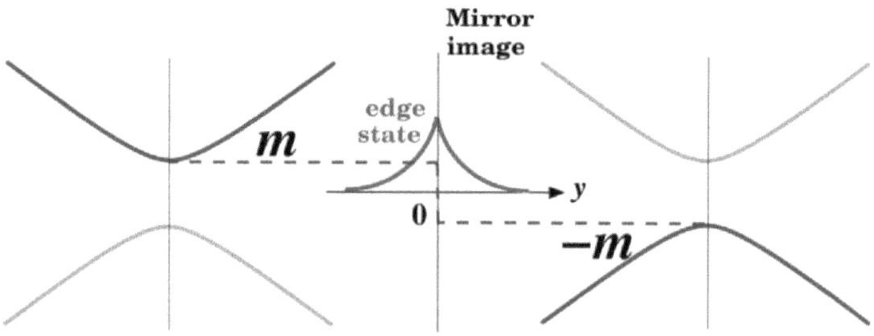

Figure 2.2 Illustration of the band inversion and localized mode at the edge of a Chern insulator. The edge is equivalent to a domain boundary between the region with the "mass" m and its mirror image with the "mass" $-m$ (see also text).

edge accompanied by the gap closing and, as a consequence, the appearance of a gapless mode. Our model is equivalent to a sharp, step-like band inversion at the edge. Similar results can be obtained for a smoothly varying odd-parity $m(y)$ [75].

(iv) The Chern insulator discussed in this section has the TR-partner. It can be described by the Hamiltonian

$$\hat{h}_{\boldsymbol{k}} = \boldsymbol{d}_{\boldsymbol{k}} \cdot \boldsymbol{\sigma}, \qquad \boldsymbol{d}_{\boldsymbol{k}} = (\hbar v k_x, -\hbar v k_y, m), \tag{2.17}$$

where we change the relative sign of the d_x and d_y components. In the TR-partner, the chirality of the edge mode flips to $-\mathrm{sgn}(m)$ (see end-of-chapter problems).

The existence of the gapless edge state distinguishes Chern insulators from ordinary band insulators. As we shall see, this distinction is topological as it is based on a topological Chern number. Before going to that point, we need to introduce the notions of Berry's phase and Berry's curvature.

2.1.2 Berry's Phase and Curvature

Berry's phase is a quantum phase factor that appears in the course of an adiabatic evolution of a quantum state [76]. Let us consider an evolution of a bulk state $|\psi_{\boldsymbol{k}s}\rangle$ (2.5) under an adiabatic rotation of the wave vector, which we model by a time-dependent function $\boldsymbol{k}(t)$ as sketched in Fig. 2.3. It is assumed that at any time t the evolving state

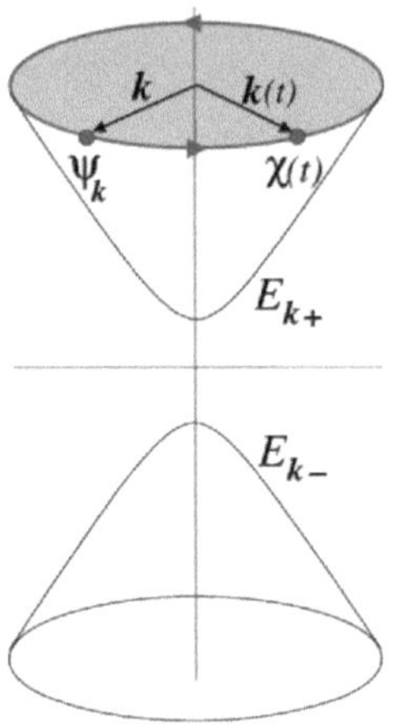

Figure 2.3 Illustration of an evolution of a bulk state in a CI under an adiabatic rotation of the wave vector modeled by a time-dependent function $\boldsymbol{k}(t)$ (see also text).

$|\chi(t)\rangle$ remains an eigenstate of the Hamiltonian $\hat{h}(\boldsymbol{k}(t))$ that depends on time only through the function $\boldsymbol{k}(t)$. Then, the equation of motion for $|\chi(t)\rangle$ reduces to

$$i\hbar\partial_t|\chi(t)\rangle = E_{\boldsymbol{k}(t)s}\,|\chi(t)\rangle. \tag{2.18}$$

Its solution must be proportional to $|\psi_{\boldsymbol{k}(t)s}\rangle$, so that we use the ansatz

$$|\chi(t)\rangle = e^{-i\theta(t)}|\psi_{\boldsymbol{k}(t)s}\rangle, \qquad \langle\psi_{\boldsymbol{k}(t)s}|\psi_{\boldsymbol{k}(t)s}\rangle = 1, \tag{2.19}$$

where $\theta(t)$ is the unknown phase acquired by time t. Inserting Eq. (2.19) into Eq. (2.18) yields

$$i\hbar\partial_t|\psi_{\boldsymbol{k}(t)s}\rangle + \hbar\dot{\theta}(t)|\psi_{\boldsymbol{k}(t)s}\rangle = E_{\boldsymbol{k}(t)s}|\psi_{\boldsymbol{k}(t)s}\rangle. \tag{2.20}$$

Applying $\langle\psi_{\boldsymbol{k}(t)s}|$ from the left, we have

$$\dot{\theta}(t) = \frac{1}{\hbar}E_{\boldsymbol{k}(t)s} - i\langle\psi_{\boldsymbol{k}(t)s}|\partial_t|\psi_{\boldsymbol{k}(t)s}\rangle, \tag{2.21}$$

and, finally, the solution for the phase

$$\theta(t) = \frac{1}{\hbar}\int_0^t E_{\boldsymbol{k}(t')s}dt' - i\int_0^t \langle\psi_{\boldsymbol{k}(t')s}|\partial_{t'}|\psi_{\boldsymbol{k}(t')s}\rangle dt'. \tag{2.22}$$

Here the first term is the ordinary dynamic phase, whereas the second one originates from the adiabatic evolution of $|\psi_{\boldsymbol{k}(t)s}\rangle$. Upon completing the full cycle at time t_{cycle}, the second contribution becomes

$$\gamma_s = -i\int_0^{t_{cycle}} \langle \psi_{\boldsymbol{k}(t')s}|\partial_{t'}|\psi_{\boldsymbol{k}(t')s}\rangle dt' = -i\oint \langle \psi_{\boldsymbol{k}s}|\nabla|\psi_{\boldsymbol{k}s}\rangle \cdot d\boldsymbol{k}, \quad (2.23)$$

where the time integration is converted into a path integral around $\boldsymbol{k} = 0$ in momentum space. This geometrical phase is the analogue of Berry's phase for our system. Defining the geometrical vector potential (Berry's connection) by

$$\boldsymbol{A}_{\boldsymbol{k}s} = -i\langle \psi_{\boldsymbol{k}s}|\nabla|\psi_{\boldsymbol{k}s}\rangle, \quad (2.24)$$

one can express the Berry phase in the form analogous to the magnetic flux,

$$\gamma_s = \oint \boldsymbol{A}_{\boldsymbol{k}s} \cdot d\boldsymbol{k} = \int (\nabla \times \boldsymbol{A}_{\boldsymbol{k}s})_z \, dk_x dk_y, \quad (2.25)$$

where Berry's curvature $(\nabla \times \boldsymbol{A}_{\boldsymbol{k}s})_z$ plays the role of the magnetic field perpendicular to the area swept out by the integration path.

Next, we take advantage of the axial symmetry of the problem and write the Berry connection in the polar coordinate system

$$\boldsymbol{A}_{\boldsymbol{k}s} = A^{(k)}_{\boldsymbol{k}s}\boldsymbol{e}_k + A^{(\phi)}_{\boldsymbol{k}s}\boldsymbol{e}_\phi, \quad (2.26)$$

where $\boldsymbol{e}_k$ and $\boldsymbol{e}_\phi$ are the radial and azimuthal basis unit vectors, respectively. With the wave functions given by Eq. (2.5), the radial and azimuthal projections of $\boldsymbol{A}_{\boldsymbol{k}s}$ are

$$A^{(k)}_{\boldsymbol{k}s} = -i\langle \psi_{\boldsymbol{k}s}|\frac{\partial}{\partial k}|\psi_{\boldsymbol{k}s}\rangle = 0, \quad (2.27)$$

$$A^{(\phi)}_{\boldsymbol{k}s} = -i\langle \psi_{\boldsymbol{k}s}|\frac{\partial}{k\partial \phi}|\psi_{\boldsymbol{k}s}\rangle = \frac{1 - su_z}{2k}. \quad (2.28)$$

By symmetry, only the azimuthal projection is nonzero, which yields the Berry phase

$$\gamma_s = \oint \boldsymbol{A}_{\boldsymbol{k}s} \cdot d\boldsymbol{k} = 2\pi k A^{(\phi)}_{\boldsymbol{k}s} = \pi(1 - su_z) \quad (2.29)$$

$$= \pi\left(1 - \frac{sm}{\sqrt{\hbar^2 v^2 \boldsymbol{k}^2 + m^2}}\right), \quad (2.30)$$

and the Berry curvature

$$(\nabla \times \boldsymbol{A}_{\boldsymbol{k}s})_z = \frac{1}{k}\frac{\partial}{\partial k}\left(kA^{(\phi)}_{\boldsymbol{k}s}\right) = -\frac{s}{2k}\frac{\partial u_z}{\partial k} \quad (2.31)$$

$$= \frac{s(\hbar v)^2}{2}\frac{m}{(\hbar^2 v^2 \boldsymbol{k}^2 + m^2)^{3/2}} = \frac{s(\hbar v)^2}{2}\frac{\boldsymbol{d}_{\boldsymbol{k}} \cdot \boldsymbol{z}}{|\boldsymbol{d}_{\boldsymbol{k}}|^3}. \quad (2.32)$$

We note that for a gapless Dirac cone the Berry phase is the universal number of π. The nonuniversal contribution is proportional to the mass term m and has opposite signs for the conduction and valence bands. Interestingly, Berry's curvature (2.32) is analogous to the field of a monopole positioned at the point $\boldsymbol{k} = 0$ with the fractional strength $s/2$ in units of $(\hbar v)^2$. This observation is intimately related to the quantum Hall effect and topological invariants discussed below.

It is important to pay attention to the signs of the Berry connection, phase and curvature above. For the TR-partner in Eq. (2.17), these quantities should be taken with the opposite signs (see end-of-chapter problems).

2.1.3 Thouless–Kohmoto–Nightingale–den Nijs Formula for the Hall Conductivity. Topological Invariants

Our next goal is to introduce the notion of a topological invariant. It allows a rigorous classification of topological phases of an electronic system and is insensitive to continuous deformations of the electronic band structure. For systems with broken TRS, the topological invariant is directly related to an observable—the quantized Hall conductivity. In order to illustrate all these points we shall derive an important formula, first proposed by TKNN [3], that expresses the quantum Hall conductivity in terms of the Berry curvature of all occupied electronic bands of the system.

To calculate the electric conductivity we employ the linear response theory with respect to an external uniform electric field $\mathcal{E}e^{-i\omega t}$ with the frequency ω. The details of this standard approach are well described in literature (see e.g. Refs. [77, 78]). The conductivity is a 2×2 tensor whose components are given by the Kubo formula

$$\sigma_{\alpha\beta}(\omega) = -\frac{ie^2\hbar}{a}\sum_{l,m}\frac{f(E_l) - f(E_m)}{E_l - E_m}\frac{\langle l|v_\alpha|m\rangle\langle m|v_\beta|l\rangle}{\hbar\omega + E_l - E_m}, \tag{2.33}$$

where $\langle l|v_\alpha|m\rangle$ is the matrix element of the α-projection of the velocity operator

$$v_\alpha = (x_\alpha\hat{h} - \hat{h}x_\alpha)/i\hbar \tag{2.34}$$

between quantum states $|m\rangle$ and $|l\rangle$ with energies E_m and E_l:

$$\hat{h}|m, l\rangle = E_{m,l}|m, l\rangle. \quad (2.35)$$

The indices m and l include the full sets of quantum numbers, while α and β refer to the spatial coordinates x and y; $f(E)$ is the Fermi-Dirac distribution function, e is the electron charge, and a is the area of the system.

We are interested in the transverse (Hall) conductivity σ_{xy} in the limit of zero frequency and temperature

$$\sigma_{xy} = -\frac{ie^2\hbar}{a}\sum_{l,m}[\Theta(\mu - E_l) - \Theta(\mu - E_m)]\frac{\langle l|v_x|m\rangle\langle m|v_y|l\rangle}{(E_l - E_m)^2}, \quad (2.36)$$

and with the Fermi level, μ, lying between two arbitrary energy levels E_m and E_l [$\Theta(E)$ is the unit-step function]. Two level configurations are possible:

$$E_l < \mu < E_m, \quad (2.37)$$

and

$$E_m < \mu < E_l. \quad (2.38)$$

For the former the difference of the Θ functions in Eq. (2.36) is $+1$, whereas for the latter this difference is -1. Summing up the two possibilities, we can write

$$\sigma_{xy} = -\frac{ie^2\hbar}{a}\sum_{E_l<\mu,\, E_m>\mu}\frac{\langle l|v_x|m\rangle\langle m|v_y|l\rangle - \langle l|v_y|m\rangle\langle m|v_x|l\rangle}{(E_l - E_m)^2}. \quad (2.39)$$

Next, we write the velocity matrix elements in terms of the matrix elements of the coordinate operator:

$$\langle m|v_\alpha|l\rangle = \frac{\langle m|x_\alpha\,\hat{h}|l\rangle - \langle m|\hat{h}\,x_\alpha|l\rangle}{i\hbar} = \frac{E_l - E_m}{i\hbar}\langle m|x_\alpha|l\rangle \quad (2.40)$$

and

$$\langle l|v_\alpha|m\rangle = \frac{\langle l|x_\alpha\,\hat{h}|m\rangle - \langle l|\hat{h}\,x_\alpha|m\rangle}{i\hbar} = \frac{E_m - E_l}{i\hbar}\langle l|x_\alpha|m\rangle, \quad (2.41)$$

where we used Eq. (2.35) and its conjugate $\langle m, l|\hat{h} = E_{m,l}\langle m, l|$. Now, the Hall conductivity is expressed in terms of the eigenfunctions only

$$\sigma_{xy} = -\frac{ie^2}{\hbar a}\sum_{E_l<\mu,\, E_m>\mu}[\langle l|x|m\rangle\langle m|y|l\rangle - \langle l|y|m\rangle\langle m|x|l\rangle]. \quad (2.42)$$

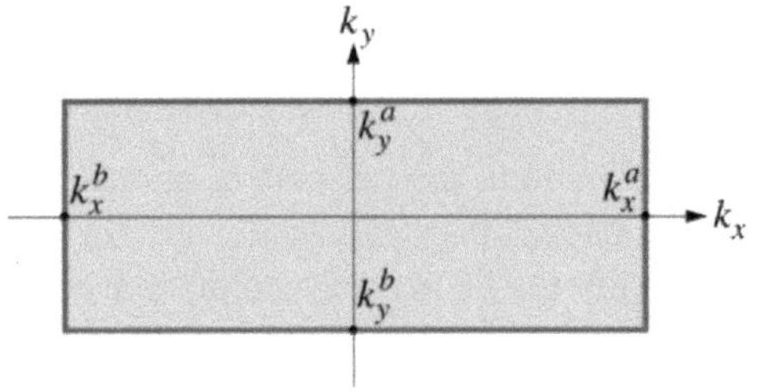

Figure 2.4 Schematic of the Brillouin zone assumed in the derivation of the TKNN formula. The electronic wave function is periodic in both k_x- and k_y-directions.

Note that the unoccupied states with $E_m > \mu$ can be eliminated, using the completeness condition

$$\sum_{E_m>\mu} |m\rangle\langle m| = 1 - \sum_{E_m<\mu} |m\rangle\langle m|, \tag{2.43}$$

which yields

$$\begin{aligned}\sigma_{xy} = &-\frac{ie^2}{\hbar a}\sum_{E_l<\mu}\langle l|xy - yx|l\rangle \\ &+\frac{ie^2}{\hbar a}\sum_{E_l<\mu,\,E_m<\mu}\left[\langle l|x|m\rangle\langle m|y|l\rangle - \langle m|x|l\rangle\langle l|y|m\rangle\right].\end{aligned} \tag{2.44}$$

Up to the summation index names the two terms in the last line are identical and cancel each other, so the result is

$$\sigma_{xy} = -\frac{ie^2}{\hbar a}\sum_{E_l<\mu}\langle l|xy - yx|l\rangle. \tag{2.45}$$

The matrix elements are taken over the first Brillouin zone. For concreteness, we assume a rectangular Brillouin zone as sketched in Fig. 2.4, with periodic boundary conditions for the electronic wave function $\psi_{\mathbf{k}l}$ in both k_x- and k_y-directions. For example, the matrix element $\langle l|xy|l\rangle$ is given by

$$\begin{aligned}\langle l|xy|l\rangle &= \sum_{\mathbf{k}}\langle\psi_{\mathbf{k}l}|i\partial_{k_x} i\partial_{k_y}\psi_{\mathbf{k}l}\rangle = -a\int_{BZ}\frac{dk_x dk_y}{(2\pi)^2}\langle\psi_{\mathbf{k}l}|\partial_{k_x}\partial_{k_y}\psi_{\mathbf{k}l}\rangle \\ &= a\int_{BZ}\frac{dk_x dk_y}{(2\pi)^2}\langle\partial_{k_x}\psi_{\mathbf{k}l}|\partial_{k_y}\psi_{\mathbf{k}l}\rangle.\end{aligned} \tag{2.46}$$

The last line is obtained by partial integration over k_x under periodic boundary conditions at the Brillouin zone edges $k_x^{a,b}$. This yields the Hall conductivity as an integral in $\boldsymbol{k}$ space

$$\sigma_{xy} = -\frac{ie^2}{h}\int_{BZ}\frac{dk_x dk_y}{2\pi}\sum_{E_l<\mu}\left[\langle\partial_{k_x}\psi_{\boldsymbol{k}l}|\partial_{k_y}\psi_{\boldsymbol{k}l}\rangle - \langle\partial_{k_y}\psi_{\boldsymbol{k}l}|\partial_{k_x}\psi_{\boldsymbol{k}l}\rangle\right]. \tag{2.47}$$

Define the Berry connection of the occupied bands as

$$A_{\boldsymbol{k}}^{(x)} = -i\sum_{E_l<\mu}\langle\psi_{\boldsymbol{k}l}|\partial_{k_x}\psi_{\boldsymbol{k}l}\rangle, \qquad A_{\boldsymbol{k}}^{(y)} = -i\sum_{E_l<\mu}\langle\psi_{\boldsymbol{k}l}|\partial_{k_y}\psi_{\boldsymbol{k}l}\rangle. \tag{2.48}$$

This brings us to the TKNN formula

$$\begin{aligned}\sigma_{xy} &= \frac{e^2}{h}\int_{BZ}\frac{dk_x dk_y}{2\pi}\left(\partial_{k_x}A_{\boldsymbol{k}}^{(y)} - \partial_{k_y}A_{\boldsymbol{k}}^{(x)}\right)\\ &= \frac{e^2}{h}\int_{BZ}\frac{dk_x dk_y}{2\pi}(\nabla\times\boldsymbol{A}_{\boldsymbol{k}})_z.\end{aligned} \tag{2.49}$$

It establishes a general link between the Hall conductivity and the topology of the electronic bands in the first Brillouin zone.

Two aspects of Eq. (2.49) merit special attention:

(i) The Hall conductivity is nonzero due to broken TRS and is linked to chiral gapless edge states. To illustrate this point, take the model of the Chern insulator introduced in Section 2.1.1. Since the Fermi level must lie in the band gap, there is only one occupied band—the valence one. The Berry connection in Eq. (2.48) is then identified as $\boldsymbol{A}_{\boldsymbol{k}s}$ with $s = -1$ from Section 2.1.2; the Berry curvature in Eq. (2.49) is given by Eq. (2.31), and the $\boldsymbol{k}$ integral can be done in the polar coordinates as follows:

$$\begin{aligned}\sigma_{xy} &= \frac{e^2}{h}\int_0^\infty\frac{2\pi k dk}{2\pi}\frac{1}{2k}\frac{\partial u_z(k)}{\partial k}\\ &= \frac{e^2}{h}\frac{u_z(\infty)-u_z(0)}{2} = -\frac{e^2}{h}\frac{\mathrm{sgn}(m)}{2}.\end{aligned} \tag{2.50}$$

We used Eq. (2.3) for $u_z(k)$ with limits $u_z(\infty) = 0$ and $u_z(0) = \mathrm{sgn}(m)$. As we see, the Hall conductivity is induced by the TRS-breaking mass term m. If we set $m = 0$ before evaluating the integral, the result is $\sigma_{xy} = 0$. This type of

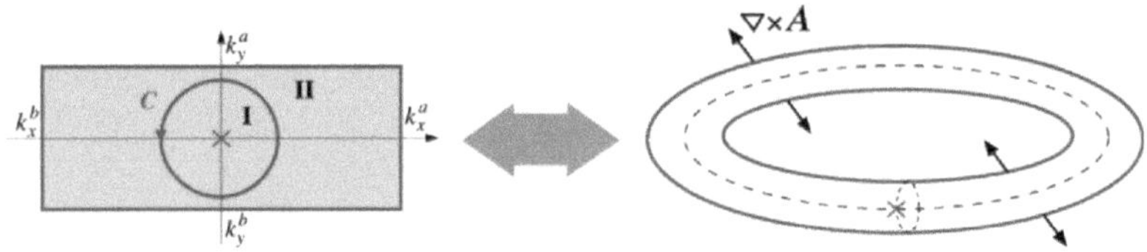

Figure 2.5 Mapping the Brillouin zone to the surface of a torus. The cross indicates a vortex-like singularity of the Berry connection resulting in the nontrivial winding of the phase (see also text).

the quantum Hall effect is called anomalous as it occurs without an external magnetic field and the Landau levels [6]. We shall address it in more detail in the forthcoming sections. Although the Hall conductivity (2.50) is calculated from the bulk band model, σ_{xy} depends on the chirality sgn(m) of the edge state in a Chern insulator [cf. Eq. (2.14)]. This dependence establishes the link between the bulk Hall conductivity and the edge mode. It is a concrete manifestation of the bulk-edge correspondence principle of TIs.

(ii) The Hall conductivity is quantized in integer units of the conductance quantum e^2/h. We cannot see this in our Chern insulator model for the reasons that will become clear later. Fortunately, the proof rests only on topological properties of the TKNN formula and is free of any model specifics [3, 79]. In view of the periodic boundary conditions we can connect the edges of the Brillouin zone in both k_x- and k_y-directions, so it maps to the surface of a torus and, as such, has no boundary (see also Fig. 2.5). This allows us to divide the Brillouin zone into two closed manifolds, denoted below as I and II, and write Eq. (2.49) as

$$\sigma_{xy} = \frac{e^2}{h}\left[\int_I \frac{dk_x dk_y}{2\pi}(\nabla \times \boldsymbol{A_k})_z + \int_{II} \frac{dk_x dk_y}{2\pi}(\nabla \times \boldsymbol{A_k})_z\right]. \tag{2.51}$$

The area integrals can be converted into the line ones along the common boundary C, but with the opposite signs because the circulation direction of C is opposite for manifolds I and II:

$$\sigma_{xy} = \frac{e^2}{h}\frac{1}{2\pi}\left[\oint_C \boldsymbol{A}_{\boldsymbol{k}}^{I} \cdot d\boldsymbol{k} - \oint_C \boldsymbol{A}_{\boldsymbol{k}}^{II} \cdot d\boldsymbol{k}\right]. \tag{2.52}$$

The gauge invariance allows the Berry connections $\boldsymbol{A}_{\boldsymbol{k}}^{I}$ and $\boldsymbol{A}_{\boldsymbol{k}}^{II}$ to differ only by a gradient of an arbitrary phase of the wave function,

$$\boldsymbol{A}_{\boldsymbol{k}}^{I} - \boldsymbol{A}_{\boldsymbol{k}}^{II} = \nabla\theta, \tag{2.53}$$

so that the Hall conductivity is proportional to the winding number of the phase of the electronic wave function, n:

$$\sigma_{xy} = \frac{e^2}{h}n, \qquad n = \frac{1}{2\pi}\oint_C \nabla\theta \cdot d\boldsymbol{k}. \tag{2.54}$$

For a uniquely defined wave function, n is an integer and, hence, the Hall conductivity is quantized in integer units of e^2/h. This result has parallels with the quantization of the superfluid velocity and the magnetic flux in superconductors. In those systems, the nontrivial phase winding numbers $n \neq 0$ are associated with vortices—spatial singularities of the condensate order parameter. In our case, the role of the vortices is played by the singularities of the Berry connection $\boldsymbol{A}_{\boldsymbol{k}}$. To illustrate this point let us consider the Berry connection for the Chern insulator in Eq. (2.28). For $\boldsymbol{k} \to 0$, the Berry connection behaves as

$$A_{\boldsymbol{k}s}^{(\phi)} \approx \frac{1 - s\cdot \mathrm{sgn}(m)}{2k} = \begin{cases} 0, & s\cdot\mathrm{sgn}(m) = 1, \\ \infty, & s\cdot\mathrm{sgn}(m) = -1. \end{cases} \tag{2.55}$$

Assuming $m > 0$, we see that for the valence band ($s = -1$) the Berry connection is singular at the origin $\boldsymbol{k} = 0$, while for the conduction band ($s = 1$) it vanishes there. Furthermore, according to Eq. (2.29), the singularity of the Berry connection carries a quantized flux (Berry's phase):

$$\gamma_s = \pi\left[1 - s\cdot\mathrm{sgn}(m)\right] = \begin{cases} 0, & s\cdot\mathrm{sgn}(m) = 1, \\ 2\pi, & s\cdot\mathrm{sgn}(m) = -1. \end{cases} \tag{2.56}$$

This indicates a quantum vortex at the origin $\boldsymbol{k} = 0$. Thus, the quantized Hall conductivity is proportional to the total vorticity of the Berry connection in the Brillouin zone under broken TRS [3, 79].

The TKNN number n is determined solely by the electronic band structure:

$$
\begin{aligned}
n &= \int_{BZ} \frac{dk_x dk_y}{2\pi} (\nabla \times \boldsymbol{A_k})_z \\
&= -i \int_{BZ} \frac{dk_x dk_y}{2\pi} \sum_{E_l < \mu} \left[\langle \partial_{k_x} \psi_{\boldsymbol{k}l} | \partial_{k_y} \psi_{\boldsymbol{k}l} \rangle - \langle \partial_{k_y} \psi_{\boldsymbol{k}l} | \partial_{k_x} \psi_{\boldsymbol{k}l} \rangle \right].
\end{aligned}
\tag{2.57}
$$

It is a topological invariant since it is robust against continuous deformations of the band structure that leave the number of the occupied bands and, consequently, the number of the singularities of the Berry connection unchanged. In particular, Eq. (2.57) shows that the transition from the $n = 0$ to $n \neq 0$ state is a topological one since the topology of the Brillouin zone changes from singly connected to multiply connected if $\boldsymbol{A_k}$ acquires a vortex-like singularity in the Brillouin zone. There is also an insightful parallel with differential geometry in which n in Eq. (2.57) is identified as the first Chern number of a principal $U(1)$ bundle over a torus.

To conclude this section, we need to rationalize why our Chern insulator model predicts a half-integer Hall conductivity [see Eq. (2.50)] instead of the integer one (2.54). The reason for this discrepancy is that a single two-component Dirac species (see Section 2.1.1) is insufficient to represent a crystalline solid. According to the general theorem proved by Nielsen and Ninomiya [80], in lattice crystal structures Dirac fermions always come in pairs, which is referred to as fermion doubling. In other words, in order to represent all relevant degrees of freedom the fermion content of our model must be at least twice as large, which would yield the correct, integer-quantized Hall conductivity. This has been shown explicitly by Haldane for a 2D honeycomb lattice where a pair of fermion species emerges in two inequivalent valleys at the corners of the Brillouin zone [6]. We shall obtain analogous results within quite different approaches which use square lattice and thin-film models.

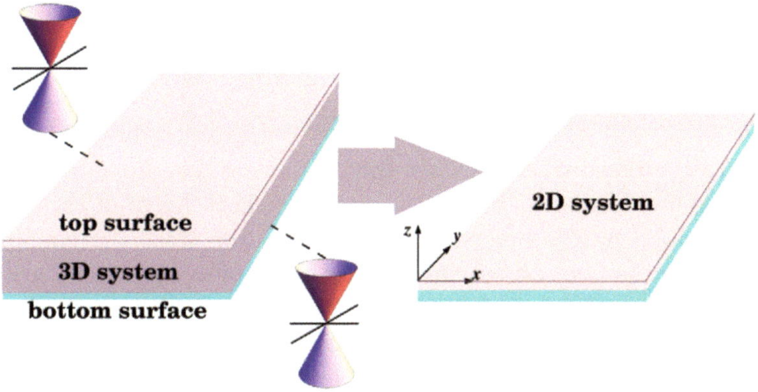

Figure 2.6 Transformation of a 3D slab with top and bottom Dirac surfaces (left) into a thin film (right) as a model for a QSHI.

2.2 Quantum Spin Hall Insulators I

2.2.1 Dirac Model

Similar to Chern insulators, quantum spin Hall insulators (QSHIs) must have a band gap in the bulk, but the gap opening should occur without breaking the TRS. The model satisfying these requirements can be introduced by a rather simple reasoning. Consider two 2D electronic systems at the top and bottom surfaces of a 3D slab with an insulating interior [see Fig. 2.6 (left)]. Each system is described by a gapless TR-invariant Dirac Hamiltonian. As the thickness of the slab is reduced, the two systems become coupled by tunneling through the insulating layer between them. In the quasi-2D limit, when the top and bottom states overlap [see Fig. 2.6 (right)], the whole system can be described by an equation

$$\hat{H}\,\psi(\boldsymbol{r}) = E\,\psi(\boldsymbol{r}), \qquad \boldsymbol{r} = (x, y, 0), \tag{2.58}$$

for a four-component wave function, $\psi(\boldsymbol{r})$, in the basis of the original top (t) and bottom (b) states each having two spin components $\uparrow$ and $\downarrow$:

$$\psi(\boldsymbol{r}) = \begin{bmatrix} \psi_{t\uparrow}(\boldsymbol{r}) \\ \psi_{t\downarrow}(\boldsymbol{r}) \\ \psi_{b\uparrow}(\boldsymbol{r}) \\ \psi_{b\downarrow}(\boldsymbol{r}) \end{bmatrix}. \tag{2.59}$$

The Hamiltonian $\hat{H}$ in Eq. (2.58) is a 4×4 matrix given by

$$\hat{H} = \begin{bmatrix} -i\hbar v(\sigma_x \partial_x + \sigma_y \partial_y) & \Delta\sigma_0 \\ \Delta\sigma_0 & +i\hbar v(\sigma_x \partial_x + \sigma_y \partial_y) \end{bmatrix}. \quad (2.60)$$

Here the diagonal entries are the real-space Hamiltonians of the originally decoupled t and b subsystems. The choice of the signs is dictated by the inversion symmetry of the slab. The coupling between the two subsystems is described by the off-diagonal entry $\Delta\sigma_0$, where σ_0 is a unit matrix in spin space, which corresponds to spin-independent tunneling. For a real parameter Δ, the whole system is manifestly TR invariant since its Hamiltonian does not change under the $\mathbb{T}$ operation

$$\mathbb{T}\hat{H}\mathbb{T}^{-1} = \hat{H}, \quad \mathbb{T} = \begin{bmatrix} i\sigma_y K & 0 \\ 0 & i\sigma_y K \end{bmatrix}. \quad (2.61)$$

At the same time, as we shall see later, the coupling Δ induces an energy gap in the electron spectrum. Thus, our model is analogous to a massive Dirac model, and Hamiltonian (2.60) is a good starting point to search for the TR-invariant topological states. It is worth noting that this Hamiltonian can be derived from the band structure theory of concrete TI materials, e.g. HgTe/CdTe quantum wells [13] and thin films of Bi_2Se_3 [81, 82]. In Section 2.4, we shall discuss how to obtain Hamiltonian (2.60) from a tight-binding model on a square lattice.

We are interested in the solutions that decay exponentially from the edge of the system, $y = 0$, into its interior in the 2D semi-space $-\infty < x < \infty$ and $0 \leq y < \infty$ [see also Fig. 2.6 (left)]. The spinor $\psi(\boldsymbol{r})$ must satisfy a boundary condition that ensures the vanishing of the normal component of the particle current density, $j_y(\boldsymbol{r})$, at the edge $y = 0$. In order to derive such a boundary condition, we start with the expression for the particle current density, $\boldsymbol{j}(\boldsymbol{r})$, in terms of an expectation value of the velocity operator, $\hat{\boldsymbol{v}}$:

$$\boldsymbol{j}(\boldsymbol{r}) = \tilde{\psi}(\boldsymbol{r})\hat{\boldsymbol{v}}\psi(\boldsymbol{r}), \qquad \hat{\boldsymbol{v}} = (\boldsymbol{r}\hat{H} - \hat{H}\boldsymbol{r})/i\hbar, \quad (2.62)$$

where $\tilde{\psi}(\boldsymbol{r})$ denotes the conjugated spinor,

$$\tilde{\psi}(\boldsymbol{r}) = [\psi^*_{t\uparrow}(\boldsymbol{r}), \psi^*_{t\downarrow}(\boldsymbol{r}), \psi^*_{b\uparrow}(\boldsymbol{r}), \psi^*_{b\downarrow}(\boldsymbol{r})]. \quad (2.63)$$

From Eqs. (2.60) and (2.62) we find

$$j_y(\mathbf{r}) = v\tilde{\psi}(\mathbf{r}) \begin{bmatrix} \sigma_y & 0 \\ 0 & -\sigma_y \end{bmatrix} \psi(\mathbf{r})$$
$$= iv[\psi^*_{t\downarrow}(\mathbf{r})\psi_{t\uparrow}(\mathbf{r}) - \psi^*_{t\uparrow}(\mathbf{r})\psi_{t\downarrow}(\mathbf{r})]$$
$$+ iv[\psi^*_{b\uparrow}(\mathbf{r})\psi_{b\downarrow}(\mathbf{r}) - \psi^*_{b\downarrow}(\mathbf{r})\psi_{b\uparrow}(\mathbf{r})]. \quad (2.64)$$

To fulfill the requirement $j_y(x, y = 0) = 0$ we impose the following boundary conditions:

$$\psi_{b\uparrow}(x, y = 0) = \psi_{t\downarrow}(x, y = 0), \quad (2.65)$$
$$\psi_{b\downarrow}(x, y = 0) = -\psi_{t\uparrow}(x, y = 0), \quad (2.66)$$

Importantly, this is the only choice compatible with the TRS. In order to see this we write Eqs. (2.65) and (2.66) in a spinor form

$$\psi(x, y = 0) = \hat{M}\psi(x, y = 0), \qquad \hat{M} = \begin{bmatrix} 0 & -i\sigma_y \\ i\sigma_y & 0 \end{bmatrix}. \quad (2.67)$$

Since $i\sigma_y$ commutes with the TR operator $i\sigma_y K$, matrix $\hat{M}$ in Eq. (2.67) is invariant under operation (2.61),

$$\mathbb{T}\hat{M}\mathbb{T}^{-1} = \hat{M}, \quad (2.68)$$

i.e. the transformed wave function $\mathbb{T}\psi(\mathbf{r})$ satisfies the same boundary condition as $\psi(\mathbf{r})$ in Eq. (2.67).

2.2.2 Bulk Band Structure

Before approaching the problem of edge states, it is instructive to discuss the bulk band structure of an unconfined 2D Dirac matter. For that purpose, we are going to diagonalize the Hamiltonian $\hat{H}$ (2.60) by an appropriate unitary transformation, $\hat{U}$, of Eq. (2.58),

$$\hat{H}'\,\psi'(\mathbf{r}) = E\,\psi'(\mathbf{r}), \qquad \mathbf{r} = (x, y, 0), \quad (2.69)$$

where the new spinor $\psi'(\mathbf{r})$ and Hamiltonian $\hat{H}'$ are

$$\psi'(\mathbf{r}) = \hat{U}\,\psi(\mathbf{r}), \quad (2.70)$$

$$\hat{H}' = \hat{U}\,\hat{H}\,U^{-1}. \quad (2.71)$$

Choosing the unitary matrix in the form

$$\hat{U} = \frac{1}{\sqrt{2}} \begin{bmatrix} \sigma_0 & \sigma_z \\ \sigma_z & -\sigma_0 \end{bmatrix}, \qquad \hat{U}^{-1} = \hat{U}, \quad (2.72)$$

we write Eqs. (2.70) and (2.71) explicitly as

$$\psi'(\boldsymbol{r}) = \frac{1}{\sqrt{2}} \begin{bmatrix} \psi_{t\uparrow}(\boldsymbol{r}) + \psi_{b\uparrow}(\boldsymbol{r}) \\ \psi_{t\downarrow}(\boldsymbol{r}) - \psi_{b\downarrow}(\boldsymbol{r}) \\ \psi_{t\uparrow}(\boldsymbol{r}) - \psi_{b\uparrow}(\boldsymbol{r}) \\ -\psi_{t\downarrow}(\boldsymbol{r}) - \psi_{b\downarrow}(\boldsymbol{r}) \end{bmatrix} \equiv \begin{bmatrix} \psi_{s\uparrow}(\boldsymbol{r}) \\ \psi_{a\downarrow}(\boldsymbol{r}) \\ \psi_{a\uparrow}(\boldsymbol{r}) \\ \psi_{s\downarrow}(\boldsymbol{r}) \end{bmatrix}, \tag{2.73}$$

$$\hat{H}' = \begin{bmatrix} -i\hbar v(\sigma_x \partial_x + \sigma_y \partial_y) + \Delta\sigma_z & 0 \\ 0 & +i\hbar v(\sigma_x \partial_x + \sigma_y \partial_y) - \Delta\sigma_z \end{bmatrix}. \tag{2.74}$$

The new basis in Eq. (2.73) has a simple meaning. It consists of all symmetric (s) and anti-symmetric (a) combinations of the original t and b states for the same spin projection. This basis has the advantage that the new Hamiltonian (2.74) is block diagonal. The coupling Δ enters similar to the mass term in a 4×4 Dirac model.

Ignoring, for a moment, the boundaries, we expand $\psi'(\boldsymbol{r})$ in plane waves in 2D wave vector ($\boldsymbol{k}$) space,

$$\psi'(\boldsymbol{r}) = \sum_{\boldsymbol{k}} \psi'_{\boldsymbol{k}} e^{i\boldsymbol{k}\boldsymbol{r}}, \tag{2.75}$$

where spinor $\psi'_{\boldsymbol{k}}$ satisfies the equation

$$\hat{H}'_{\boldsymbol{k}} \psi'_{\boldsymbol{k}} = E\, \psi'_{\boldsymbol{k}}, \tag{2.76}$$

with Hamiltonian $\hat{H}'_{\boldsymbol{k}}$ given by

$$\hat{H}'_{\boldsymbol{k}} = \begin{bmatrix} \hbar v \boldsymbol{\sigma} \cdot \boldsymbol{k} + \Delta\sigma_z & 0 \\ 0 & -\hbar v \boldsymbol{\sigma} \cdot \boldsymbol{k} - \Delta\sigma_z \end{bmatrix}. \tag{2.77}$$

For $\boldsymbol{k} = 0$ Eq. (2.76) assumes a simple form,

$$\Delta\sigma_z \begin{bmatrix} \psi_{s\uparrow\boldsymbol{k}=0} \\ \psi_{a\downarrow\boldsymbol{k}=0} \end{bmatrix} = E \begin{bmatrix} \psi_{s\uparrow\boldsymbol{k}=0} \\ \psi_{a\downarrow\boldsymbol{k}=0} \end{bmatrix}, \tag{2.78}$$

$$-\Delta\sigma_z \begin{bmatrix} \psi_{a\uparrow\boldsymbol{k}=0} \\ \psi_{s\downarrow\boldsymbol{k}=0} \end{bmatrix} = E \begin{bmatrix} \psi_{a\uparrow\boldsymbol{k}=0} \\ \psi_{s\downarrow\boldsymbol{k}=0} \end{bmatrix}. \tag{2.79}$$

Equations (2.78) and (2.79) each have two solutions being the eigenstates of σ_z. In total, there are four solutions:

$$\begin{bmatrix} \psi_{s\uparrow \boldsymbol{k}=0} \\ 0 \\ 0 \\ 0 \end{bmatrix} \quad \text{with} \quad E_{s\uparrow} = \Delta, \tag{2.80}$$

$$\begin{bmatrix} 0 \\ \psi_{a\downarrow \boldsymbol{k}=0} \\ 0 \\ 0 \end{bmatrix} \quad \text{with} \quad E_{a\downarrow} = -\Delta, \tag{2.81}$$

$$\begin{bmatrix} 0 \\ 0 \\ \psi_{a\uparrow \boldsymbol{k}=0} \\ 0 \end{bmatrix} \quad \text{with} \quad E_{a\uparrow} = -\Delta, \tag{2.82}$$

$$\begin{bmatrix} 0 \\ 0 \\ 0 \\ \psi_{s\downarrow \boldsymbol{k}=0} \end{bmatrix} \quad \text{with} \quad E_{s\downarrow} = \Delta. \tag{2.83}$$

The energies of these states correspond to the bottom of the conduction band and top of the valence band at $\boldsymbol{k} = 0$. Compared to the gapless spectrum in Fig. 1.8 there is an energy gap $2|\Delta|$ between the conduction and valence bands. The sign of Δ is of particular importance. Upon the sign change $\Delta \rightarrow -\Delta$ the band structure transforms from a normal one to an inverted one or vice versa, which is an essential ingredient of TI materials and will be explained in detail in Section 2.3.

In order to derive the band structure for $\boldsymbol{k} \neq 0$ we note that Eq. (2.76) decouples into two equations,

$$(\hbar v \boldsymbol{\sigma} \cdot \boldsymbol{k} + \Delta \sigma_z) \begin{bmatrix} \psi_{s\uparrow \boldsymbol{k}} \\ \psi_{a\downarrow \boldsymbol{k}} \end{bmatrix} = E \begin{bmatrix} \psi_{s\uparrow \boldsymbol{k}} \\ \psi_{a\downarrow \boldsymbol{k}} \end{bmatrix}, \tag{2.84}$$

$$-(\hbar v \boldsymbol{\sigma} \cdot \boldsymbol{k} + \Delta \sigma_z) \begin{bmatrix} \psi_{a\uparrow \boldsymbol{k}} \\ \psi_{s\downarrow \boldsymbol{k}} \end{bmatrix} = E \begin{bmatrix} \psi_{a\uparrow \boldsymbol{k}} \\ \psi_{s\downarrow \boldsymbol{k}} \end{bmatrix}, \tag{2.85}$$

which we can also write as

$$\mathrm{sgn}(\Delta)\sqrt{(\hbar v \boldsymbol{k})^2 + \Delta^2}\,(\boldsymbol{u}_{\boldsymbol{k}} \cdot \boldsymbol{\sigma}) \begin{bmatrix} \psi_{s\uparrow \boldsymbol{k}} \\ \psi_{a\downarrow \boldsymbol{k}} \end{bmatrix} = E \begin{bmatrix} \psi_{s\uparrow \boldsymbol{k}} \\ \psi_{a\downarrow \boldsymbol{k}} \end{bmatrix}, \tag{2.86}$$

$$-\mathrm{sgn}(\Delta)\sqrt{(\hbar v \boldsymbol{k})^2 + \Delta^2}\,(\boldsymbol{u}_{\boldsymbol{k}} \cdot \boldsymbol{\sigma}) \begin{bmatrix} \psi_{a\uparrow \boldsymbol{k}} \\ \psi_{s\downarrow \boldsymbol{k}} \end{bmatrix} = E \begin{bmatrix} \psi_{a\uparrow \boldsymbol{k}} \\ \psi_{s\downarrow \boldsymbol{k}} \end{bmatrix}, \tag{2.87}$$

where $\boldsymbol{u_k}$ is a unit vector defined by

$$\boldsymbol{u_k} = \frac{[\mathrm{sgn}(\Delta)\hbar v k_x,\ \mathrm{sgn}(\Delta)\hbar v k_x,\ |\Delta|]}{\sqrt{(\hbar v \boldsymbol{k})^2 + \Delta^2}}. \tag{2.88}$$

Each of Eqs. (2.86) and (2.87) have again two solutions being the eigenstates of helicity matrix $\boldsymbol{u_k} \cdot \boldsymbol{\sigma}$. The latter is a generalization of σ_z in Eqs. (2.78) and (2.79) and, as before, has the eigenvalues equal to ± 1. Instead of Eqs. (2.80)–(2.83), we have the following four solutions:

$$\begin{bmatrix} \psi_{s\uparrow \boldsymbol{k}} \\ \psi_{a\downarrow \boldsymbol{k}} \\ 0 \\ 0 \end{bmatrix}_{1,2} \quad \text{with} \quad E_{\boldsymbol{k}}^{(1,2)} = \pm \mathrm{sgn}(\Delta)\sqrt{(\hbar v \boldsymbol{k})^2 + \Delta^2}, \tag{2.89}$$

$$\begin{bmatrix} 0 \\ 0 \\ \psi_{a\uparrow \boldsymbol{k}} \\ \psi_{s\downarrow \boldsymbol{k}} \end{bmatrix}_{3,4} \quad \text{with} \quad E_{\boldsymbol{k}}^{(3,4)} = \mp \mathrm{sgn}(\Delta)\sqrt{(\hbar v \boldsymbol{k})^2 + \Delta^2}, \tag{2.90}$$

where labels 1, 2 and 3, 4 correspond to the signs $\pm$ and $\mp$, respectively. We see that for $\boldsymbol{k} \neq 0$ the symmetric and anti-symmetric states are mixed. For instance, solution 1 coincides at $\boldsymbol{k} = 0$ with the symmetric state $s \uparrow$ [see Eq. (2.80)], acquiring for $\boldsymbol{k} \neq 0$ an admixture of the anti-symmetric state $a \downarrow$ [see Eq. (2.81)]. The corresponding band dispersions are shown in Figs. 2.7 and 2.8.

2.2.3 Edge States. Spin Helicity

We are now in a position to examine the topological properties of the system by looking at its edge excitations. The boundary problem that we intend to solve is given by Eqs. (2.58), (2.60) and (2.67). To write them in a more compact form, we introduce another set of Pauli matrices $\tau_{x,y}$ and τ_z that operate in the space of the t and b subsystems. Then, the boundary problem assumes the form

$$\left[-i\hbar v \tau_z(\sigma_x \partial_x + \sigma_y \partial_y) + \Delta \tau_x \sigma_0\right] \psi(\boldsymbol{r}) = E\psi(\boldsymbol{r}), \tag{2.91}$$

$$\psi(x, y = 0) = \tau_y \sigma_y \psi(x, y = 0). \tag{2.92}$$

We again seek the solutions that propagate along the edge with a wave vector k and decay exponentially in the bulk on a length λ:

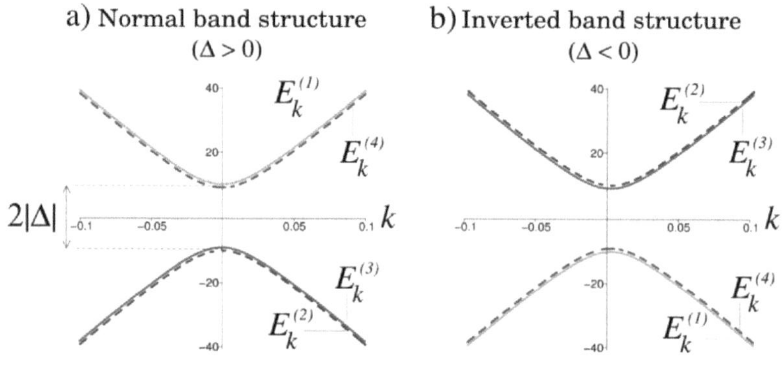

Figure 2.7 Energy bands (in meV) versus absolute value of wave vector $k = |\boldsymbol{k}|$ (in nm^{-1}) [see Eqs. (2.89) and (2.90)]. We assume that $\Delta > 0$ and $\Delta < 0$ correspond to normal (a) and inverted (b) band structures; $\hbar v = 380$ meV nm and $|\Delta| = 10$ meV.

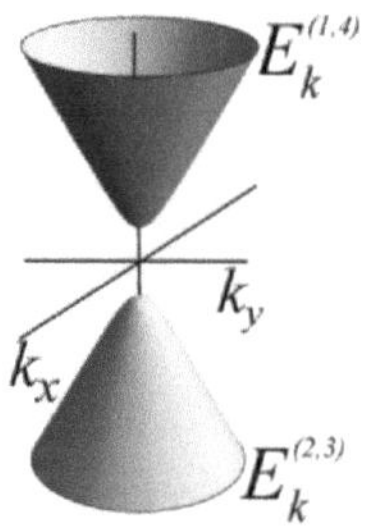

Figure 2.8 3D plot of energy bands in Eqs. (2.89) and (2.90) for $\hbar v = 380$ meV nm and $\Delta = 10$ meV. Note that conduction and valence bands consist of degenerate branches 1, 4 and 2, 3, respectively.

$$\psi_k(x, y) = \Phi\, e^{ikx - y/\lambda}, \tag{2.93}$$

$$\Phi = \tau_y \sigma_y \Phi. \tag{2.94}$$

The constraint (2.94) on the spinor Φ follows from the boundary condition (2.92). Inserting Eq. (2.93) into Eq. (2.91) and using the identity $\tau_z \sigma_y \Phi = -i\tau_x \sigma_0 \Phi$, we have

$$\hbar v k \tau_z \sigma_x \Phi + (\Delta + \hbar v/\lambda)\, \tau_x \sigma_0 \Phi = E\Phi. \tag{2.95}$$

Setting then the expression in the brackets to zero, we find the decay length

$$\lambda = -\hbar v/\Delta, \tag{2.96}$$

while Eq. (2.95) yields

$$\hbar v k \tau_z \sigma_x \Phi = E \Phi. \tag{2.97}$$

We are about to make an important observation: At the edge the number of independent degrees of freedom is reduced to the electron spin only. Indeed, the constraint (2.92) restricts the form of the spinor Φ to

$$\Phi = \begin{bmatrix} \chi \\ i\sigma_y \chi \end{bmatrix}, \tag{2.98}$$

due to which Eq. (2.97) reduces to an equation for the two-component spinor χ,

$$\hbar v k \sigma_x \chi = E \chi. \tag{2.99}$$

Its solutions are the eigenstates of the spin projection σ_x,

$$\sigma_x \chi_\sigma = \sigma \chi_\sigma, \quad \sigma = \pm 1, \tag{2.100}$$

with the energies

$$E_{k\sigma} = \hbar v k \sigma. \tag{2.101}$$

Finally, the normalized edge states (2.93) are

$$\psi_{k+} = \begin{bmatrix} 1 \\ 1 \\ 1 \\ -1 \end{bmatrix} \sqrt{\frac{|\Delta|}{2\hbar v}}\, e^{ikx - |\Delta| y/\hbar v}, \tag{2.102}$$

and

$$\psi_{k-} = \begin{bmatrix} 1 \\ -1 \\ -1 \\ -1 \end{bmatrix} \sqrt{\frac{|\Delta|}{2\hbar v}}\, e^{ikx - |\Delta| y/\hbar v}. \tag{2.103}$$

These results illustrate all the hallmarks of the TR-invariant topological insulators, as summarized below:

(i) The edge states are gapless, which can again be attributed to the inversion of the band structure at the edge. Indeed, the system can be extended to the other semispace ($y < 0$) by the parity transformation,

$$\psi'(x, y) = \tau_y \sigma_y \psi(x, -y), \tag{2.104}$$

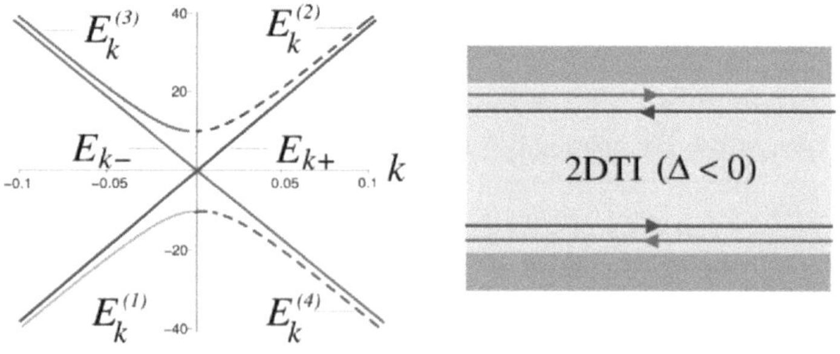

Figure 2.9 Band structure of a QSHI with gapless helical edge states, $E_{k\pm}$, [see Eq. (2.101)] for $\hbar v = 380$ meV nm and $\Delta = -10$ meV (left) and schematic of helical (counter-propagating) edge states at opposite boundaries of a QSHI (right).

that inverts the sign of the gap term in Eq. (2.91)

$$[-i\hbar v\tau_z(\sigma_x\partial_x + \sigma_y\partial_y) - \Delta\tau_x\sigma_0]\psi'(x,y) = E\,\psi'(x,y). \tag{2.105}$$

Notice the difference to the Chern insulator in Section 2.1.1: The edge states exist only for a negative Δ inside the material. The origin of this restriction can be traced back to boundary condition (2.92). The latter is suitable for TI materials with an inverted order of the conduction and valence bands such as HgTe quantum wells.

(ii) The bulk degrees of freedom are reduced at the edge to the electron spin only. Unlike the Chern insulator, there is a pair of counter-propagating edge states with the opposite spin projections σ_x (see also Fig. 2.9). That is, the propagation direction of the electron is tied to the orientation of its spin. This property is known as helicity, due to which the edge states in QSHIs are frequently called helical, as opposed to the chiral edge states in the quantum Hall systems. The helical edge states are related to each other by the TRS,

$$\psi_{-k,-} = \mathbb{T}\,\psi_{k,+}, \tag{2.106}$$

where $\mathbb{T}$ is defined in Eq. (2.61). Also, in accord with Kramers' theorem, edge states (2.102) and (2.103) are orthogonal and, therefore, protected against local perturbations that do not break the TRS. One of the most common types of such perturbations in realistic systems is potential disorder created by randomly distributed ionized impurities. If the impurities do not flip the spin of the edge state, its propagation direction remains unchanged, which means that potential disorder cannot cause scattering between the helical edge channels and their localization. Such robustness of the edge states has important implications for electron transport in QSHIs.

2.3 Chern Insulators II

In the Dirac model above, we from the start adopted a linear dependence on the momentum. This can be justified if the band structure has a single valley (e.g. at $\boldsymbol{k} = [0, 0]$) and only its vicinity matters. This is to some extent the case for the surface states of a 3DTI (examined in Chapter 4). If we wish to define a TI on a periodic lattice, the situation is quite different. In lattice models, the band structure has multiple valleys at different points of the Brillouin zone depending on the lattice symmetry. The band behavior between the valleys reveals another key attribute of TIs, namely, their inverted band structure. Furthermore, this feature is intimately related to the concept of the fermion doubling and Nielsen–Ninomiya theorem [80]. To illustrate these ideas, we start with a tight-binding model for a Chern insulator on a square lattice, as proposed in Ref. [12].

2.3.1 Tight-Binding Model on a Square Lattice

The model assumes one s-like orbital (even under space inversion) and one p-like orbital (odd under space inversion) at each site of the lattice (see also Fig. 2.10). Each lattice site has four nearest neighbors at the positions given by the vectors $\pm\boldsymbol{X}$ and $\pm\boldsymbol{Y}$, where

$$\boldsymbol{X} = [a, 0], \qquad \boldsymbol{Y} = [0, a]. \tag{2.107}$$

In the following, we set the lattice constant a to 1. The tight-binding Hamiltonian, including the on-site terms and the nearest-neighbor

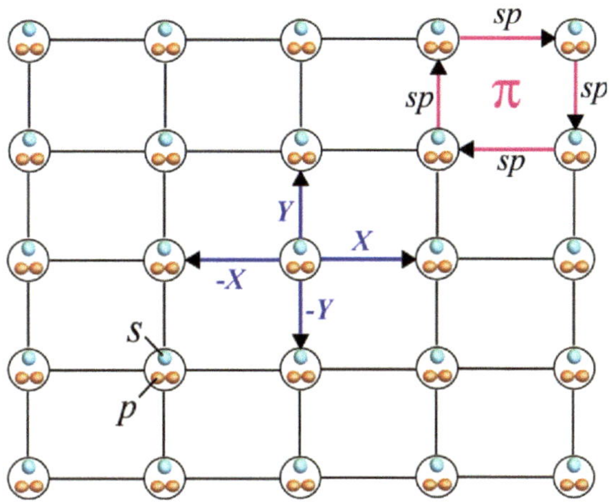

Figure 2.10 Schematic of a square lattice with one s-like and one p-like orbital per site. Each site has four nearest neighbors at the positions given by the vectors $\pm\boldsymbol{X}$ and $\pm\boldsymbol{Y}$ (see also text).

hopping, can be written as

$$\hat{h}_{t.b.} = \sum_{\boldsymbol{r}} (V_s c_{\boldsymbol{r}}^\dagger c_{\boldsymbol{r}} + V_p b_{\boldsymbol{r}}^\dagger b_{\boldsymbol{r}})$$
$$+ \sum_{\boldsymbol{r},\boldsymbol{\delta}=\pm\boldsymbol{X},\pm\boldsymbol{Y}} [t_s c_{\boldsymbol{r}}^\dagger c_{\boldsymbol{r}+\boldsymbol{\delta}} + t_p b_{\boldsymbol{r}}^\dagger b_{\boldsymbol{r}+\boldsymbol{\delta}} + t_{sp}(\boldsymbol{\delta}) c_{\boldsymbol{r}}^\dagger b_{\boldsymbol{r}+\boldsymbol{\delta}} + t_{sp}^*(\boldsymbol{\delta}) b_{\boldsymbol{r}+\boldsymbol{\delta}}^\dagger c_{\boldsymbol{r}}], \tag{2.108}$$

where the operators $c_{\boldsymbol{r}}^\dagger$ ($c_{\boldsymbol{r}}$) and $b_{\boldsymbol{r}}^\dagger$ ($b_{\boldsymbol{r}}$) create (annihilate) a spinless electron in the s- and p-orbital state, respectively; V_s and V_p are the electron on-site energies in the s- and p- states, t_s and t_p are the hopping energies between the same-parity (s- and p-) orbitals, while $t_{sp}(\boldsymbol{\delta})$ is the hopping amplitude between the opposite-parity orbitals. The latter is an odd function of the nearest-neighbor position $\boldsymbol{\delta}$,

$$t_{sp}(-\boldsymbol{\delta}) = -t_{sp}(\boldsymbol{\delta}), \tag{2.109}$$

which excludes the possibility of an on-site overlap between the s- and p-states (for $\boldsymbol{\delta} = 0$).

To diagonalize $\hat{h}_{t.b.}$ we use the Fourier transformation

$$c_{\boldsymbol{r}} = \frac{1}{\sqrt{N}} \sum_{\boldsymbol{k}} c_{\boldsymbol{k}} e^{i\boldsymbol{kr}}, \qquad b_{\boldsymbol{r}} = \frac{1}{\sqrt{N}} \sum_{\boldsymbol{k}} b_{\boldsymbol{k}} e^{i\boldsymbol{kr}} \tag{2.110}$$

where the vector $\boldsymbol{r}$ specifies the position of a given unit cell in the lattice, N is the number of the unit cells, and the sum runs over the points $\boldsymbol{k}$ in the reciprocal space. In view of the orthogonality and completeness of the plane wave basis, we have

$$\begin{aligned}\hat{h}_{t.b.} &= \sum_{\boldsymbol{k}} (V_s c_{\boldsymbol{k}}^{\dagger} c_{\boldsymbol{k}} + V_p b_{\boldsymbol{k}}^{\dagger} b_{\boldsymbol{k}}) \\ &+ \sum_{\boldsymbol{k}} [t_s(\boldsymbol{k}) c_{\boldsymbol{k}}^{\dagger} c_{\boldsymbol{k}} + t_p(\boldsymbol{k}) b_{\boldsymbol{k}}^{\dagger} b_{\boldsymbol{k}} + t_{sp}(\boldsymbol{k}) c_{\boldsymbol{k}}^{\dagger} b_{\boldsymbol{k}} + t_{sp}^{*}(\boldsymbol{k}) b_{\boldsymbol{k}}^{\dagger} c_{\boldsymbol{k}}],\end{aligned} \tag{2.111}$$

or

$$\hat{h}_{t.b.} = \sum_{\boldsymbol{k}} [c_{\boldsymbol{k}}^{\dagger} \; b_{\boldsymbol{k}}^{\dagger}] \begin{bmatrix} V_s + t_s(\boldsymbol{k}) & t_{sp}(\boldsymbol{k}) \\ t_{sp}^{*}(\boldsymbol{k}) & V_p + t_p(\boldsymbol{k}) \end{bmatrix} \begin{bmatrix} c_{\boldsymbol{k}} \\ b_{\boldsymbol{k}} \end{bmatrix}. \tag{2.112}$$

The Fourier transforms of the hopping amplitudes are given by

$$t_s(\boldsymbol{k}) = t_s \sum_{\boldsymbol{\delta}=\pm\boldsymbol{X},\pm\boldsymbol{Y}} \mathrm{e}^{i\boldsymbol{k}\boldsymbol{\delta}} = 2t_s(\cos k_x + \cos k_y), \tag{2.113}$$

$$t_p(\boldsymbol{k}) = t_p \sum_{\boldsymbol{\delta}=\pm\boldsymbol{X},\pm\boldsymbol{Y}} \mathrm{e}^{i\boldsymbol{k}\boldsymbol{\delta}} = 2t_p(\cos k_x + \cos k_y), \tag{2.114}$$

$$\begin{aligned}t_{sp}(\boldsymbol{k}) &= \sum_{\boldsymbol{\delta}=\pm\boldsymbol{X},\pm\boldsymbol{Y}} t_{sp}(\boldsymbol{\delta}) \mathrm{e}^{i\boldsymbol{k}\boldsymbol{\delta}} \\ &= 2i\,[t_{sp}(\boldsymbol{X}) \sin k_x + t_{sp}(\boldsymbol{Y}) \sin k_y].\end{aligned} \tag{2.115}$$

Here, we used symmetry relation (2.109). Furthermore, since $t_{sp}(\boldsymbol{k})$ is generally neither imaginary nor real, we can choose $t_{sp}(\boldsymbol{Y}) = i\,t_{sp}(\boldsymbol{X}) = t_{sp}$, which yields a complex-valued hopping amplitude

$$t_{sp}(\boldsymbol{k}) = 2t_{sp}(\sin k_x + i \sin k_y). \tag{2.116}$$

In real space, the hopping amplitudes $t_{sp}(\boldsymbol{X}) = -i\,t_{sp}$ and $t_{sp}(\boldsymbol{Y}) = t_{sp}$ result in a total phase of π around a square plaque, as sketched in Fig. 2.10.

From the above equations we obtain the first-quantized Hamiltonian as the 2×2 matrix

$$\hat{h}_{\boldsymbol{k}} = \begin{bmatrix} V_s + 2t_s(\cos k_x + \cos k_y) & 2t_{sp}(\sin k_x + i \sin k_y) \\ 2t_{sp}(\sin k_x - i \sin k_y) & V_p + 2t_p(\cos k_x + \cos k_y) \end{bmatrix}.$$

It is convenient to write it in terms of Pauli matrices τ_x, τ_y, τ_z and unit matrix τ_0:

$$\hat{h}_{\boldsymbol{k}} = \epsilon_{\boldsymbol{k}} \tau_0 + \boldsymbol{d}_{\boldsymbol{k}} \cdot \boldsymbol{\tau}, \qquad \boldsymbol{\tau} = [\tau_x, \tau_y, \tau_z], \tag{2.117}$$

where vector $\boldsymbol{d_k}$ and scalar $\epsilon_{\boldsymbol{k}}$ are defined as follows:

$$\boldsymbol{d_k} = [\mathcal{A}\sin k_x, -\mathcal{A}\sin k_y, \mathcal{M}_{\boldsymbol{k}}], \tag{2.118}$$

$$\mathcal{M}_{\boldsymbol{k}} = \Delta + 2\mathcal{B}(2 - \cos k_x - \cos k_y), \tag{2.119}$$

$$\epsilon_{\boldsymbol{k}} = \mathcal{C} + 2\mathcal{D}(2 - \cos k_x - \cos k_y). \tag{2.120}$$

The constants, Δ, $\mathcal{A}$, $\mathcal{B}$, $\mathcal{C}$ and $\mathcal{D}$, are expressed in terms of the parameters of the tight-binding model:

$$\Delta = \frac{V_s - V_p}{2} + 2(t_s - t_p), \quad \mathcal{A} = 2t_{sp}, \quad \mathcal{B} = -\frac{t_s - t_p}{2}, \tag{2.121}$$

$$\mathcal{C} = \frac{V_s + V_p}{2} + 2(t_s + t_p), \qquad \mathcal{D} = -\frac{t_s + t_p}{2}. \tag{2.122}$$

The main results of the model are summarized by Eqs. (2.117) – (2.120). In particular, its energy bands

$$E_{\boldsymbol{k}s} = \epsilon_{\boldsymbol{k}} + s\sqrt{\mathcal{A}^2\sin^2 k_x + \mathcal{A}^2\sin^2 k_y + \mathcal{M}_{\boldsymbol{k}}^2}, \qquad s = \pm 1, \tag{2.123}$$

are defined in a square Brillouin zone $|k_{x,y}| \le \pi$. The band extrema lie in two inequivalent valleys, at Γ-point $\boldsymbol{k} = [0, 0]$ and at M-point $\boldsymbol{k} = [\pi, \pi]$. The corresponding gap terms, $\mathcal{M}_{[0,0]} = \Delta$ and $\mathcal{M}_{[\pi,\pi]} = \Delta + 8\mathcal{B}$, characterize an insulating state.

2.3.2 Normal and Inverted Band Structure. Chern Number and Edge Mode

Equation (2.117) resembles the Hamiltonian of a ferromagnet, with vector $\boldsymbol{d_k}$ playing the role of the spin magnetization. In this case, the d-vector defines the texture of pseudospin $\boldsymbol{\tau}/2$ in the Brillouin zone. It is instructive to approximate $\boldsymbol{d_k}$ (2.118) by an expansion up to quadratic terms around $\boldsymbol{k} = 0$:

$$\boldsymbol{d_k} \approx [\mathcal{A}k_x, -\mathcal{A}k_y, \mathcal{M}_{\boldsymbol{k}}], \qquad \mathcal{M}_{\boldsymbol{k}} = \Delta + \mathcal{B}k^2. \tag{2.124}$$

Depending on the relative sign of parameters Δ and $\mathcal{B}$, two topologically distinct types of the texture are possible, as depicted in Figs. 2.11 and 2.12. Their topologies can be characterized by winding

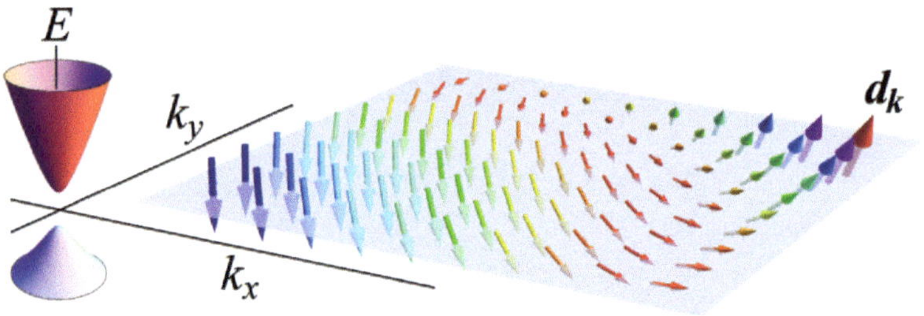

Figure 2.11 Schematic of the pseudospin texture in a Chern insulator [$\mathcal{B}\Delta < 0$ in Eq. (2.124)]. The texture has the skyrmion topology with a nontrivial winding (Chern) number $n = \pm 1$ characteristic of an inverted band structure. Adapted from [60].

number

$$n = \frac{1}{4\pi} \int \boldsymbol{u}_k \cdot \left(\frac{\partial \boldsymbol{u}_k}{\partial k_x} \times \frac{\partial \boldsymbol{u}_k}{\partial k_y} \right) dk_x dk_y, \quad \boldsymbol{u}_k = \frac{\boldsymbol{d}_k}{|\boldsymbol{d}_k|}. \tag{2.125}$$

It is just Chern invariant (2.57) recast as a surface integral divided by the area of a unit sphere. Geometrically, $|n|$ is the number of times the unit vector $\boldsymbol{u}_k$ sweeps a unit sphere as $\boldsymbol{k}$ covers the entire momentum space. The Chern invariant acquires nontrivial values $n = \pm 1$ under condition $\mathcal{B}\Delta < 0$, when the configuration of vector $\boldsymbol{d}_k$ defines a skyrmion, a topological defect that compactifies the physical space on a sphere (see Fig. 2.11). In this case, $\boldsymbol{d}_k$ takes all possible directions, accompanied by an inversion of the sign of the gap term $\mathcal{M}_k = \Delta + \mathcal{B}\boldsymbol{k}^2$. Such type of the band structure is called inverted, and the corresponding insulating state is topologically nontrivial (Chern insulator) as opposed to the case $\mathcal{B}\Delta > 0$ (Fig. 2.12) in which no $\boldsymbol{d}_k$ winding takes place, hence $n = 0$. The latter case corresponds to an ordinary insulator with a normal (non-topological) band structure.

The exact dependence of $\boldsymbol{d}_k$ confirms the winding in the Brillouin zone. Indeed, at the Γ- and M-points, we have

$$\boldsymbol{d}_{[0,0]} = [0, 0, \Delta], \qquad \boldsymbol{d}_{[\pi,\pi]} = [0, 0, \Delta + 8\mathcal{B}]. \tag{2.126}$$

A band inversion occurs between the Γ- and M-points if gaps Δ and $\Delta + 8\mathcal{B}$ have opposite signs, i.e. if

$$\mathcal{B}\Delta < 0, \qquad |\Delta| < 8|\mathcal{B}|. \tag{2.127}$$

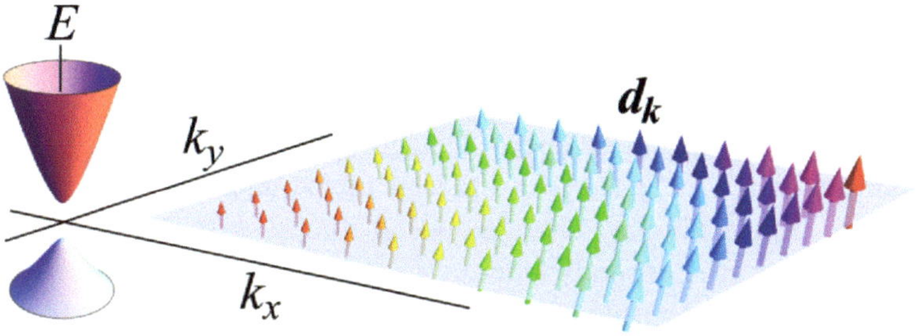

Figure 2.12 Schematic of the pseudospin texture in an ordinary insulator [$\mathcal{B}\Delta > 0$ in Eq. (2.124)]. In this case, $\boldsymbol{d_k}$ shows no winding. Hence, Chern number $n = 0$ for a normal band structure. Adapted from [60].

Our geometric considerations are corroborated by an explicit calculation of n, using Eq. (2.57). We note, however, that here the Chern insulator is not identical to that in Section 2.1.1, even if we neglect all higher-order terms. The sign $'-'$ in Eq. (2.124) indicates that this Chern insulator has the opposite chirality and Berry's curvature, i.e. it is the TR-partner to that in Section 2.1.1. In order to correctly calculate its Chern number, we should use Berry's curvature in Eq. (2.204) (see end-of-chapter problems). We also need the z-component of unit vector $\boldsymbol{u_k}$, which is given here by

$$u_z(k) = \frac{\mathcal{M}_{\boldsymbol{k}}}{\sqrt{\mathcal{A}^2\boldsymbol{k}^2 + \mathcal{M}_{\boldsymbol{k}}^2}}. \tag{2.128}$$

Otherwise, the calculation is fully analogous to that in Eq. (2.50):

$$\begin{aligned} n &= -\frac{1}{2}\int_0^\infty \frac{\partial u_z(k)}{\partial k} dk = -\frac{u_z(\infty) - u_z(0)}{2} \\ &= \frac{\operatorname{sgn}(\Delta) - \operatorname{sgn}(\mathcal{B})}{2} \\ &= \begin{cases} \operatorname{sgn}(\Delta), & \mathcal{B}\Delta < 0, \\ 0, & \mathcal{B}\Delta > 0. \end{cases} \end{aligned} \tag{2.129}$$

This analytical result confirms the winding of $\boldsymbol{d_k}$ in momentum space shown in Figs. 2.11 and 2.12. As expected, we have $n = \pm 1$ and $n = 0$ for inverted and normal band structures, respectively. In agreement with TKNN formula (2.54), Chern number (2.129) is integer. This is because all inequivalent valleys of our lattice model are taken into account, e.g. factors sgn(Δ) and sgn($\mathcal{B}$) characterize the gaps at the Γ- and M-points.

As in Section 2.1, nontrivial number n indicates a chiral gapless state at the edge of a Chern insulator. Here, we prove this for Hamiltonian (2.117), keeping all terms up to quadratic ones in $\boldsymbol{k}$. For the sake of learning, we change the coordinate system such that the x and y axes are perpendicular and parallel to the edge, respectively, and the system occupies 2D half-space $x \geq 0$. In this geometry, k_y remains a good quantum number, while k_x needs to be replaced by operator $-i\partial_x$. Hamiltonian (2.117) can be written as

$$\hat{h}_{k_y}(x) = \hat{h}^{A}_{k_y}(x) + \hat{h}^{S}_{k_y}(x), \tag{2.130}$$

where $\hat{h}^{A}_{k_y}(x)$ and $\hat{h}^{S}_{k_y}(x)$ are the particle-hole asymmetric and symmetric parts of the Hamiltonian given by

$$\hat{h}^{A}_{k_y}(x) = (\mathcal{C} + \mathcal{D}k_y^2 - \mathcal{D}\partial_x^2)\tau_0 \tag{2.131}$$

and

$$\hat{h}^{S}_{k_y}(x) = -i\mathcal{A}\partial_x\tau_x - \mathcal{A}k_y\tau_y + (\Delta + \mathcal{B}k_y^2 - \mathcal{B}\partial_x^2)\tau_z. \tag{2.132}$$

This yields the eigenvalue equation

$$\hat{h}^{A}_{k_y}(x)\psi(x) + \hat{h}^{S}_{k_y}(x)\psi(x) = E\psi(x) \tag{2.133}$$

for energy E and two-component wave function $\psi(x)$. Everywhere outside the system and at the edge the wave function is identically zero, $\psi(x) = 0$ for $x \leq 0$, while in the interior of the material $\psi(x) \to 0$ for $x \to +\infty$.

To determine E, we integrate Eq. (2.133) across the edge,

$$\int_{-\infty}^{\infty} \hat{h}^{A}_{k_y}(x)\psi(x)dx + \int_{-\infty}^{\infty} \hat{h}^{S}_{k_y}(x)\psi(x)dx = E\int_{-\infty}^{\infty} \psi(x)dx, \tag{2.134}$$

and evaluate the first integral, using $\partial_x\psi(x) = 0$ at $\pm\infty$, which yields $(\mathcal{C} + \mathcal{D}k_y^2)\int_{-\infty}^{\infty}\psi(x)dx$. The second term can be cast into a similar form by choosing $\psi(x)$ to be an eigenstate of the particle-hole symmetric Hamiltonian $\hat{h}^{S}_{k_y}(x)$,

$$\hat{h}^{S}_{k_y}(x)\psi(x) = E^{S}_{k_y}\psi(x), \tag{2.135}$$

with an eigenvalue $E^{S}_{k_y}$. Then, collecting the pre-integral factors in Eq. (2.134), we can write the edge eigenenergy as

$$E_{k_y} = \mathcal{C} + \mathcal{D}k_y^2 + E^S_{k_y}. \tag{2.136}$$

Now, the problem reduces to solving particle-hole symmetric Eq. (2.135). The substitution

$$\psi(x) = \Phi e^{-\varkappa x} \tag{2.137}$$

yields

$$[i\mathcal{A}\varkappa\tau_x - \mathcal{A}k_y\tau_y + (\Delta + \mathcal{B}k_y^2 - \mathcal{B}\varkappa^2)\tau_z]\Phi = E^S_{k_y}\Phi. \tag{2.138}$$

Since there are two unknowns ($\varkappa$ and $E^S_{k_y}$), we split Eq. (2.138) into two equations:

$$[\mathcal{A}\varkappa\tau_0 - (\Delta + \mathcal{B}k_y^2 - \mathcal{B}\varkappa^2)\tau_y]\Phi = 0, \tag{2.139}$$

and

$$-\mathcal{A}k_y\tau_y\Phi = E^S_{k_y}\Phi. \tag{2.140}$$

In the former we used $\tau_z = -i\tau_x\tau_y$. Clearly, Φ is an eigenstate of τ_y, so the equations above become algebraic and have the following solutions:

$$\varkappa_{1,2} = -\tau\frac{\mathcal{A}}{2\mathcal{B}} \pm \sqrt{\left(\frac{\mathcal{A}}{2\mathcal{B}}\right)^2 + \frac{\Delta}{\mathcal{B}} + k_y^2}, \tag{2.141}$$

and

$$E^S_{k_y} = -\mathcal{A}k_y\tau, \tag{2.142}$$

with τ being an eigenvalue of τ_y.

Further, eigenvalue τ must be chosen such that the decay constants $\varkappa_1$ and $\varkappa_2$ have positive real parts. For a normal band structure with $\mathcal{B}\Delta > 0$, this is impossible. We can readily see this by inspecting the product $\varkappa_1\varkappa_2 = -\left(\frac{\Delta}{\mathcal{B}} + k_y^2\right)$. For an inverted band structure with $\mathcal{B}\Delta < 0$, to the contrary, both $\varkappa_1$ and $\varkappa_2$ can have a positive real part in the segment

$$-k_0 \le k_y \le k_0, \qquad k_0 = \sqrt{-\Delta/\mathcal{B}}. \tag{2.143}$$

Indeed, choosing the eigenvalue

$$\tau = -\mathrm{sgn}(\mathcal{A}\mathcal{B}), \tag{2.144}$$

we obtain

$$\varkappa_{1,2} = \left|\frac{\mathcal{A}}{2\mathcal{B}}\right| \pm \sqrt{\left(\frac{\mathcal{A}}{2\mathcal{B}}\right)^2 + \frac{\Delta}{\mathcal{B}} + k_y^2}, \tag{2.145}$$

with $\mathrm{Re}(\varkappa_{1,2}) > 0$ for any wave number in segment (2.143). In order for the wave function to vanish at the boundary, $x = 0$, the two fundamental solutions $\mathrm{e}^{-\varkappa_1 x}$ and $\mathrm{e}^{-\varkappa_2 x}$ must be taken with the opposite signs. Hence, the edge solution

$$\psi(x) = \Phi(\mathrm{e}^{-\varkappa_1 x} - \mathrm{e}^{-\varkappa_2 x}). \tag{2.146}$$

As before, Φ is the eigenstate of τ_y with eigenvalue (2.144). Combining Eqs. (2.136), (2.142) and (2.144), we arrive at the energy dispersion of the edge mode

$$E_{k_y} = \mathcal{C} + \mathcal{D}k_y^2 + |\mathcal{A}|\mathrm{sgn}(\mathcal{B})k_y. \tag{2.147}$$

As in Section 2.1.1, we obtain a single edge mode. The direction of its velocity—the chirality—is given by $\mathrm{sgn}(\mathcal{B})$, i.e. by the sign of Chern number (2.129). Unlike the Dirac model in Section 2.1.1, we have taken into account quadratic $\boldsymbol{k}$-terms. In particular, $\mathcal{D}k_y^2$ makes the edge dispersion particle-hole asymmetric (cf. Fig. 1.6 of Chapter 1), while $\mathcal{B}k_y^2$ restricts the allowed k_y-values to segment (2.143).

2.4 Quantum Spin Hall Insulators II

2.4.1 Tight-Binding, Bernevig–Hughes–Zhang and Related Models

Our next goal is to deduce a lattice Hamiltonian for a QSHI. This can be done solely on the basis of the TR-invariance of the QSHI state. First of all, our Chern insulator model is spinless. Therefore, TR is just the complex conjugation combined with reflection $\boldsymbol{k} \to -\boldsymbol{k}$. The operator $\hat{h}^*_{-\boldsymbol{k}}$ is thus the Kramers partner to $\hat{h}_{\boldsymbol{k}}$ (2.117). In the simplest case, the QSHI Hamiltonian is a Kramers doublet

$$\hat{H}_{\boldsymbol{k}} = \begin{bmatrix} \hat{h}_{\boldsymbol{k}} & 0 \\ 0 & \hat{h}^*_{-\boldsymbol{k}} \end{bmatrix}, \tag{2.148}$$

or, explicitly,

$$\hat{H}_{\boldsymbol{k}} = \begin{bmatrix} \epsilon_{\boldsymbol{k}}\tau_0 + d^x_{\boldsymbol{k}}\tau_x + d^y_{\boldsymbol{k}}\tau_y + d^z_{\boldsymbol{k}}\tau_z & 0 \\ 0 & \epsilon_{\boldsymbol{k}}\tau_0 - d^x_{\boldsymbol{k}}\tau_x + d^y_{\boldsymbol{k}}\tau_y + d^z_{\boldsymbol{k}}\tau_z \end{bmatrix}, \tag{2.149}$$

where we used $\tau_y^* = -\tau_y$, $d_{-\boldsymbol{k}}^{x,y} = -d_{\boldsymbol{k}}^{x,y}$ and $d_{-\boldsymbol{k}}^z = d_{\boldsymbol{k}}^z$. Equation (2.149) takes a compact form

$$\hat{H}_{\boldsymbol{k}} = \epsilon_{\boldsymbol{k}}\tau_0\sigma_0 + d_{\boldsymbol{k}}^x\tau_x\sigma_z + d_{\boldsymbol{k}}^y\tau_y\sigma_0 + d_{\boldsymbol{k}}^z\tau_z\sigma_0 \tag{2.150}$$

$$= \epsilon_{\boldsymbol{k}}\tau_0\sigma_0 + \mathcal{A}\sin k_x\tau_x\sigma_z - \mathcal{A}\sin k_y\tau_y\sigma_0 + \mathcal{M}_{\boldsymbol{k}}\tau_z\sigma_0, \tag{2.151}$$

with σ_z and σ_0 being additional Pauli and unit matrices. This Hamiltonian has the TRS represented by an antiunitary operation $\mathbb{T} = i\sigma_y\tau_0 K$. We can therefore identify vector $\frac{\hbar}{2}\boldsymbol{\sigma} = \frac{\hbar}{2}[\sigma_x, \sigma_y, \sigma_z]$ as the electron spin. Its z-component is conserved, so that its two projections on the z-axis label the Kramers partners in Eq. (2.148).

Lattice Hamiltonian (2.151) describes the band structure of a QSHI in the entire Brillouin zone. Often, it suffices to expand Eq. (2.151) up to quadratic terms in $\boldsymbol{k}$:

$$\hat{H}_{\boldsymbol{k}} \approx (\mathcal{C}+\mathcal{D}\boldsymbol{k}^2)\tau_0\sigma_0 + \mathcal{A}k_x\tau_x\sigma_z - \mathcal{A}k_y\tau_y\sigma_0 + (\Delta+\mathcal{B}\boldsymbol{k}^2)\tau_z\sigma_0. \tag{2.152}$$

The resulting Hamiltonian coincides with that derived by Bernevig, Hughes and Zhang for HgTe/CdTe quantum wells in Ref. [13]. This is not accidental. HgTe and CdTe both have a crystal structure of the zinc-blend type, and our square lattice model captures the essential symmetries of the lattice structure and atomic orbitals in such materials.

Hamiltonian (2.152) is an important generalization of the Dirac-like models introduced earlier. We shall first establish the connection to the Chern insulator studied in Section 2.1. To that end, we write Eq. (2.152) as

$$\hat{H}_{\boldsymbol{k}} = \epsilon_{\boldsymbol{k}}\tau_0\sigma_0 + \begin{bmatrix} \mathcal{A}k_x\tau_x - \mathcal{A}k_y\tau_y + \mathcal{M}_{\boldsymbol{k}}\tau_z & 0 \\ 0 & -\mathcal{A}k_x\tau_x - \mathcal{A}k_y\tau_y + \mathcal{M}_{\boldsymbol{k}}\tau_z \end{bmatrix}, \tag{2.153}$$

and make the unitary transformation

$$\hat{H}_{\boldsymbol{k}} \to \hat{U}\hat{H}_{\boldsymbol{k}}\hat{U}^\dagger, \qquad \hat{U} = \begin{bmatrix} 0 & \tau_z \\ -i\tau_y & 0 \end{bmatrix}. \tag{2.154}$$

The matrix multiplication yields

$$\hat{H}_{\boldsymbol{k}} = \epsilon_{\boldsymbol{k}}\tau_0\sigma_0 + \begin{bmatrix} \mathcal{A}k_x\tau_x + \mathcal{A}k_y\tau_y + \mathcal{M}_{\boldsymbol{k}}\tau_z & 0 \\ 0 & -\mathcal{A}k_x\tau_x - \mathcal{A}k_y\tau_y - \mathcal{M}_{\boldsymbol{k}}\tau_z \end{bmatrix}. \tag{2.155}$$

Another unitary transformation

$$\hat{H}_k \to \hat{U}\,\hat{H}_k\hat{U}^{\dagger}, \qquad \hat{U} = \begin{bmatrix} \tau_0 & 0 \\ 0 & \tau_z \end{bmatrix}, \tag{2.156}$$

brings the Hamiltonian to the form

$$\hat{H}_k = \epsilon_k \tau_0 \sigma_0 + \begin{bmatrix} \boldsymbol{d}_k^+ \cdot \boldsymbol{\tau} & 0 \\ 0 & \boldsymbol{d}_k^- \cdot \boldsymbol{\tau} \end{bmatrix},$$
$$\boldsymbol{d}_k^{\pm} = [\mathcal{A}k_x, \mathcal{A}k_y, \pm\mathcal{M}_k]. \tag{2.157}$$

We have obtained two species of a Chern insulator specified by vectors $\boldsymbol{d}_k^{\pm}$. The difference between them is only in the sign of the gap term.

Next, we can map Hamiltonian (2.155) to the thin-film model. To that end, we apply the unitary transformation

$$\hat{H}_k \to \hat{U}\,\hat{H}_k\hat{U}^{\dagger}, \qquad \hat{U} = \frac{1}{\sqrt{2}}\begin{bmatrix} \tau_0 & \tau_z \\ \tau_z & -\tau_0 \end{bmatrix}, \tag{2.158}$$

after which one finds the Hamiltonian

$$\hat{H}_k = \begin{bmatrix} (\mathcal{C} + \mathcal{D}\boldsymbol{k}^2)\tau_0 + \mathcal{A}\boldsymbol{k}\cdot\boldsymbol{\tau} & (\Delta + \mathcal{B}\boldsymbol{k}^2)\tau_0 \\ (\Delta + \mathcal{B}\boldsymbol{k}^2)\tau_0 & (\mathcal{C} + \mathcal{D}\boldsymbol{k}^2)\tau_0 - \mathcal{A}\boldsymbol{k}\cdot\boldsymbol{\tau} \end{bmatrix}. \tag{2.159}$$

It reproduces the structure of the thin-film Hamiltonian in Eq. (2.60) and additionally includes the higher order $\boldsymbol{k}$ terms, most notably in the (off-diagonal) coupling terms. We see therefore that the thin-film model can formally be obtained from the square lattice one with an s-like and a p-like orbital per site.

2.4.2 Z_2 Invariant and Helical Edge States

In view of the TRS, the Hall conductivity in QSHIs is zero. This means that the topological TKNN invariant (2.57) vanishes identically. Still, it is possible to define another topological number, the so-called Z_2 invariant [8], that distinguishes between topologically trivial and nontrivial phases of a TR-invariant insulator.

For Hamiltonian (2.157), TKNN invariant (2.57) is the sum of the Chern numbers of the two blocks:

$$n = \sum_{\nu=\pm 1} \int_{BZ} \frac{dk_x dk_y}{2\pi} (\nabla \times \boldsymbol{A}_{k\nu})_z, \tag{2.160}$$

each characterized by the Berry connection of the occupied (valence) band, $\boldsymbol{A}_{k\nu}$, where index $\nu = \pm 1$ labels the two Chern species. For an explicit calculation, we refer to Eq. (2.31), setting $s = -1$ for the valence band and using

$$u_{z\nu}(k) = \frac{\nu \mathcal{M}_{\boldsymbol{k}}}{\sqrt{\mathcal{A}^2 \boldsymbol{k}^2 + \mathcal{M}_{\boldsymbol{k}}^2}} \tag{2.161}$$

for the Chern species with flavor ν. Extending the $\boldsymbol{k}$ integral to the whole plane [cf. Eqs. (2.50) and (2.129)], we find

$$n = \frac{1}{2} \sum_{\nu=\pm 1} \int_0^\infty \frac{\partial u_{z\nu}(k)}{\partial k} dk = \sum_{\nu=\pm 1} \frac{u_{z\nu}(\infty) - u_{z\nu}(0)}{2} - n_+ + n_-, \tag{2.162}$$

where

$$n_\pm = \pm \frac{\mathrm{sgn}(\mathcal{B}) - \mathrm{sgn}(\Delta)}{2}. \tag{2.163}$$

As the consequence of TRS, the Chern numbers of the two subsystems have the opposite signs, $n_- = -n_+$, so the TKNN number is zero independently of the band structure details. On the other hand, difference $n_+ - n_-$ does not have to vanish, and we can define the topological number

$$\frac{n_+ - n_-}{2} = \frac{\mathrm{sgn}(\mathcal{B}) - \mathrm{sgn}(\Delta)}{2} = \begin{cases} \mathrm{sgn}(\mathcal{B}), & \mathcal{B}\Delta < 0, \\ 0, & \mathcal{B}\Delta > 0. \end{cases} \tag{2.164}$$

Its nonzero value indicates inverted band structures of the Chern subsystems, with a skyrmion and an antiskyrmion topology in $\boldsymbol{k}$ space, respectively. The number (2.164) can be associated with the quantized spin-Hall conductivity [7] and is called the Z_2 invariant. The above definition invokes the conservation of the z-component of the spin. A more general approach to the Z_2 invariant exploits the symmetry of the matrix elements of the TR operator in the Brillouin zone [8]. The details of this theoretical framework have been explained in Refs. [27, 28, 83].

The nontrivial Z_2 invariant in the bulk comes along with a pair of counter-propagating gapless modes at the edge of a QSHI. Qualitatively, such helical modes have been discussed in Section 2.2.3. There are, however, a few new details because here the edge modes are the eigenstates of BHZ Hamiltonian (2.152). We already have the edge mode for spin projection $\sigma =\uparrow$ [see Eq. (2.147)]. A similar calculation yields the edge state for $\sigma =\downarrow$. Hence, the edge spectrum of a QSHI

$$E_{k_y\sigma} = \mathcal{C} + \mathcal{D}k_y^2 + |\mathcal{A}|\mathrm{sgn}(\mathcal{B})k_y\sigma. \tag{2.165}$$

The full energy spectrum of a QSHI in the BHZ model is shown in Fig. 1.6 of the introductory chapter. In the plot,[1] the two edge modes cross in the band gap, each merging into the bulk conduction and valence bands at higher energies. The lattice and BHZ models both allow for a strong particle-hole asymmetry and for this reason are more suitable for describing realistic 2DTI materials. We also note in passing that Z_2 number (2.164) can be linked to the total spin accumulated by a Kramers pair at the edge [7].

2.4.3 Nonlocal Edge Transport

Helical edge states of QSHIs generate nonlocal transport effects that have no analogues in conventional 2D systems. This has been observed in experiments on HgTe quantum wells [84, 85]. HgTe has a zinc blende lattice type, shown in Fig. 2.13, and the first Brillouin zone in the form of an octahedron. The conduction and valence bands meet at the center of the Brillouin zone and have the s- and p-type symmetries inherited from the atomic orbitals of Hg and Te, respectively. In the presence of the crystal potential, the bands are split by the $\boldsymbol{k}\cdot\boldsymbol{p}$ and spin–orbit couplings (see e.g. Ref. [86]), resulting in a Dirac-like dispersion in HgTe quantum wells [13], similar to that in Fig. 2.11. Moreover, because of the atomically heavy mercury, the relativistic corrections to the band structure are large, leading to an inverted band structure where the s-type band (usually acting as the conduction band) has a lower energy than the p-type ones. Thus, the conditions needed for the occurrence of the helical edge states can be met in appropriately designed HgTe quantum well structures. A sketch of such a structure is shown in Fig. 2.14.

In order to illustrate the nonlocal transport through the helical edge states we follow Ref. [84] and consider a four-terminal Hall bar

[1] The parameters are $\Delta = -0.1$ eV, $\mathcal{A} = 1.5$ eVÅ, $\mathcal{B} = 12$ eVÅ^2, $\mathcal{C} = 0.1$ eV and $\mathcal{D} = 9$ eVÅ^2.

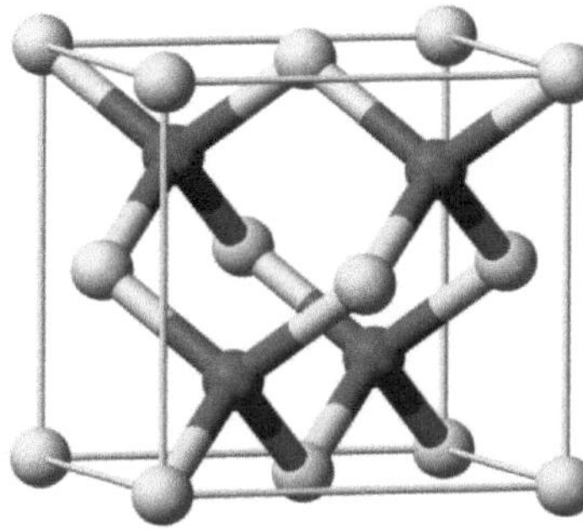

Figure 2.13 Zinc blende lattice structure of HgTe. Hg and Te atoms are depicted as dark blue and yellow spheres, respectively.

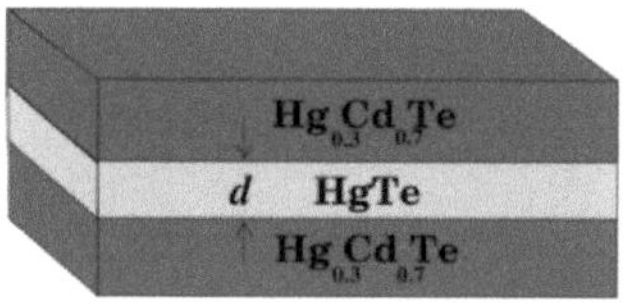

Figure 2.14 Schematic of HgTe/CdTe quantum well structure. The HgTe layer hosts a QSHI system.

depicted in Fig. 2.15. For comparison, panel a of Fig. 2.15 shows a chiral edge state in a strong perpendicular magnetic field (cf. Fig. 1.1 of Chapter 1), whereas panel b shows the helical edge states in a QSHI in a zero magnetic field. The Hall bar terminals serve to inject electric currents and (or) to measure induced voltages. We are interested in the relation between current I_i injected through contact i and voltages V_j induced on all four contacts (i, j = 1 . . . 4). For that purpose, one commonly uses the Landauer–Buetikker formula,

$$I_i = \frac{e^2}{h}\sum_{j=1}^{N}(T_{ji}V_i - T_{ij}V_j), \tag{2.166}$$

where T_{ji} is the transmission probability from contact i to contact j. Consider the chiral edge state in Fig. 2.15a. Since it connects the neighboring contacts in one propagation direction, the corresponding transmission coefficient, T_{ji}^{chi}, can be written as

$$T_{i+1,i}^{chi} = 1, \qquad i = 1, \ldots, N, \tag{2.167}$$

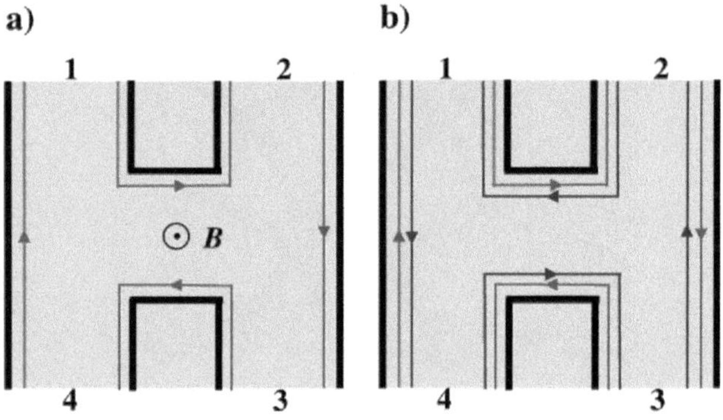

Figure 2.15 Schematic of edge transport in four-terminal device: (a) chiral edge state in a strong magnetic field $\boldsymbol{B}$ and (b) helical edge states in a QSHI in zero magnetic field.

where N is the number of the terminals (i.e. $N = 4$ in Fig. 2.15), with the convention that $T_{N+1,N} = T_{1,N}$ describes the transmission from terminal N to terminal 1. Assuming, for concreteness, that the current flows from terminal 1 to terminal 4, while leads 2 and 3 are used as voltage probes, we have from Eq. (2.166)

$$I_1 \equiv I_{14} = \frac{e^2}{h}(V_1 - V_4), \ V_2 = V_1, \ V_3 = V_2, \ I_4 = -I_1. \quad (2.168)$$

This yields a finite two-terminal resistance

$$R_{14,14} = \frac{V_1 - V_4}{I_{14}} = \frac{h}{e^2}, \quad (2.169)$$

and zero four-terminal (nonlocal) resistances

$$R_{14,12} = \frac{V_1 - V_2}{I_{14}} = 0, \quad (2.170)$$

$$R_{14,23} = \frac{V_2 - V_3}{I_{14}} = 0, \quad (2.171)$$

$$R_{14,13} = \frac{V_1 - V_3}{I_{14}} = 0. \quad (2.172)$$

In contrast, the helical edge channels connect the neighboring contacts in both propagation directions with the transmission coefficients, T^{hel}_{ji}, given by

$$T^{hel}_{i+1,i} = T^{hel}_{i,i+1} = 1, \qquad i = 1, \dots, N, \quad (2.173)$$

with the conventions $T_{N+1,N} = T_{1,N}$ and $T_{N,N+1} = T_{N,1}$. As the helical edge states are protected by the TRS, the transmission between the contacts is perfect even in the presence of potential disorder (e.g. ionized impurities). For the same current contacts (1 and 4) and with $T^{hel}_{i+1,i}$ given by Eq. (2.173), we find from Eq. (2.166)

$$I_1 \equiv I_{14} = \frac{e^2}{h}(2V_1 - V_4 - V_2), \qquad I_4 = -I_1, \tag{2.174}$$

$$2V_2 - V_1 - V_3 = 0, \qquad 2V_3 - V_2 - V_4 = 0. \tag{2.175}$$

This, in turn, yields the two-terminal resistance

$$R_{14,14} = \frac{V_1 - V_4}{I_{14}} = \frac{3}{4}\frac{h}{e^2}, \tag{2.176}$$

and the four-terminal resistances

$$R_{14,12} = \frac{V_1 - V_2}{I_{14}} = R_{14,23} = \frac{V_2 - V_3}{I_{14}} = \frac{1}{4}\frac{h}{e^2}, \tag{2.177}$$

$$R_{14,13} = \frac{V_1 - V_3}{I_{14}} = \frac{1}{2}\frac{h}{e^2}. \tag{2.178}$$

We see that the helical transport is characterized by the lower two-terminal resistance $R_{14,14}$ (2.176). However, the most striking feature of the helical edge states is the nonzero nonlocal resistances (2.177) and (2.178). They are the direct consequence of the TRS and unique to QSHIs.

2.5 Towards Room-Temperature Quantum Spin Hall Insulators: 2D Transition Metal Dichalcogenides

2.5.1 Monolayer Transition Metal Dichalcogenides as Quantum Spin Hall Insulators

Recently, monolayer transition metal dichalcogenides (TMDs) in the distorted octahedral 1T′ phase have been predicted to be QSHIs with an intrinsic inverted band structure [17]. Experimentally, the topological edge states have been reported in single-layers of WTe_2 and WSe_2 (see e.g. [19, 20]). The topological states typically come as boundary states at a crystallographically aligned interface between a 1T′ phase domain and a semiconducting (1H) domain. Such crystalline phase boundaries are well ordered, accessible to high-resolution scanned probe microscopy and offer other opportunities for testing predictions regarding topological edge states. In quest for topological electronics, transport properties of 1T′ - TMDs deserve particular attention as they pave the way for room-temperature QSHIs.

There is also a general theoretical reason for taking a closer look at the boundary states in 2D TMDs. In typical QSHIs, the edge states resemble massless Dirac fermions in the sense that they exhibit a linear level crossing over a substantial energy range. In this energy range, the essential properties of the edge states, such as their electric transport, can be successfully explained by a Dirac-like model. In contrast, in 2D TMDs, the crossing of the boundary states is highly nonlinear (see e.g. [19]), rendering the picture of the Dirac fermions invalid in this case. Consequently, transport calculations based on Dirac-like models cannot be directly applied to the boundary states in 2D TMDs.

In this section, we examine the boundary states in 2D TMDs, using an effective Hamiltonian theory for 1T′ - 1H phase boundaries. The effective Hamiltonian operates in a reduced Hilbert space spanned by the conduction and valence bands, including the spin, as adopted in [87]. We find a strongly nonlinear boundary spectrum reminiscent of a SO-split parabolic band in a 1D conductor. Its nonlinearity and particle-hole asymmetry are consistent with the *ab initio* calculations of Ugeda *et al* [19].

The solution for the boundary states is implemented to calculate their electric conductance and thermopower in the ballistic regime. A subtlety is that the ballistic transport depends on how the boundary spectrum merges into the bulk bands. This happens at special points on the bulk conduction and valence bands at which the bound state ceases to exist. The implications of such spectrum termination points for electron transport have not been fully understood yet. This question is clarified here for 1T′ - TMDs in the context of the recent experimental and *ab initio* study of Ugeda *et al* [19]. Notably, the temperature and chemical potential dependences of both conductance and thermopower are found to be sensitive to the termination points of the boundary spectrum. Furthermore, through the spectrum termination points the thermoelectric coefficients depend on a structural inversion asymmetry, providing extra information on the material properties. These results establish a link between the bulk band structure of 1T′ - TMDs and the boundary electron transport and complement earlier transport studies of QSHI systems. The following subsections explain the details of the calculations and provide an extended discussion of the results.

2.5.2 Effective Model for 1T′ - TMDs

To set the scene, we define the Hamiltonian for a plane-wave state with wave vector $\boldsymbol{k} = [k_x, k_y, 0]$ in a homogeneous 2D TMD,

$$\hat{H}(\boldsymbol{k}) = \hat{H}_0(\boldsymbol{k}) + \hat{H}_{SO}(\boldsymbol{k}). \tag{2.179}$$

Here, the first term describes a monolayer without SOC, while the second term accounts for SOC due to a structural $z \to -z$ reflection asymmetry. As long as only the properties of the conduction and valence bands are of concern, we can work in the reduced Hilbert space in which a state vector $|\boldsymbol{k}\rangle$ has four components, $|\boldsymbol{k}\rangle = [\Psi_{c\uparrow}(\boldsymbol{k}), \Psi_{v\uparrow}(\boldsymbol{k}), \Psi_{c\downarrow}(\boldsymbol{k}), \Psi_{v\downarrow}(\boldsymbol{k})]^T$, where subscripts c and v refer to the conduction and valence bands, while $\uparrow$ and $\downarrow$ to the spin states. A Hamiltonian acting in this reduced space—an effective Hamiltonian—can be represented by a 4×4 matrix [17, 87]. In particular (see [87]),

$$\hat{H}_0(\boldsymbol{k}) = \begin{bmatrix} \epsilon_{\boldsymbol{k}} + m_{\boldsymbol{k}} & v_x k_x + i v_y k_y & 0 & 0 \\ v_x k_x - i v_y k_y & \epsilon_{\boldsymbol{k}} - m_{\boldsymbol{k}} & 0 & 0 \\ 0 & 0 & \epsilon_{\boldsymbol{k}} + m_{\boldsymbol{k}} & -v_x k_x + i v_y k_y \\ 0 & 0 & -v_x k_x - i v_y k_y & \epsilon_{\boldsymbol{k}} - m_{\boldsymbol{k}} \end{bmatrix}$$
$$= \epsilon_{\boldsymbol{k}} \sigma_0 \tau_0 + v_x k_x \sigma_z \tau_1 - v_y k_y \sigma_0 \tau_2 + m_{\boldsymbol{k}} \sigma_0 \tau_3, \tag{2.180}$$

where $\epsilon_{\boldsymbol{k}}$ and $m_{\boldsymbol{k}}$ are quadratic functions of the wave vector,

$$\epsilon_{\boldsymbol{k}} = \epsilon_0 + \epsilon_x k_x^2 + \epsilon_y k_y^2, \qquad m_{\boldsymbol{k}} = m_0 + m_x k_x^2 + m_y k_y^2, \tag{2.181}$$

and ϵ_0, m_0, $\epsilon_{x,y}$, $m_{x,y}$ and $v_{x,y}$ are the band structure constants of the effective model. In particular, $\epsilon_0 \pm m_0$ are the energies of

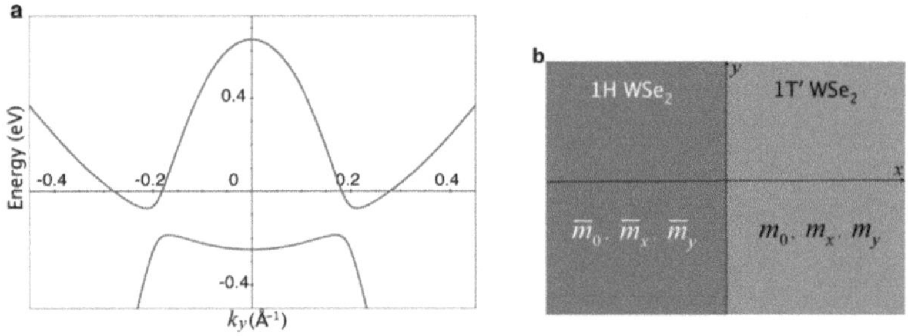

Figure 2.16 (a) Conduction and valence bands of 1T′ - TMDs from the effective Hamiltonian model. The plot shows the band dispersion in equation (2.183) along the k_y direction for $\epsilon_0 = 0.2$ eV, $\epsilon_x = -10$ eVÅ^2, $\epsilon_y = -9$ eVÅ^2, $v_x = 0.8$ eVÅ, $v_y = 0.45$ eVÅ, $m_0 = 0.45$ eV, $m_x = -13$ eVÅ^2, $m_y = -12$ eVÅ^2 and $\lambda = \delta = 0$. The effective model is tailored to qualitatively reproduce the *ab initio* band structure calculations along the $Y - \Gamma - Y$ direction in the Brillouin zone of 1T′ - WSe_2, e.g. its band gap $E_g \approx 120$ meV. Along the k_x direction, the band dispersion is similar, with a somewhat larger band gap. (b) Schematic of a 1T′ - 1H phase boundary in the effective model of 2D TMDs (e.g. single-layer WSe_2). Adapted from [21].

the conduction and valence bands at the Γ point; $\epsilon_{x,y}$ and $m_{x,y}$ characterize the band curvature, while $v_{x,y}$ the atomic SOC. We use the Pauli matrices in the band (τ_1, τ_2 and τ_3) and spin (σ_z) subspaces along with the corresponding unit matrices τ_0 and σ_0.

The lack of the $z \to -z$ symmetry allows for $\hat{H}_{\text{SO}}(\boldsymbol{k})$ of different types. We consider a particular one

$$\hat{H}_{\text{SO}}(\boldsymbol{k}) = \begin{bmatrix} \lambda k_y & i\delta & 0 & 0 \\ -i\delta & \lambda k_y & 0 & 0 \\ 0 & 0 & -\lambda k_y & -i\delta \\ 0 & 0 & i\delta & -\lambda k_y \end{bmatrix} = (\lambda k_y \tau_0 - \delta \tau_2)\sigma_z, \qquad (2.182)$$

where λ and δ are the structural SOC constants. For the purpose of this study, it is sufficient that block-diagonal $\hat{H}_{\text{SO}}(\boldsymbol{k})$ lifts the spin degeneracy of $\hat{H}_0(\boldsymbol{k})$. The inclusion of the off-diagonal SOC would result in a more involving boundary problem later on.

The band structure of the effective model is given by the eigenvalues of the total Hamiltonian in Eq. (2.179). It is instructive to look at the dispersion of the conduction and valence bands:

$$E_\sigma^\pm(\boldsymbol{k}) = \epsilon_0 + \epsilon_x k_x^2 + \epsilon_y k_y^2 + \lambda\sigma k_y \pm\sqrt{v_x^2 k_x^2 + (v_y k_y + \delta\sigma)^2 + (m_0 + m_x k_x^2 + m_y k_y^2)^2}, \tag{2.183}$$

where $\sigma = \pm 1$ is the eigenvalue of σ_z. With appropriately chosen parameters, Eq. (2.183) qualitatively reproduces the conduction and valence bands of 1T′ - TMDs. For example, Fig. 2.16(a) shows the plot for 1T′ - WSe_2. The positions of the bands, their profiles and the energy gap between them overall agree with the *ab initio* calculations along the $Y - \Gamma - Y$ direction in the Brillouin zone (cf. [19]). The effective model is tailored to have the band gap $E_g \approx 120$ meV as calculated in [19]. Notably, the band curvature at $k_y = 0$ (Γ point) indicates an inverted band ordering (cf. [17]), which is a necessary prerequisite for the occurrence of the topological boundary modes. Similar to the BHZ model, an intrinsic band inversion is realized under conditions $m_0 m_x < 0$ and $m_0 m_y < 0$ for the coefficients in the gap term $m_{\boldsymbol{k}}$ in Eq. (2.181), see also the caption for Fig. 2.16(a).

2.5.3 1T′ - 1H Phase Interface and Boundary States

We consider a crystallographically well-defined interface between a 1T′ phase domain and a semiconducting 1H domain in contiguous single layers of TMD. The two phases are separated by a straight boundary; the x and y axes are perpendicular and parallel to it, as shown in Fig. 2.16(b). In this geometry, k_y remains a good quantum number, while k_x needs to be replaced by operator $-i\partial_x$. Compared to the 1T′ phase, the 1H one has a larger band gap and a normal band ordering. This difference can be accounted for by an appropriately generalized gap term $m_{\boldsymbol{k}}$. To model the 1T′ - 1H interface, we use the position-dependent gap term

$$m_{\boldsymbol{k}} \to m_0(x) - m_x(x)\partial_x^2 + m_y(x)k_y^2, \tag{2.184}$$

where the coefficients $m_0(x)$ and $m_{x,y}(x)$ coincide with m_0 and $m_{x,y}$ in the 1T′ domain ($x \geq 0$), while taking different values $\overline{m}_0$ and $\overline{m}_{x,y}$ on the 1H side ($x \leq 0$). The relative sign of $\overline{m}_0$ and $\overline{m}_{x,y}$ is positive, meaning a normal band ordering in the 1H domain.

The next steps are analogous to Section 2.3.2. Using Eqs. (2.179)–(2.182) and (2.184), we can write the interface Hamiltonian as

$$\hat{H}_{\sigma k_y}(x) = \hat{H}^A_{\sigma k_y}(x) + \hat{H}^S_{\sigma k_y}(x),$$

where $\hat{H}^S_{\sigma k_y}(x)$ and $\hat{H}^A_{\sigma k_y}(x)$ are the particle-hole symmetric and asymmetric parts of the Hamiltonian. For given spin projection $\sigma = \pm 1$ (resp. $\uparrow$ and $\downarrow$), these are 2×2 matrices given by

$$\hat{H}^A_{\sigma k_y}(x) = (\epsilon_0 + \epsilon_y k_y^2 + \lambda \sigma k_y - \epsilon_x \partial_x^2)\tau_0 \tag{2.185}$$

and

$$\begin{aligned}\hat{H}^S_{\sigma k_y}(x) &= -i\sigma v_x \partial_x \tau_1 - (v_y k_y + \delta\sigma)\tau_2 \\ &\quad + [m_0(x) + m_y(x)k_y^2 - m_x(x)\partial_x^2]\tau_3.\end{aligned} \tag{2.186}$$

This yields the eigenvalue equation

$$\hat{H}^A_{\sigma k_y}(x)\psi(x) + \hat{H}^S_{\sigma k_y}(x)\psi(x) = E\psi(x)$$

for energy E and two-component wave function $\psi(x)$. The latter is assumed to vanish away from the interface: $\psi(x) \to 0$ for $x \to \pm\infty$, while being continuous at $x = 0$.

The solution of the above boundary problem is pretty straightforward (see Section 2.3.2). Here we just quote the result:

$$\psi(x) = \Phi \begin{cases} -\frac{\varkappa_3+\varkappa_2}{\varkappa_1-\varkappa_2} e^{-\varkappa_1 x} + \frac{\varkappa_3+\varkappa_1}{\varkappa_1-\varkappa_2} e^{-\varkappa_2 x}, & x \geq 0, \\ e^{\varkappa_3 x}, & x \leq 0. \end{cases} \tag{2.187}$$

It describes a bound state localized on length-scales $\varkappa_{1,2}^{-1}$ and $\varkappa_3^{-1}$ in the 1T′ and 1H domains, respectively, where

$$\varkappa_{1,2} = \left|\frac{v_x}{2m_x}\right| \pm \sqrt{\left(\frac{v_x}{2m_x}\right)^2 + \frac{m_0 + m_y k_y^2}{m_x}}$$

and

$$\varkappa_3 = \mathrm{sgn}(m_0 \overline{m}_0)\left|\frac{v_x}{2\overline{m}_x}\right| + \sqrt{\left(\frac{v_x}{2\overline{m}_x}\right)^2 + \frac{\overline{m}_0 + \overline{m}_y k_y^2}{\overline{m}_x}}.$$

The inverted 1T′ band structure allows the k_y values in the segment

$$-k_0 \leq k_y \leq k_0, \qquad k_0 = \sqrt{-m_0/m_y}, \tag{2.188}$$

where endpoints $\pm k_0$ are the zeros of the gap term $m_0 + m_y k_y^2$. Further, it can be shown that Φ is an eigenstate of Pauli matrix τ_2 defined by

$$\tau_2 \Phi = -\sigma\, \mathrm{sgn}(v_x m_x)\Phi,$$

i.e. the choice of the eigenstate depends on the spin projection as well as on the relative sign of the band structure parameters v_x and m_x. For $\sigma = \pm 1$, there are two orthogonal boundary modes. Their energy dispersion is given by

$$E_{\sigma k_y} = \epsilon + \epsilon_y k_y^2 + v\sigma k_y, \tag{2.189}$$

with

$$v = v_y \,\mathrm{sgn}(v_x m_x) + \lambda, \qquad \epsilon = \epsilon_0 + \delta \,\mathrm{sgn}(v_x m_x). \tag{2.190}$$

The parameters v and ϵ absorb the structural SOC constants and account for the signs of other involved parameters.

Overall, the above boundary solution is analogous to the edge states of a QSHI in the BHZ model. There are a few new details, though. The solution of the BHZ model (2.146) is a "hard-wall" one, i.e. the electronic wave function vanishes upon approaching the boundary of a QSHI. In contrast, Eq. (2.187) accounts for the leakage of the wave function into the semiconducting (1H) region, which was observed in the scanning tunneling experiment of Ugeda *et al* [19]. Further, the SOC in Eq. (2.182) makes the bulk bands asymmetric with respect to $k_y \to -k_y$. In this case, we find a specific dependence of the boundary spectrum on the SOC constants λ and δ. The implications of this finding for the thermoelectric coefficients will be discussed in the next subsection. Finally, the dependence on the signs of the model parameters is generic, allowing for different types of band structures.

The energy spectrum of a 1T′ - TMD is depicted in Fig. 1.8 of the introductory chapter. The plot shows boundary states (2.189) along with bulk bands (2.183). The parameters are the same as in Fig. 2.16(a). The two boundary modes with opposite spins $\uparrow$ and $\downarrow$ connect the bulk conduction and valence bands, crossing at the Γ point. Their dispersion resembles a SOC—split parabolic band in a one-dimensional conductor. However, the boundary modes terminate on the conduction and valence bands, so only one Kramers pair occurs in the band gap. The termination points of the boundary spectrum are the endpoints of the allowed k_y segment in Eq. (2.188). In Fig. 1.8, $\pm k_0$ are approximately $\pm 0.2\,\text{Å}^{-1}$. At these points, bound state (2.187) gets delocalized, spreading into the bulk of the 1T′ domain. In energy, the spectrum termination points lie at

$$E_{\mathrm{c,v}} = \epsilon + \epsilon_y k_0^2 \mp v k_0 = \epsilon - \frac{\epsilon_y m_0}{m_y} \mp v\sqrt{-\frac{m_0}{m_y}} \tag{2.191}$$

in the conduction ("$-$") and valence ("$+$") band, respectively. In Fig. 1.8, these energies are $E_c \approx -50$ meV and $E_v \approx -225$ meV. The spectrum termination points $\pm k_0$ do not coincide with the positions of the local extrema of the bulk bands, so the energy difference $E_c - E_v \approx 175$ meV is somewhat larger than the band gap $E_g \approx 120$ meV, although the scale is the same.

Beside the energy spectrum, the above results provide an estimate for the distance over which the boundary states decay from the interface. In the 1T$'$ domain, the decay length is of order of $|2m_x/v_x|$, which for the chosen parameters is about 3 nm. For comparison, the estimate of Ugeda *et al* [19] is 2 nm.

2.5.4 Boundary Conductance and Thermopower

When the Fermi level is adjusted in the band gap of the 1T$'$ domain (see dashed line in Fig. 1.8), the phase boundary acts as a quasi-1D conductor with a Kramers pair of propagating modes. We discuss first the equilibrium case. For one mode, say the $\uparrow$ one, the electric current can be calculated as the equilibrium expectation value in k space:

$$
\begin{aligned}
j_\uparrow &= e \int_{-k_0}^{k_0} v_\uparrow(k_y) f[E_\uparrow(k_y)] \frac{dk_y}{2\pi} \\
&= e \int_{E_v}^{E_c} N_\uparrow(E) v_\uparrow(E) f(E) dE,
\end{aligned} \tag{2.192}
$$

where we trace over all k_y values of the boundary mode [see Eq. (2.188)], $v_\uparrow(k_y) = \hbar^{-1}\partial E_\uparrow(k_y)/\partial k_y$ is the mode velocity, and $f(E) = \left[e^{(E-\mu)/(k_B T)} + 1\right]^{-1}$ is the Fermi occupation number (μ and k_B are the chemical potential and Boltzmann constant). The k_y integration is replaced by the energy integral with the 1D density of states $N_\uparrow(E) = h^{-1}|v_\uparrow(E)|^{-1}$ and velocity

$$
v_\uparrow(E) = \frac{\text{sgn}(\epsilon_y)}{\hbar} \sqrt{v^2 - 4\epsilon_y(\epsilon - E)}, \tag{2.193}
$$

obtained from Eq. (2.189). Only the energies between the spectrum termination points E_v and E_c (2.191) contribute to the current because this energy window corresponds to a chiral (one-way moving) state. The sign of its velocity, $\text{sgn}(v_\uparrow) = \text{sgn}(\epsilon_y)$, determines the direction of the current:

$$j_\uparrow = \frac{e}{h}\int_{E_\mathrm{v}}^{E_\mathrm{c}} \mathrm{sgn}(\nu_\uparrow) f(E) dE$$
$$= \mathrm{sgn}(\epsilon_y)\frac{e}{h}\int_{E_\mathrm{v}}^{E_\mathrm{c}} f(E) dE. \tag{2.194}$$

At zero temperature, the current $j_\uparrow = \mathrm{sgn}(\epsilon_y)(e/h)(\mu - E_\mathrm{v})$ is carried by all occupied states from E_v to μ. Likewise, the current $j_\downarrow$ depends on the sign of the velocity of the $\downarrow$ mode, $\mathrm{sgn}(\nu_\downarrow) = -\mathrm{sgn}(\epsilon_y)$, which is opposite to that in Eq. (2.194). At equilibrium, the two modes are equally occupied, rendering the net electric current $j = j_\uparrow + j_\downarrow$ null.

We now turn to the non-equilibrium transport. It can be realized by attaching a boundary channel to two electronic reservoirs, each being in equilibrium with its own chemical potential and temperature. We assume a ballistic boundary channel, which is justified if it is shorter than both elastic and inelastic mean free paths. Now, the counter-propagating $\uparrow$ and $\downarrow$ states come from different reservoirs with unequal occupation numbers. Say, the $\uparrow$ occupation number is still $f(E)$, while that of the $\downarrow$ mode is $f'(E) = \left[\mathrm{e}^{(E-\mu')/(k_B T')} + 1\right]^{-1}$, with chemical potential $\mu' \neq \mu$ and temperature $T' \neq T$. The net electric current $j = j_\uparrow + j_\downarrow$ can be written as

$$j = \mathrm{sgn}(\epsilon_y)\frac{e}{h}\int_{E_\mathrm{v}}^{E_\mathrm{c}} [f(E) - f'(E)] dE$$
$$\approx \mathrm{sgn}(\epsilon_y)\left[G\,\frac{\mu - \mu'}{e} + GS\,(T - T')\right],$$

where we linearized j with respect to the differences $\mu - \mu'$ and $T - T'$ (both assumed small enough), introducing the electric conductance, G, and the Seebeck coefficient (thermopower), S [77]:

$$G = \frac{e^2}{h}\int_{E_\mathrm{v}}^{E_\mathrm{c}} \left(-\frac{\partial f}{\partial E}\right) dE = \frac{e^2}{h}[f(E_\mathrm{v}) - f(E_\mathrm{c})], \tag{2.195}$$

$$S = \frac{1/e}{f(E_\mathrm{v}) - f(E_\mathrm{c})}\int_{E_\mathrm{v}}^{E_\mathrm{c}} \frac{E-\mu}{T}\left(-\frac{\partial f}{\partial E}\right) dE$$
$$= \frac{k_B/e}{f(E_\mathrm{v}) - f(E_\mathrm{c})}\int_{\frac{E_\mathrm{v}-\mu}{2k_B T}}^{\frac{E_\mathrm{c}-\mu}{2k_B T}} \frac{\eta\, d\eta}{\cosh^2\eta}. \tag{2.196}$$

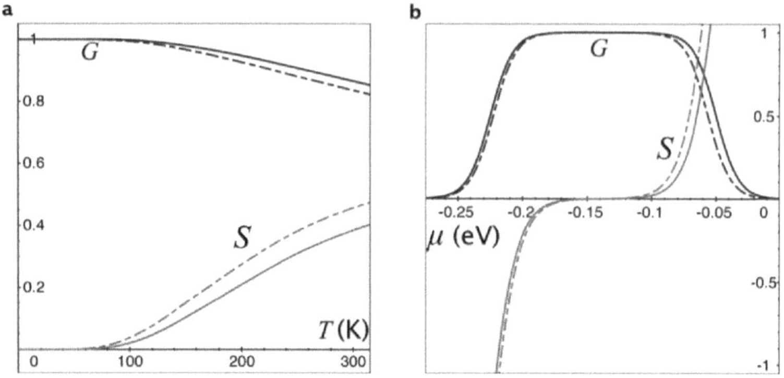

Figure 2.17 Electric conductance G (in units of e^2/h) and thermopower S (in units of $k_B/|e|$) along a 1T′ - 1H phase boundary in WSe_2 [see Eqs. (2.195) and (2.196)]: (a) temperature dependence for $\mu = -0.1$ eV in the band gap and (b) dependence on chemical potential inside the band gap for $T =$ 100 K. Solid and dashed curves are for vanishing and finite structural SOC, with $\lambda = \delta = 0$ and $\lambda = 0.02$ eVÅ, $\delta = 0.002$ eV, respectively. Other band structure parameters are the same as in Fig. 2.16(a). Adapted from [21].

Here, the energy bounds E_c and E_v (2.191) contain the details of the band structure of a 1T′ - TMD. In particular, the transport coefficients reflect the particle-hole asymmetry as well as the atomic and structural SOC. The overall sign of current j depends on the curvature of the particle-hole asymmetric dispersion along the k_y direction. This sign determines which of the two reservoirs acts as the electron source and which as the sink. The calculation of the transport coefficients is restricted to the subgap states, implying $k_B T < \frac{1}{2} E_g$, which holds well up to the room temperatures for 1T′ - WSe_2 with $E_g \approx 120$ meV.

Figure 2.17(a) shows the temperature dependence of Eqs. (2.195) and (2.196) for single-layer WSe_2. The large band gap of this material manifests itself as the conductance plateau at e^2/h up to $\approx$ 100 K. The deviation from e^2/h remains less than 20% up to the room temperatures. The thermopower is exponentially suppressed, but grows faster than the conductance deviation from e^2/h. These observations are corroborated by the asymptotic formulae

$$G \approx \frac{e^2}{h}\left(1 - e^{-\frac{\mu - E_v}{k_B T}} - e^{-\frac{E_c - \mu}{k_B T}}\right), \qquad (2.197)$$

$$S \approx \frac{k_B}{e}\left(\frac{\mu - E_{\rm v}}{k_B T}\mathrm{e}^{-\frac{\mu - E_{\rm v}}{k_B T}} - \frac{E_{\rm c} - \mu}{k_B T}\mathrm{e}^{-\frac{E_{\rm c} - \mu}{k_B T}}\right), \qquad (2.198)$$

for $k_B T < |E_{\rm c,v} - \mu|$.

As a function of the chemical potential, $G(\mu)$ has a plateau-like maximum close to e^2/h [see Fig. 2.17(b)]. We recall that G does not include the contribution of the bulk states. The total conductance of the boundary and bulk states is expected to have a plateau-like minimum at e^2/h [13]. Away from the centre of the plateau, $G(\mu)$ drops exponentially. This behavior is band-structure-dependent, and the knowledge of the spectrum termination points $E_{\rm c}$ and $E_{\rm v}$ is the minimal information needed to understand it. A specific feature of the thermopower $S(\mu)$ is a sign reversal inside the band gap of the material. The zero of $S(\mu)$ is given by the average of the energy bounds $E_{\rm c}$ and $E_{\rm v}$

$$\mu_0 = \frac{E_{\rm c} + E_{\rm v}}{2} = \epsilon - \frac{\epsilon_y m_0}{m_y}, \qquad (2.199)$$

which is estimated to be about -0.14 eV for 1T′ - WSe_2. The point of the sign reversal μ_0 lies at the centre of the conductance plateau and reflects the particle-hole asymmetry of the band structure. Away from the plateau centre, the function $S(\mu)$ shows an exponential increase, depending on the spectrum termination points $E_{\rm c}$ and $E_{\rm v}$.

Also noteworthy is the effect of the structural inversion asymmetry on the thermoelectric coefficients (compare solid and dashed curves in Fig. 2.17). It is caused by the shifts of the energy bounds

$$\Delta E_{\rm c,v} = \delta\, \mathrm{sgn}(v_x m_x) \mp \lambda k_0, \qquad (2.200)$$

due to the structural SOC [see Eqs. (2.190) and (2.191)]. The effect is well visible already for small values of the SOC constants λ and δ, such as in typical 2D semiconductor heterostructures. This can be explained by the exponential sensitivity of G and S to the changes in energy bounds $E_{\rm c}$ and $E_{\rm v}$. It is worth reminding that we consider the SOC without mixing the spin states. The Rashba-like SOC [87] requires a separate treatment. It can be included in Eq. (2.185), while the particle-hole symmetric Hamiltonian in Eq. (2.186) remains unchanged. We can therefore still use the approach of Section 2.5.3 and expect similar results to those in Fig. 2.17.

The temperature dependence of the boundary conductance has been measured for 1T′ - WTe_2 [20]. A conductance plateau followed by a decrease in $G(T)$ has been seen in a few experiments, e.g. in [20] the conductance plateau persisted up to 100K. The behavior of $G(T)$ in Fig. 2.17(a) is quite similar, indicating that our ballistic model captures the essential features of the boundary transport. Especially, the spectrum termination points of the boundary states are important for modeling the temperature and chemical potential dependences of the ballistic conductance and thermopower.

Because of their large band gap, the 1T′ materials should be particularly suitable for measurements of the boundary thermopower, at least conceptually. In the model studied above, the thermopower is small for temperatures below the band gap, but it can be detected by the sign reversal as an applied gate voltage shifts the Fermi level between the conduction and valence bands. The position of the zero of the thermopower is determined solely by the spectrum termination points.

2.6 Problems

(1) Consider a Chern insulator described by the Hamiltonian in Eq. (2.17). Using the results of Section 2.1.1 and 2.1.2, show that this Chern insulator has the edge mode

$$E(k) = -\mathrm{sgn}(m)\,\hbar v, \tag{2.201}$$

Berry's connection components

$$A^{(k)}_{\mathbf{k}s} = 0, \qquad A^{(\phi)}_{\mathbf{k}s} = -\frac{1 - su_z}{2k}, \tag{2.202}$$

Berry's phase

$$\gamma_s = \oint \boldsymbol{A}_{\mathbf{k}s} \cdot d\boldsymbol{k} = -\pi(1 - su_z), \tag{2.203}$$

and Berry's curvature

$$(\nabla \times \boldsymbol{A}_{\mathbf{k}s})_z = \frac{s}{2k}\frac{\partial u_z}{\partial k}. \tag{2.204}$$

What is the relation between this Chern insulator and that introduced in Section 2.1.1?

(2) Prove that in the transformed basis (2.73) the boundary condition (2.92) takes the form

$$\psi'(x, y = 0) = \hat{M}'\psi'(x, y = 0), \qquad \hat{M}' = \begin{bmatrix} \sigma_x & 0 \\ 0 & \sigma_x \end{bmatrix}. \tag{2.205}$$

(3) Consider the boundary problem given by Eqs. (2.69), (2.74) and (2.205). Find the edge solutions and show that they are helical with the spectrum given by Eq. (2.101). Identify further properties of these edge states that are independent of the basis.

(4) In Ref. [13], Bernevig, Hughes and Zhang (BHZ) derived the following Hamiltonian for HgTe/CdTe quantum wells:

$$\hat{H}_{BHZ} = \begin{bmatrix} \hat{h}_{\boldsymbol{k}} & 0 \\ 0 & \hat{h}^*_{-\boldsymbol{k}} \end{bmatrix}, \tag{2.206}$$

with

$$\hat{h}_{\boldsymbol{k}} = \mathcal{A}(\sigma_x k_x - \sigma_y k_y) + \mathcal{M}_{\boldsymbol{k}}\sigma_z + \mathcal{D}\boldsymbol{k}^2\sigma_0, \tag{2.207}$$

and

$$\mathcal{M}_{\mathbf{k}} = \mathcal{M} + \mathcal{B}\mathbf{k}^2, \tag{2.208}$$

where $\mathcal{A}$, $\mathcal{B}$, $\mathcal{D}$, and $\mathcal{M}$ are constants characterizing the band structure. Using the unitary transformation

$$U = \begin{bmatrix} 0 & \sigma_z \\ -i\sigma_y & 0 \end{bmatrix}, \tag{2.209}$$

obtain the Hamiltonian $\hat{H}'_{BHZ} = U\hat{H}_{BHZ}U^{\dagger}$. Show that for $\mathcal{D} = \mathcal{B} = 0$ the new Hamiltonian coincides with $\hat{H}'$ in Eq. (2.74).

Chapter 3

Two-Dimensional Topological Insulators with Broken Time-Reversal Symmetry

We have just explored QSHIs—the class of 2DTIs restricted by TRS. Spin magnetization or a strong magnetic field removes TRS. What happens to QSHIs in that case? And should we expect new 2DTI phases? In this chapter, we address these questions, which are of both theoretical and practical interest. We first examine the effect of spin magnetization, using both Dirac and BHZ models and building upon the results of the preceding chapter. Our goal is to introduce phenomenological theories of quantum anomalous Hall insulators (QAHIs) and recover the main findings of Haldane's [6] and more recent [22, 23] theories. We then focus on the orbital effect of a perpendicular magnetic field on a QSHI. As the calculation is rather involved, we do it for the Dirac model only. The results turn out to be rather counter-intuitive: A QSHI can exist in strong quantizing magnetic fields. This robustness can be explained by the fact that the helical edge states are bound to an effective domain wall that is not destroyed by the magnetic field. However, as we shall see, both spectral and transport properties of the helical edge states are modified by a strong quantizing magnetic field.

Topological Quantum Materials: Concepts, Models, and Phenomena
Grigory Tkachov

ISBN 978-981-4968-12-6 (Hardcover), 978-1-003-26641-9 (eBook)
www.jennystanford.com

3.1 Quantum Anomalous Hall Insulators I

3.1.1 Dirac Model

The simplest model for a QAHI is a Dirac-like one. It can be deduced by analogy with the QSHI model from Section 2.2 (see also Fig. 2.6). We again begin with an insulating 3D slab with two Dirac surfaces, gradually reducing the thickness of the slab until the two surfaces overlap, forming a thin, quasi-2D film. As before, the film is described by 2D Eq. (2.58) for a four-component wave function (2.59), but the film Hamiltonian is generalized as follows:

$$\hat{H}_{\boldsymbol{k}} = \begin{bmatrix} \hbar v\boldsymbol{\sigma}\cdot\boldsymbol{k} + m\sigma_z & \Delta\sigma_0 \\ \Delta\sigma_0 & -\hbar v\boldsymbol{\sigma}\cdot\boldsymbol{k} + m\sigma_z \end{bmatrix}. \tag{3.1}$$

Here we add a phenomenological TR-breaking term $m\sigma_z$ at each surface. It represents the surface magnetization due to either an intrinsic exchange interaction in the film or the proximity effect of ferromagnetic insulators. Other ingredients of the model remain the same, e.g. the off-diagonal entry $\Delta\sigma_0$ accounts for the coupling between the surfaces, and $\boldsymbol{k} = [k_x, k_y, 0]$ lies in the plane of the film.

Hamiltonian (3.1) describes a thin-film magnetic insulator. More insight can be gained by making a unitary transformation

$$\hat{H}_{\boldsymbol{k}} \to \hat{U}\,\hat{H}_{\boldsymbol{k}} U^{\dagger}, \qquad \hat{U} = \frac{1}{\sqrt{2}}\begin{bmatrix} \sigma_0 & \sigma_z \\ -\sigma_z & \sigma_0 \end{bmatrix}, \tag{3.2}$$

which brings the Hamiltonian to the block-diagonal form

$$\hat{H}_{\boldsymbol{k}} = \begin{bmatrix} \hbar v\boldsymbol{\sigma}\cdot\boldsymbol{k} + (m+\Delta)\sigma_z & 0 \\ 0 & -\hbar v\boldsymbol{\sigma}\cdot\boldsymbol{k} + (m-\Delta)\sigma_z \end{bmatrix}. \tag{3.3}$$

Unlike the magnetization m, the coupling energy Δ enters the two blocks with the opposite signs, providing a TR-invariant mechanism for band gap generation. The two blocks in Eq. (3.3) correspond to a pair of Chern insulators. A precise map to the models introduced earlier can be achieved by making another unitary transformation

$$\hat{H}_{\boldsymbol{k}} \to \hat{U}\,\hat{H}_{\boldsymbol{k}}\hat{U}^{-1}, \qquad \hat{U} = \begin{bmatrix} \sigma_0 & 0 \\ 0 & \sigma_z \end{bmatrix}. \tag{3.4}$$

This yields

$$\hat{H}_{\boldsymbol{k}} = \begin{bmatrix} \boldsymbol{d}^{+}_{\boldsymbol{k}}\cdot\boldsymbol{\sigma} & 0 \\ 0 & \boldsymbol{d}^{-}_{\boldsymbol{k}}\cdot\boldsymbol{\sigma} \end{bmatrix}, \qquad \boldsymbol{d}^{\pm}_{\boldsymbol{k}} = [\hbar v k_x, \hbar v k_y, m \pm \Delta]. \tag{3.5}$$

We have obtained two copies of a Chern insulator with different "masses" $m \pm \Delta$ due to the competition of the TR-breaking and TR-preserving gap terms. It is worth noting that Hamiltonian (3.5) is equivalent to the long-wave-length limit of Haldane's model [6], which was originally set up on a 2D hexagonal lattice. Besides, this Hamiltonian is related to the BHZ model for magnetically doped semiconductor materials such as $Hg_{1-y}Mn_yTe$ quantum wells [22].

3.1.2 Phase Classification by TKNN Invariant

The topological phases of Hamiltonian (3.5) can be identified by inspecting the TKNN invariant

$$n = \sum_{\nu=\pm 1} \int_{BZ} \frac{dk_x dk_y}{2\pi} (\nabla \times \boldsymbol{A}_{k\nu})_z,$$

where $\boldsymbol{A}_{k\nu}$ is the Berry connection for the Chern species with flavour $\nu = \pm 1$ in Eq. (3.5). We just need to repeat the calculation in Eqs. (2.162) and (2.163) for the function

$$u_{z\nu}(k) = \frac{m + \nu\Delta}{\sqrt{(\hbar v)^2 \boldsymbol{k}^2 + (m + \nu\Delta)^2}}. \tag{3.6}$$

The result is

$$\begin{aligned} n &= \sum_{\nu=\pm 1} \frac{u_{z\nu}(\infty) - u_{z\nu}(0)}{2} = -\frac{1}{2} \sum_{\nu=\pm 1} \text{sgn}(m + \nu\Delta) \\ &= \frac{1}{2} \left[\text{sgn}(\Delta - m) - \text{sgn}(\Delta + m) \right]. \end{aligned} \tag{3.7}$$

Clearly, the number n can only take the integer values 0, +1 or −1. We also note that, up to the notations, Eq. (3.7) is identical to Haldane's result [6]. The value of n depends critically on the strength of the TR-breaking, which is quantified by the ratio of the two gaps m/Δ. There are only two topologically distinct phases:

$$n = \begin{cases} \pm 1, & |m| > \Delta \quad \text{(QAHI)}, \\ 0, & |m| < \Delta \quad \text{(QSHI)}. \end{cases} \tag{3.8}$$

Here, $n = \pm 1$ correspond to QAHIs with the quantized Hall conductivities

$$\sigma_{xy} = \pm \frac{e^2}{h}, \tag{3.9}$$

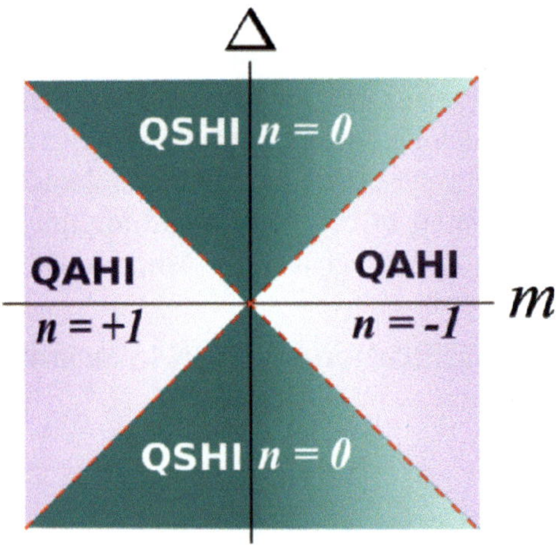

Figure 3.1 Phase diagram for a 2D topological insulator [see also Eq. (3.8)]. There are two topologically distinct phases: the quantum anomalous Hall insulator (QAHI) and the quantum spin Hall insulator (QSHI). Such type of the phase diagram was first obtained by Haldane.

while $n = 0$ to a QSHI in which $\sigma_{xy} = 0$. The necessary requirement for the QAHI is the existence of a TR breaking mechanism (here mass term m) generating the band gaps of the same sign, $m \pm \Delta > 0$ or < 0, in the two Chern subsystems. This has first been shown in Ref. [6] for a 2D honeycomb lattice model. As to the QSHI, it occurs when the band gaps $m + \Delta$ and $m - \Delta$ have opposite signs, i.e. when the hybridization of the surfaces is stronger than the TR breaking. These results are summarized in the phase diagram shown in Fig. 3.1 (cf. Ref. [6]). Crossing the phase boundaries $\Delta = \pm m$ causes a topological transition between a QAHI and a QSHI.

3.1.3 Phase Classification by Edge States

Another way to characterize the 2D topological phases is to look at the edge states of the two Chern subsystems in Hamiltonian (3.5). The problem reduces to the one solved in Section 2.1. Using Eqs. (2.13) and (2.14), we find the localization lengths and spectra of the

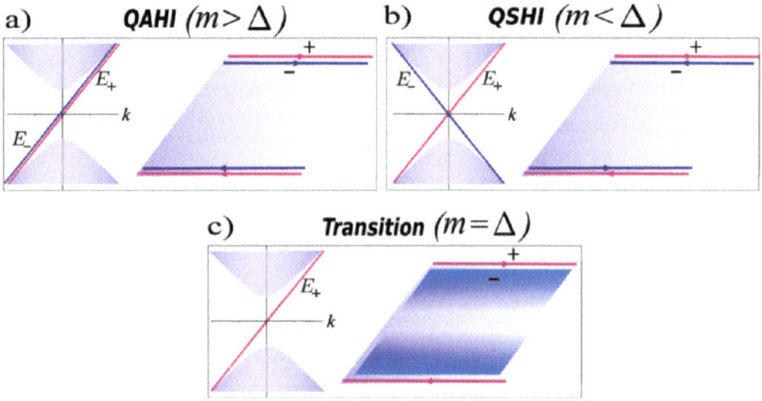

Figure 3.2 Schematic of edge states in (a) QAHI, (b) QSHI and (c) at the transition between QAHI and QSHI [see also Eqs. (3.11)–(3.13)].

edge states for the upper (+) and lower (−) blocks of Eq. (3.5):

$$\lambda_{\pm} = \frac{\hbar v}{|m \pm \Delta|}, \qquad E_{\pm}(k) = \mathrm{sgn}\,(m \pm \Delta)\; v\hbar k. \tag{3.10}$$

Taking, for concreteness, positive m and Δ, we can write the possible edge spectra as (see also Fig. 3.2)

$$E_{\pm}(k) = v\hbar k, \quad m > \Delta \quad \text{(QAHI)}, \tag{3.11}$$

$$E_{\pm}(k) = \pm v\hbar k, \quad m < \Delta \quad \text{(QSHI)}, \tag{3.12}$$

$$E_{+}(k) = v\hbar k, \quad m = \Delta \quad \text{(Transition)}. \tag{3.13}$$

In a QAHI, the edge states of the two Chern subsystems propagate in the same direction, forming a single chiral edge mode. In a QSHI, the edge states of the Chern subsystems are helical, i.e. propagating in the opposite directions. At the transition between the QAHI and QSHI, one of the Chern species loses its edge mode, as it gets spread into the bulk, e.g. for $\lambda_- \to \infty$. Such a critical state is interesting because it features no fermion doubling at the edge. This observation is also in agreement with Haldane's microscopic lattice model [6]. Thus, Dirac-like Hamiltonian (3.5) proves to be an efficient basic tool for understanding the new 2DTI state—the QAHI—and topological QAHI-QSHI transitions.

3.2 Quantum Anomalous Hall Insulators II

The above results for the TKNN invariant and edge modes are specific to the linear momentum dependence of the Dirac model. How do they change if we include higher order terms in momentum? In the following, we address this question, starting from the square-lattice and BHZ models for QAHIs.

3.2.1 Tight-Binding and Bernevig–Hughes–Zhang Models

Our goal here is to appropriately extend the results of Section 2.4 to QAHIs. We start from the square-lattice Hamiltonian in $\boldsymbol{k}$-space [see Eq. (2.148)] and introduce a phenomenological term $J\,\tau_z$ in it:

$$\hat{H}_{\boldsymbol{k}} = \begin{bmatrix} \hat{h}_{\boldsymbol{k}} - J\,\tau_z & 0 \\ 0 & \hat{h}^*_{-\boldsymbol{k}} + J\,\tau_z \end{bmatrix}, \tag{3.14}$$

or, explicitly,

$$\hat{H}_{\boldsymbol{k}} = \begin{bmatrix} \epsilon_{\boldsymbol{k}}\tau_0 & 0 \\ 0 & \epsilon_{\boldsymbol{k}}\tau_0 \end{bmatrix}$$

$$+ \begin{bmatrix} d^x_{\boldsymbol{k}}\tau_x + d^y_{\boldsymbol{k}}\tau_y + (d^z_{\boldsymbol{k}} - J\,)\tau_z & 0 \\ 0 & -d^x_{\boldsymbol{k}}\tau_x + d^y_{\boldsymbol{k}}\tau_y + (d^z_{\boldsymbol{k}} + J\,)\tau_z \end{bmatrix}, \tag{3.15}$$

Using spin matrices σ_z and σ_0, we cast the Hamiltonian into a compact form

$$\hat{H}_{\boldsymbol{k}} = \epsilon_{\boldsymbol{k}}\sigma_0\tau_0 + d^x_{\boldsymbol{k}}\sigma_z\tau_x + d^y_{\boldsymbol{k}}\sigma_0\tau_y + (d^z_{\boldsymbol{k}}\sigma_0 - J\,\sigma_z)\tau_z \tag{3.16}$$

$$= \epsilon_{\boldsymbol{k}}\sigma_0\tau_0 + \mathcal{A}\sin k_x\sigma_z\tau_x - \mathcal{A}\sin k_y\sigma_0\tau_y + (\mathcal{M}_{\boldsymbol{k}}\sigma_0 - J\,\sigma_z)\tau_z. \tag{3.17}$$

The generalized Hamiltonian has a simple meaning. The usual gap term $\mathcal{M}_{\boldsymbol{k}}\sigma_0$ acquires the spin-dependent contribution $-J\,\sigma_z$, which breaks the TRS. That is, similar to the Dirac model above, there is a competition between TR-symmetric and TR-breaking mechanisms of band gap generation. Furthermore, the TR-breaking term $-J\,\sigma_z$ acts akin to spin magnetization.

In practice, it suffices to expand Eq. (3.17) up to quadratic terms in $\boldsymbol{k}$:

$$\hat{H}_{\boldsymbol{k}} \approx (\mathcal{C} + \mathcal{D}\boldsymbol{k}^2)\sigma_0\tau_0 + \mathcal{A}k_x\sigma_z\tau_x - \mathcal{A}k_y\sigma_0\tau_y + [(\Delta + \mathcal{B}\boldsymbol{k}^2)\sigma_0 - J\,\sigma_z]\tau_z. \tag{3.18}$$

The resulting Hamiltonian captures the essential features of the generalized BHZ model for magnetically doped zinc blende semiconductors, e.g. $Hg_{1-y}Mn_yTe$ quantum wells [22].

We again seek the connection between the BHZ Hamiltonian and the Chern insulator from Section 2.1. To that end, we write Eq. (3.18) as

$$\hat{H}_{\boldsymbol{k}} = \epsilon_{\boldsymbol{k}}\sigma_0\tau_0 + \begin{bmatrix} \mathcal{A}k_x\tau_x - \mathcal{A}k_y\tau_y + (\mathcal{M}_{\boldsymbol{k}} - J)\tau_z & 0 \\ 0 & -\mathcal{A}k_x\tau_x - \mathcal{A}k_y\tau_y + (\mathcal{M}_{\boldsymbol{k}} + J)\tau_z \end{bmatrix}, \tag{3.19}$$

and make the unitary transformation

$$\hat{H}_{\boldsymbol{k}} \to \hat{U}\hat{H}_{\boldsymbol{k}}\hat{U}^{\dagger}, \qquad \hat{U} = \begin{bmatrix} 0 & \tau_z \\ -i\tau_y & 0 \end{bmatrix}.$$

After the matrix multiplication, we arrive at

$$\hat{H}_{\boldsymbol{k}} = \epsilon_{\boldsymbol{k}}\sigma_0\tau_0 + \begin{bmatrix} \mathcal{A}k_x\tau_x + \mathcal{A}k_y\tau_y + (J + \mathcal{M}_{\boldsymbol{k}})\tau_z & 0 \\ 0 & -\mathcal{A}k_x\tau_x - \mathcal{A}k_y\tau_y + (J - \mathcal{M}_{\boldsymbol{k}})\tau_z \end{bmatrix}. \tag{3.20}$$

Another unitary transformation

$$\hat{H}_{\boldsymbol{k}} \to \hat{U}\hat{H}_{\boldsymbol{k}}\hat{U}^{\dagger}, \qquad \hat{U} = \begin{bmatrix} \tau_0 & 0 \\ 0 & \tau_z \end{bmatrix},$$

brings the Hamiltonian to the final form

$$\hat{H}_{\boldsymbol{k}} = \epsilon_{\boldsymbol{k}}\sigma_0\tau_0 + \begin{bmatrix} \boldsymbol{d}_{\boldsymbol{k}}^{+}\cdot\boldsymbol{\tau} & 0 \\ 0 & \boldsymbol{d}_{\boldsymbol{k}}^{-}\cdot\boldsymbol{\tau} \end{bmatrix},$$
$$\boldsymbol{d}_{\boldsymbol{k}}^{\pm} = [\mathcal{A}k_x, \mathcal{A}k_y, J \pm \mathcal{M}_{\boldsymbol{k}}]. \tag{3.21}$$

The two Chern species are again specified by vectors $\boldsymbol{d}_{\boldsymbol{k}}^{\pm}$ and have different gap terms $J \pm \mathcal{M}_{\boldsymbol{k}}$. We see the one-to-one correspondence between the pair J and $\mathcal{M}_{\boldsymbol{k}}$ and parameters m and Δ of the Dirac model in Section 3.1.

3.2.2 Phase Classification by TKNN Invariant

For the sake of comparison with the Dirac model, here we examine TKNN invariant (2.160) for the generalized BHZ model with Hamiltonian (3.21). The difference to the Dirac model lies in the function

$$u_{z\nu}(k) = \frac{J + \nu\mathcal{M}_{k}}{\sqrt{\mathcal{A}^2\boldsymbol{k}^2 + (J + \nu\mathcal{M}_{k})^2}}. \tag{3.22}$$

Otherwise, the calculation is analogous to that in Section 3.1.2:

$$\begin{aligned} n &= \sum_{\nu=\pm1} \frac{u_{z\nu}(\infty) - u_{z\nu}(0)}{2} \\ &= \sum_{\nu=\pm1} \frac{\mathrm{sgn}(\nu\mathcal{B}) - \mathrm{sgn}(J + \nu\Delta)}{2} \\ &= \frac{1}{2}\left[\mathrm{sgn}(\Delta - J) - \mathrm{sgn}(\Delta + J)\right]. \end{aligned} \tag{3.23}$$

We again see that, up to the notations, Eq. (3.23) coincides with Haldane's result [6]. As long as the $\boldsymbol{k}$ integral is done over the whole plane, we obtain the same n for both the Dirac and BHZ models. That is, the quadratic $\boldsymbol{k}$ terms in Hamiltonian (3.21) do not affect TKNN number (3.7) and the phase diagram in Fig. 3.1. Still, BHZ Hamiltonian (3.21) displays such important properties as the intrinsic band inversion and particle-hole asymmetry which need to be accounted for in zinc blende QAHIs.

3.2.3 Phase Classification by Edge States

Qualitatively, the edge excitations of BHZ model (3.21) behave similar to those of the Dirac model above. However, the details are sensitive to the intrinsic band inversion and particle-hole asymmetry, and are worth a separate discussion.

We begin with the BHZ Hamiltonian in representation (3.18). Since the z-component of the spin is conserved, we can replace Pauli spin matrix σ_z with its eigenvalue $\sigma = \pm1$. The problem can then be mapped to the Chern model studied in Section 2.3.2. We just need to make the replacements

$$\partial_x \mapsto \sigma\partial_x, \qquad \Delta \mapsto \Delta - \sigma J \tag{3.24}$$

in Eq. (2.132) and proceed as before. In this way, we arrive at the following solution for the edge spectrum:

$$E_{k_y\sigma} = \mathcal{C} + \mathcal{D}k_y^2 + |\mathcal{A}|\text{sgn}(\mathcal{B})k_y\sigma, \tag{3.25}$$

and decay constants

$$\varkappa_{1,2} = \left|\frac{\mathcal{A}}{2\mathcal{B}}\right| \pm \sqrt{\left(\frac{\mathcal{A}}{2\mathcal{B}}\right)^2 + \frac{\Delta - \sigma J}{\mathcal{B}} + k_y^2}. \tag{3.26}$$

The edge modes exist only for the inverted band structure satisfying the criterion

$$(\Delta - \sigma J)\mathcal{B} < 0, \tag{3.27}$$

which involves *spin-dependent* band gap $\Delta - \sigma J$.

Let us analyze these results for a material with

$$\Delta\mathcal{B} < 0. \tag{3.28}$$

In this case, condition (3.27) holds, starting from $J = 0$. That is, the initial state is a QSHI with a Kramers pair of edge modes. The criterion for their existence implies $\text{sgn}(\mathcal{B}) = -\text{sgn}(\Delta)$, so edge spectrum (3.25) can be written as

$$E_{k_y\sigma} = \mathcal{C} + \mathcal{D}k_y^2 - |\mathcal{A}|\text{sgn}(\Delta)k_y\sigma, \tag{3.29}$$

With increasing J, the gap $\Delta - J$ for $\sigma = +1$ decreases, while the gap $\Delta + J$ for $\sigma = -1$ grows. For $J > \Delta$, the gap $\Delta - J$ flips its sign, so criterion (3.27) no longer holds for $\sigma = +1$. Consequently, the system loses its spin-$\uparrow$ edge mode, while the spin-$\downarrow$ one persists. This is the topological transition from the QSHI to a QAHI with a single chiral edge channel. Its energy spectrum can be written as

$$E_{k_y\downarrow} = \mathcal{C} + \mathcal{D}k_y^2 + |\mathcal{A}|\text{sgn}(J + \Delta)k_y, \tag{3.30}$$

where we used the existence condition $\text{sgn}(\mathcal{B}) = -\text{sgn}(J + \Delta)$ for the edge mode with spin $\sigma = -1$. While this model is more advanced, the scenario for the emergence of the QAHI is qualitatively the same as in the Dirac model [cf. Section 3.1.3].

Another scenario for the QAHI can be realized in materials with

$$\Delta\mathcal{B} > 0. \tag{3.31}$$

In this case, condition (3.27) does not hold for $J = 0$. That is, the initial state is an ordinary insulator with the normal band ordering.

When J increases and exceeds Δ, the gap $\Delta - J$ for $\sigma = +1$ again flips its sign, but now the spin-$\uparrow$ subsystem acquires an inverted band structure satisfying criterion (3.27), while the spin-$\downarrow$ subsystem remains ordinary. In this scenario, an ordinary insulator turns into a QAHI with the chiral edge mode

$$E_{k_y\uparrow} = \mathcal{C} + \mathcal{D}k_y^2 + |\mathcal{A}|\mathrm{sgn}(J - \Delta)k_y, \tag{3.32}$$

where we used the existence condition $\mathrm{sgn}(\mathcal{B}) = -\mathrm{sgn}(\Delta - J)$ for the edge state with spin $\sigma = +1$.

To conclude these two sections, we have seen that the concept of the intrinsic band inversion is also central to understanding QAHIs. We need a sufficiently strong intrinsic spin magnetization to flip the band gap for one of the spin species. This drives a topological transition to a QAHI regardless of the initial state of the material, which can be either a QSHI or an ordinary insulator.

3.3 Quantum Spin Hall Insulators under an Orbital Magnetic Field Effect

3.3.1 Boundary Problem. Green's Function Method

We adopt the Dirac model for a thin-film insulator introduced in Section 2.2.1, but now in the presence of a perpendicular magnetic field $\boldsymbol{B}$, as shown in Fig. 3.3. The gauge of vector potential $\boldsymbol{A}(\boldsymbol{r})$ is chosen to be

$$\boldsymbol{A}(\boldsymbol{r}) = (-By, 0, 0). \tag{3.33}$$

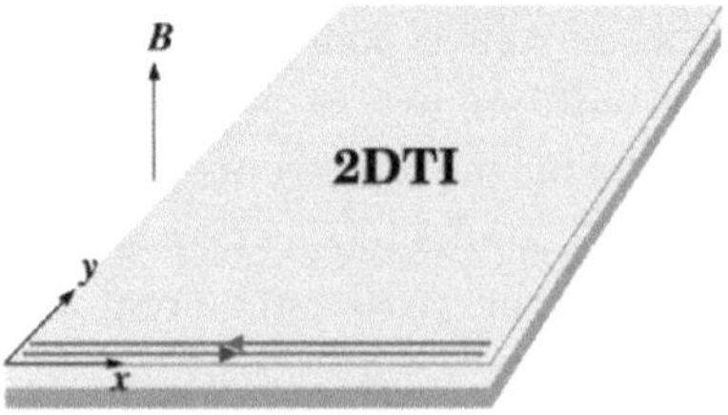

Figure 3.3 Geometry of a thin-film 2DTI with helical edge states at $y = 0$ in a perpendicular magnetic field $\boldsymbol{B}$.

The boundary problem that we intend to solve is a generalization of that considered in Chapter 2 [see also Eqs. (2.69), (2.74) and (2.205)],

$$\begin{bmatrix} E\sigma_0 - v\boldsymbol{\sigma}\cdot\boldsymbol{P} - \Delta\sigma_z & 0 \\ 0 & E\sigma_0 + v\boldsymbol{\sigma}\cdot\boldsymbol{P} + \Delta\sigma_z \end{bmatrix} \hat{G}(\boldsymbol{r}, \boldsymbol{r}') = \delta(\boldsymbol{r} - \boldsymbol{r}'), \tag{3.34}$$

$$\boldsymbol{P} = -i\hbar\nabla - \frac{e}{c}\boldsymbol{A}(\boldsymbol{r}), \tag{3.35}$$

$$\hat{G}(\boldsymbol{r}, \boldsymbol{r}')|_{y=0} = \begin{bmatrix} \sigma_x & 0 \\ 0 & \sigma_x \end{bmatrix} \hat{G}(\boldsymbol{r}, \boldsymbol{r}')|_{y=0}. \tag{3.36}$$

Instead of the spinor wave function, here we work with matrix Green's function $\hat{G}(\boldsymbol{r}, \boldsymbol{r}')$. This has an extra goal to acquaint the reader with the Green's function method, which will be needed in the subsequent chapters. Concretely, we consider the retarded Green function with a positive infinitesimally small shift of the energy, i.e. $E \to E + i0$.

Unlike the equation for the wave function, Eq. (3.34) contains a delta-functional source term, $\delta(\boldsymbol{r} - \boldsymbol{r}')$, on the right-hand side. The Green function $\hat{G}(\boldsymbol{r}, \boldsymbol{r}')$ is a 4×4 block-diagonal matrix of the form

$$\hat{G}(\boldsymbol{r}, \boldsymbol{r}') = \begin{bmatrix} \hat{G}_+(\boldsymbol{r}, \boldsymbol{r}') & 0 \\ 0 & \hat{G}_-(\boldsymbol{r}, \boldsymbol{r}') \end{bmatrix}, \tag{3.37}$$

where $\hat{G}_+(\boldsymbol{r}, \boldsymbol{r}')$ is a 2×2 matrix in the basis of spin states $s \uparrow$ and $a \downarrow$ [cf. upper two spinor components in Eq. (2.73)], whereas $\hat{G}_-(\boldsymbol{r}, \boldsymbol{r}')$ is a 2×2 matrix in the basis of spin states $a \uparrow$ and $d \downarrow$ [cf. lower two spinor components in Eq. (2.73)]. Using again the block index $\tau = \pm$ (or $\tau = \pm 1$ as a variable in algebraic expressions) we can write $\hat{G}_\pm(\boldsymbol{r}, \boldsymbol{r}')$ as

$$\hat{G}_\tau(\boldsymbol{r}, \boldsymbol{r}') = \begin{bmatrix} G_{\uparrow\uparrow,\tau}(\boldsymbol{r}, \boldsymbol{r}') & G_{\uparrow\downarrow,\tau}(\boldsymbol{r}, \boldsymbol{r}') \\ G_{\downarrow\uparrow,\tau}(\boldsymbol{r}, \boldsymbol{r}') & G_{\downarrow\downarrow,\tau}(\boldsymbol{r}, \boldsymbol{r}') \end{bmatrix}. \tag{3.38}$$

Writing Eq. (3.34) in components, it is easy to express the off-diagonal entries $G_{\uparrow\downarrow,\tau}$ and $G_{\downarrow\uparrow,\tau}$ in terms of the diagonal ones as

$$G_{\uparrow\downarrow,\tau}(\boldsymbol{r}, \boldsymbol{r}') = \frac{v(P_x - iP_y)}{\tau E - \Delta} G_{\downarrow\downarrow,\tau}(\boldsymbol{r}, \boldsymbol{r}'), \tag{3.39}$$

$$G_{\downarrow\uparrow,\tau}(\mathbf{r},\mathbf{r}') = \frac{v(P_x + iP_y)}{\tau E + \Delta} G_{\uparrow\uparrow,\tau}(\mathbf{r},\mathbf{r}'), \tag{3.40}$$

where $G_{\uparrow\uparrow,\tau}$ and $G_{\downarrow\downarrow,\tau}$ obey the equations

$$[E^2 - \Delta^2 - v^2(P_x - iP_y)(P_x + iP_y)]G_{\uparrow\uparrow,\tau} = (E + \Delta\tau)\delta(\mathbf{r} - \mathbf{r}'), \tag{3.41}$$

$$[E^2 - \Delta^2 - v^2(P_x + iP_y)(P_x - iP_y)]G_{\downarrow\downarrow,\tau} = (E - \Delta\tau)\delta(\mathbf{r} - \mathbf{r}'). \tag{3.42}$$

For a nonzero magnetic field the momentum operators P_x and P_y do not commute (see end-of-chapter problems). Taking this into account and making the plane-wave expansion,

$$\hat{G}_\tau(\mathbf{r},\mathbf{r}') = \sum_k \hat{G}_{\tau,k}(y,y')\, e^{ik(x-x')}/L, \tag{3.43}$$

where L is the system length in the x-direction, we reduce Eqs. (3.41) and (3.42) to a pair of ordinary differential equations for $G_{\uparrow\uparrow,\tau,k}(y,y')$ and $G_{\downarrow\downarrow,\tau,k}(y,y')$. It is convenient to write them in the dimensionless form

$$\left[\partial_z^2 - \frac{(z-q)^2}{4} - a\right] G_{\uparrow\uparrow,\tau,k} = \frac{(E + \Delta\tau)\ell}{(\hbar v)^2}\delta(z - z'), \tag{3.44}$$

with

$$a = \frac{(\Delta^2 - E^2)\ell^2}{(\hbar v)^2} - \frac{\mathrm{sgn}(eB)}{2}, \tag{3.45}$$

and

$$\left[\partial_z^2 - \frac{(z-q)^2}{4} - \bar{a}\right] G_{\downarrow\downarrow,\tau,k} = \frac{(E - \Delta\tau)\ell}{(\hbar v)^2}\delta(z - z'), \tag{3.46}$$

with

$$\bar{a} = \frac{(\Delta^2 - E^2)\ell^2}{(\hbar v)^2} + \frac{\mathrm{sgn}(eB)}{2}. \tag{3.47}$$

In Eqs. (3.44) and (3.46),

$$z = \frac{y}{\ell} \tag{3.48}$$

is the dimensionless position measured in units of the magnetic length,

$$\ell = \sqrt{\frac{c\hbar}{2|eB|}}, \tag{3.49}$$

and

$$q = -2k\ell\,\mathrm{sgn}(eB) \tag{3.50}$$

is the dimensionless wave vector.

The next step is to obtain the boundary conditions to Eqs. (3.44) and (3.46) from Eq. (3.36). The latter couples the off-diagonal and diagonal elements of matrix $\hat{G}_{\tau,k}$:

$$\begin{aligned} G_{\downarrow\uparrow,\tau,k}(0,z') &= G_{\uparrow\uparrow,\tau,k}(0,z'),\\ G_{\uparrow\downarrow,\tau,k}(0,z') &= G_{\downarrow\downarrow,\tau,k}(0,z'). \end{aligned} \tag{3.51}$$

Eliminating the off-diagonal matrix elements with thc help of Eqs. (3.39) and (3.40), we find the required boundary conditions for the diagonal entries:

$$\partial_z G_{\uparrow\uparrow,\tau,k}(z,z')\big|_{z=0} = \kappa\, G_{\uparrow\uparrow,\tau,k}(0,z'), \tag{3.52}$$

with

$$\kappa = \frac{(\tau E+\Delta)\ell}{\hbar v} + \frac{q\,\mathrm{sgn}(eB)}{2}, \tag{3.53}$$

and

$$\partial_z G_{\downarrow\downarrow,\tau,k}(z,z')\big|_{z=0} = \overline{\kappa}\, G_{\downarrow\downarrow,\tau,k}(0,z'), \tag{3.54}$$

with

$$\overline{\kappa} = \frac{(-\tau E+\Delta)\ell}{\hbar v} - \frac{q\,\mathrm{sgn}(eB)}{2}. \tag{3.55}$$

3.3.2 Solution in Terms of Parabolic Cylinder Functions

The boundary problem formulated above can be solved exactly in terms of the parabolic cylinder functions. It is an interesting example of such higher transcendental functions finding application in the physics of topological insulators. From this point of view, it is instructive to present here the details of the calculations. We start with Eq. (3.44) and boundary condition (3.52). For simplicity, the Green function indices $\uparrow$, τ, k are suppressed below.

The general solution of Eq. (3.44) has the form of a linear combination:

$$G(z,z') = A(z')U(a,z-q) + B(z')U(a,-z+q) \tag{3.56}$$
$$+G^{\infty}(z,z'). \tag{3.57}$$

The sum of the first and second terms is the general solution of Eq. (3.44) with the zero right-hand side, where $U(a, z-q)$ and $U(a, -z+q)$ are the parabolic cylinder functions representing two linearly independent (fundamental) solutions of the equation [88],

$$\left[\partial_z^2 - \frac{(z-q)^2}{4} - a\right] U = 0. \tag{3.58}$$

The coefficients $A(z')$ and $B(z')$ should be found from the boundary conditions at $z \to +\infty$ and $z = 0$. The function $U(a, -z+q)$ is unbounded for $z \to +\infty$, whereas $U(a, z-q)$ vanishes in this limit [88]. Therefore, $B(z')$ in Eq. (3.56) must be set to zero

$$B(z') = 0. \tag{3.59}$$

As long as the last term in Eq. (3.57) is specified, the remaining coefficient, $A(z')$, can be found from boundary condition (3.52).

The last term in Eq. (3.57) is a particular solution of the inhomogeneous Eq. (3.44), i.e. the solution generated by the nonzero right-hand side. Such a particular solution can always be chosen in the form of the Green function of the infinite system. The latter can be constructed from the two linearly independent solutions, $U(a, z-q)$ and $U(a, -z+q)$, according to the standard recipe:

$$G^{\infty}(z, z') = C\,[\,\Theta(z-z')\,U(a, z-q)U(a, -z'+q) \tag{3.60}$$

$$+\,\Theta(z'-z)\,U(a, z'-q)U(a, -z+q)\,], \tag{3.61}$$

$$C = \frac{(E+\Delta\tau)\ell}{\hbar^2 v^2 W}, \tag{3.62}$$

$$W = \partial_z U(a, \bar{z}) \cdot U(a, -\bar{z}) - U(a, \bar{z}) \cdot \partial_z U(a, -\bar{z}) \tag{3.63}$$

$$= -\frac{\sqrt{2\pi}}{\Gamma(a+1/2)}, \qquad \bar{z} = z - q, \tag{3.64}$$

where $\Theta(z)$ is the unit-step function, W is the Wronskian of the parabolic cylinder functions [88], and $\Gamma(a+1/2)$ is Euler's gamma function. We note that the constructed solution $G^{\infty}(z, z')$ vanishes for $z \to \pm\infty$.

Inserting Eq. (3.56) into boundary condition (3.52), we find the coefficient $A(z')$ as

$$A(z') = -C\,\frac{\partial_q U(a, q) + \kappa U(a, q)}{\partial_q U(a, -q) + \kappa U(a, -q)}\,U(a, z'-q). \tag{3.65}$$

Inserting this back into Eq. (3.56), we obtain the explicit solution of our boundary problem,

$$G(z, z') = G^{\infty}(z, z') - C\frac{\partial_q U(a, q) + \kappa U(a, q)}{\partial_q U(a, -q) + \kappa U(a, -q)} \tag{3.66}$$

$$\times\, U(a, z - q)U(a, z' - q). \tag{3.67}$$

The helical edge states correspond to the poles of the second term, i.e. the edge-state spectrum is obtained from the equation:

$$\partial_q U[a(E), -q] + \kappa(E)U[a(E), -q] = 0, \tag{3.68}$$

in which $a(E)$ and $\kappa(E)$ are the known functions of energy [see Eqs. (3.45) and (3.53)].

For our further discussion we only need the asymptotic formulas with energy E close to the zeros of Eq. (3.68). For that purpose, we rewrite Eqs. (3.66) and (3.67) in a slightly different form,

$$G(z, z') = G^{\infty}(z, z') \tag{3.69}$$

$$-C\frac{U(a, q)}{U(a, -q)}U(a, z - q)U(a, z' - q) \tag{3.70}$$

$$-CW\frac{U(a, z - q)U(a, z' - q)}{U(a, -q)[\partial_q U(a, -q) + \kappa U(a, -q)]}, \tag{3.71}$$

which near the poles of the last (edge) term reduces to

$$G(z, z') \approx -CW\frac{U(a, z - q)U(a, z' - q)}{U(a, -q)[\partial_q U(a, -q) + \kappa U(a, -q)]}. \tag{3.72}$$

Next, we use the recurrence relation for the parabolic cylinder functions [88],

$$\partial_q U(a, -q) = -\frac{q}{2}U(a, -q) + \left(a + \frac{1}{2}\right)U(a + 1, -q), \tag{3.73}$$

to eliminate the derivative in the denominator of Eq. (3.72). After simple algebra involving Eqs. (3.45), (3.53) and (3.63) and assuming for concreteness $eB > 0$, we arrive at the final result:

$$G_{\uparrow\uparrow,\tau,k}(z, z') \approx \frac{U(a, z - q)U(a, z' - q)}{\ell U(a, -q)U(a + 1, -q)} \tag{3.74}$$

$$\times \frac{1}{E - \Delta\tau - (\hbar v/\ell)U(a, -q)/U(a + 1, -q)}, \tag{3.75}$$

from which we obtain the equation for the edge-state spectrum as

$$E - \Delta\tau - \frac{\hbar v}{\ell} \frac{U\left[a(E), -q\right]}{U\left[a(E)+1, -q\right]} = 0. \tag{3.76}$$

In Eq. (3.75) we re-installed the indices of the Green function to distinguish it from the other diagonal Green function, $G_{\downarrow\downarrow,\tau,k}(z, z')$. It can be obtained from Eqs. (3.46) and (3.54) in a fully analogous way. Here we quote the final result

$$G_{\downarrow\downarrow,\tau,k}(z, z') \approx \frac{U(a+1, z-q)U(a+1, z'-q)U(a, -q)}{\ell\, U^3(a+1, -q)} \tag{3.77}$$

$$\times \frac{1}{E - \Delta\tau - (\hbar v/\ell)U(a, -q)/U(a+1, -q)}, \tag{3.78}$$

These results are valid for all values of the parameter $\hbar v/\ell|\Delta|$ that quantifies the relative strength of the magnetic field, B. In particular, the limits

$$\frac{\hbar v}{\ell|\Delta|} \ll 1, \tag{3.79}$$

and

$$\frac{\hbar v}{\ell|\Delta|} \gg 1, \tag{3.80}$$

correspond to the *weak*- and *strong*-field regimes, respectively. In the weak-field regime the results obtained here recover the edge-state solution discussed in Chapter 2. Below we focus on the helical edge states in strong quantizing fields.

3.3.3 Nonlinear Helical Spectrum

In order to obtain the edge-state spectrum we solve numerically Eq. (3.76) for energy E as function of dimensionless momentum q for both $\tau = \pm 1$. The numerical solutions $E_+(q)$ and $E_-(q)$ are shown in Fig. 3.4. Again, the spectrum depends crucially on the sign of the gap parameter Δ. The system with the normal band structure ($\Delta > 0$) remains an ordinary band insulator at subgap energies $|E| < \Delta$, as shown in Fig. 3.4a. Outside the band gap the conduction- and valence-band states are quantized into discrete edge-state branches evolving into flat bulk Landau levels with increasing $q > 0$. Since

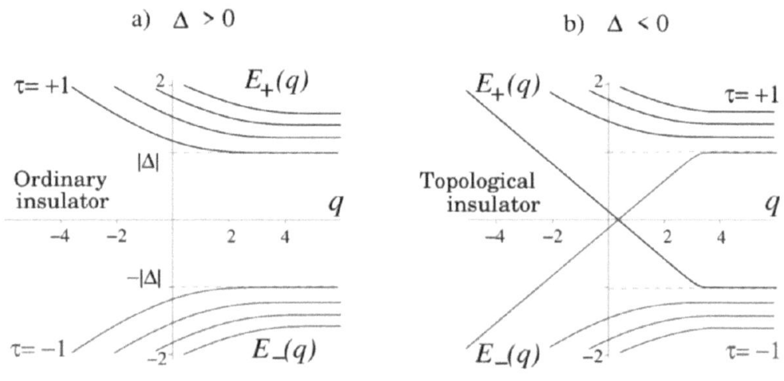

Figure 3.4 Edge-state energies $E_+(q)$ and $E_-(q)$ as function of dimensionless wave vector, q [see Eqs. (3.50) and (3.76)] for (a) normal band structure with $\Delta > 0$ and (b) inverted band structure with $\Delta < 0$; $\hbar v/\ell|\Delta| = 0.5$ and $eB > 0$. Panel b shows a 2DTI state with two helical (counter-propagating) modes crossing inside the gap. At large $q \gg 1$ helical edge states evolve into flat bulk Landau levels $\pm|\Delta|$.

these states are chiral (i.e. propagating in the same direction), the TRS is broken.

The results for the inverted band structure with $\Delta < 0$ are qualitatively different (see Fig. 3.4b). Apart from the quantized conduction- and valence-band states, there are two counter-propagating edge modes crossing in the middle of the band gap. These are the high-field counterparts of the helical edge states studied in Chapter 2. Thus, we conclude that, despite the broken TRS in the conduction and valence bands, the 2DTI state persists in strong fields. This agrees with the qualitative picture of the helical states as robust domain wall states.

In strong-field regime (3.80) an analytical formula for the edge-state spectrum can be derived. Using the asymptotic expressions (see Ref. [88])

$$U(a, z) \approx U(-1/2, z) = e^{-z^2/4}, \qquad (3.81)$$

$$U(a+1, z) \approx U(1/2, z) = e^{z^2/4}\sqrt{\pi/2}\,\mathrm{erfc}(z/\sqrt{2}), \qquad (3.82)$$

we find from Eq. (3.76)

$$E_\tau(q) = \tau\Delta + \tau\sqrt{\frac{2}{\pi}}\frac{\hbar v}{\ell}\frac{e^{-q^2/2}}{\mathrm{erfc}(-q/\sqrt{2})}, \qquad \tau = \pm 1, \qquad (3.83)$$

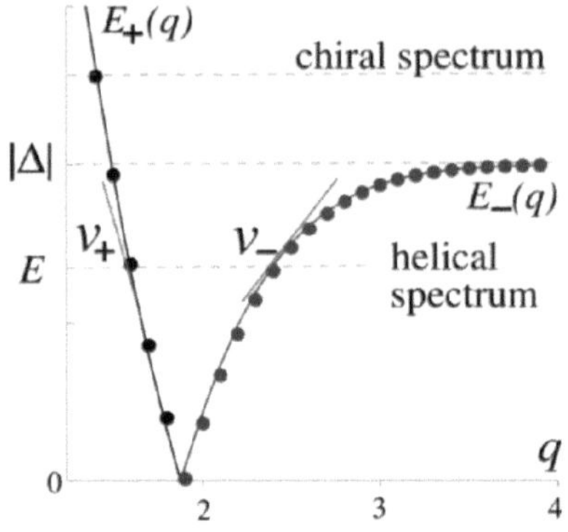

Figure 3.5 Edge-state energies $E_+(q)$ and $E_-(q)$ as function of dimensionless wave vector, q, for a strong magnetic field with $\hbar v/\ell|\Delta| = 10$. Points indicate numerical solution of Eq. (3.76), whereas solid lines are the analytical result, Eq. (3.83). Note that $E_+(q)$ and $E_-(q)$ are nonlinear and have different group velocities v_+ and v_-. The helical spectrum turns chiral at the band gap energy $E = |\Delta|$.

where erfc(z) is the complementary error function. Since it is positive for any q, the zero-energy states emerge only for $\Delta < 0$. The analytical spectrum (3.83) is plotted, along with the corresponding numerical solution of Eq. (3.76), in Fig. 3.5.

The most striking feature of the high-field spectrum (3.83) is its *nonlinearity*. In the zero field [see Eq. (2.101)] the edge-state energy is linear in wave vector, k, so that the counter-propagating states with $\tau = \pm 1$ have equal velocities, v. As shown in Fig. 3.5, in the strong-field regime the edge-state velocities are no longer equal. Moreover, at energy E close to the gap $|\Delta|$ the $\tau = -$ state merges with the bulk Landau level and has zero velocity $v_- \to 0$. Above the gap there is only one propagating edge state with $\tau = +$. Therefore, upon crossing the gap energy the spectrum changes from helical to chiral, and the 2DTI state transforms into a dissipationless quantum Hall state.

3.3.4 Spatial Distribution of Edge States

Apart from the spectral nonlinearity, the strong magnetic field modifies the spatial distribution of the helical edge states. In the zero field case [see Eqs. (2.102) and (2.103)] the maximum of the edge-state functions is exactly at the boundary $y = 0$. This is no longer the case in the strong-field regime. The complete information on the spatial dependence of the edge states is contained in the matrix Green function (3.38). Inserting Eqs. (3.39), (3.40), (3.74) and (3.77) into Eq. (3.38) and eliminating the derivatives with the help of the recurrence relations for $U(a, z)$ [88], we find the full Green function matrix:

$$\hat{G}_{\tau k}(z, z') = \begin{bmatrix} G_{\uparrow\uparrow,\tau k}(z, z') & G_{\uparrow\downarrow,\tau k}(z, z') \\ G_{\downarrow\uparrow,\tau k}(z, z') & G_{\downarrow\downarrow,\tau}(z, z') \end{bmatrix} \tag{3.84}$$

$$= \frac{\alpha(z, z') \begin{pmatrix} 1 & \beta(z') \\ \beta(z) & \beta(z)\beta(z') \end{pmatrix}}{E - \Delta\tau - (\hbar v/\ell)U(a, -q)/U(a+1, -q)}, \tag{3.85}$$

where functions $\alpha(z, z')$ and $\beta(z)$ are given by

$$\alpha(z, z') = \frac{U(a, z-q)U(a, z'-q)}{\ell U(a, -q)U(a+1, -q)}, \tag{3.86}$$

$$\beta(z) = \frac{U(a, -q)U(a+1, z-q)}{U(a+1, -q)U(a, z-q)}, \tag{3.87}$$

and parameters a and q are defined in Eqs. (3.45) and (3.50).

Figure 3.6 shows typical spatial dependences of the functions $G_{\uparrow\uparrow,\tau k}(z, z) \propto \alpha(z, z)$, $G_{\uparrow\downarrow,\tau k}(z, z) \propto \alpha(z, z)\beta(z)$ and $G_{\downarrow\downarrow,\tau k}(z, z) \propto$

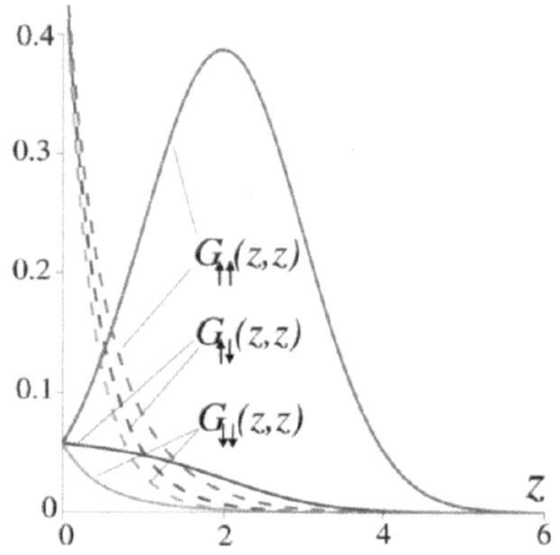

Figure 3.6 Typical spatial distributions of edge states in strong-field (solid curves) and weak-field (dashed curves) regimes defined by Eqs. (3.80) and (3.79), respectively; $E = 0$ and for $\Delta < 0$.

$\alpha(z, z)\beta^2(z)$ at zero energy $E = 0$. Interestingly, the $\uparrow$ component is no longer localized at the edge of the system ($z = 0$). Instead, it behaves almost like the lowest-Landau-level bulk wave function with the center at $z \sim q$. The other functions are small at $z \sim q$. Using Eqs. (3.81) and (3.82) and, consequently, $\beta \ll 1$, we obtain the strong-field asymptotic of Eq. (3.85) as

$$\hat{G}_{\tau k}(z, z') \approx \frac{\sigma_0 + \sigma_z}{2} G_{\tau,k}(z, z'),$$
$$G_{\tau,k}(z, z') \approx \frac{\alpha(z, z')}{E - E_\tau(q)}, \tag{3.88}$$

where edge-state spectrum $E_\tau(q)$ is given by Eq. (3.83), and function $\alpha(z, z')$ [see Eq. (3.86)] reduces to

$$\alpha(z, z') \approx \sqrt{\frac{2}{\pi}} \frac{e^{-(z-q)^2/4-(z'-q)^2/4}}{\ell \operatorname{erfc}(-q/\sqrt{2})}. \tag{3.89}$$

3.4 Backscattering, Conductance and Topological Transitions

Since the helical edge states survive in a strong magnetic field, the results of the preceding chapter for ballistic transport hold also for the finite magnetic fields. For a single ballistic edge coupled to two carrier reservoirs the longitudinal electric conductance is simply

$$g = \frac{e^2}{h}. \tag{3.90}$$

In the following, we discuss new phenomena that occur in the presence of impurities generating edge *backscattering*, whereby one edge state is scattered into the other one with the opposite chirality, as schematically depicted in Fig. 3.7. Such impurities must violate the TRS. In order to construct a theoretical model of the backscattering we first need to introduce the free edge-state propagator in real space.

3.4.1 Free Edge-State Propagator

Since in strong fields the edge-state width shrinks proportionally to the magnetic length ℓ, the free edge-state propagator can be

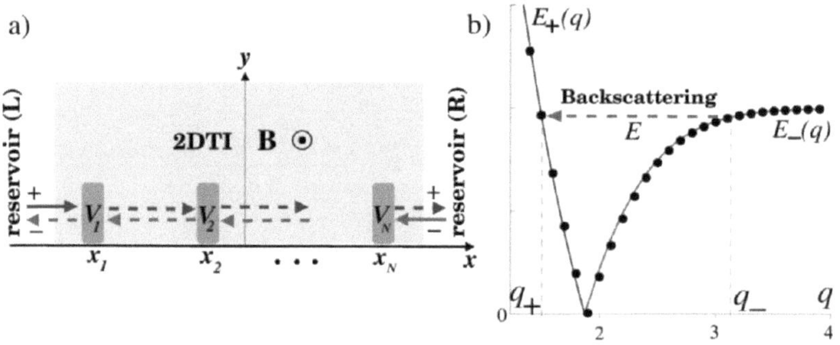

Figure 3.7 (a) 2DTI coupled to two carrier reservoirs and containing backscattering centers in the interior (shaded regions). (b) Illustration of a backscattering process between edge states with wave vectors $q_\pm$ at energy E.

described by the one-dimensional formula,

$$G_\tau(x, x') = \int \frac{dk}{2\pi}\, e^{ik(x-x')}\ell \int_0^\infty dz\, G_{\tau k}(z, z), \tag{3.91}$$

where $G_{\tau k}(z, z)$ is given by Eqs. (3.88) and (3.89). Taking into account that $\ell \int_0^\infty dz\, \alpha(z, z) = 1$ [88], we have from Eq. (3.91)

$$G_\tau(x, x') = \int \frac{dk}{2\pi} \frac{e^{ik(x-x')}}{E - E_\tau(q)}, \tag{3.92}$$

where k and q are related by Eq. (3.50), with eB chosen to be positive. The main contribution to the k-integral is expected to come from the vicinity of the point, $q_\tau(E, B)$, at which energy E crosses dispersion $E_\tau(q)$ (see also Fig. 3.7b),

$$E = E_\tau(q) \quad \to \quad q_\tau(E, B), \quad \tau = \pm. \tag{3.93}$$

We, therefore, linearize the edge dispersion [Eq. (3.83)] near fixed energy E,

$$E_\tau(q) \approx E - \frac{\hbar v_\tau}{2\ell}(q - q_\tau), \tag{3.94}$$

where velocity $v_\tau(E, B)$ is given by

$$v_\tau(E, B) = \frac{2\ell}{\hbar}(\tau|\Delta| + E)\, q_\tau(E, B). \tag{3.95}$$

This formula explicitly shows that $v_-(E, B) \to 0$ as $E \to |\Delta|$, in agreement with Fig. 3.7b. Using the linearized dispersion (3.94) in Eq. (3.92) we finally find

$$G_\tau(x, x') = \Theta([x - x']\tau)\frac{e^{ik_\tau(E,B)(x-x')}}{i\hbar|v_\tau(E, B)|},$$
$$k_\tau(E, B) = -\frac{q_\tau(E, B)}{2\ell}. \tag{3.96}$$

This equation describes a plane wave propagating with wave vector $k_\tau(\epsilon, B)$ and velocity $v_\tau(\epsilon, B)$. The nonzero values of the function $\Theta([x - x']\tau)$ determine the propagation direction, i.e. edge-state chirality. For $\tau = +1$ the coordinate x lies on the right from x', i.e. the state propagates from left to right (*right-mover*), whereas for $\tau = -1$ the coordinate x lies on the left from x', which corresponds to the *left-mover* propagating from right to left.

3.4.2 Edge Scattering Matrix and Fisher–Lee Relation

Now, we consider a disordered system connected to two carrier reservoirs on the left (L) and on the right (R), as shown in Fig. 3.7a. We introduce notations L^+ and R^- for the incoming (from the reservoirs) states, and L^- and R^+ for the states scattered towards the reservoirs (outgoing states). Transport characteristics of the disordered system can be inferred from the scattering matrix, $\hat{S}$, connecting the incoming and outgoing states,

$$\begin{bmatrix} L^- \\ R^+ \end{bmatrix}_{out} = \hat{S}\begin{bmatrix} L^+ \\ R^- \end{bmatrix}_{in}, \quad \hat{S} = \begin{bmatrix} r^{-+}_{LL} & t^{--}_{LR} \\ t^{++}_{RL} & r^{+-}_{RR} \end{bmatrix}, \tag{3.97}$$

where $r^{-+(+-)}_{LL(RR)}$ and $t^{++(--)}_{RL(LR)}$ are the all possible reflection and transmission amplitudes. Without losing generality, we assume that the current is injected into the right-moving (+) channel incoming from the L reservoir. Then, the electric conductance is proportional to the transmission probability, $|t^{++}_{RL}|^2$, from L to R

$$g = \frac{e^2}{h}\left|t^{++}_{RL}\right|^2. \tag{3.98}$$

where e^2/h is the conductance of a perfectly transparent channel. The transmission amplitude can be calculated from the Green function of the disordered system using Fisher–Lee relation [89],

$$t^{++}_{RL} = i\hbar|v_+|\mathcal{G}_{++}(x \in R, x' \in L), \tag{3.99}$$

where the Green function $\mathcal{G}_{++}(x, x')$ is the diagonal element of the Green function matrix

$$\hat{\mathcal{G}}(x, x') = \begin{bmatrix} \mathcal{G}_{++}(x, x') & \mathcal{G}_{+-}(x, x') \\ \mathcal{G}_{-+}(x, x') & \mathcal{G}_{--}(x, x') \end{bmatrix}. \tag{3.100}$$

It is a matrix in space of the right (+)- and left (−)-moving states. Its off-diagonal elements $\mathcal{G}_{+-}(x, x')$ and $\mathcal{G}_{-+}(x, x')$ originate from edge-state backscattering. We model the backscatterers by the sum of N matrix potentials localized at positions x_n ($n = 1 \ldots N$):

$$\hat{V}(x) = \sum\nolimits_{n=1\ldots N} \begin{bmatrix} 0 & V_n\delta(x - x_n) \\ V_n\delta(x - x_n) & 0 \end{bmatrix}. \tag{3.101}$$

It is off-diagonal in right-left mover space, leading to interchannel scattering with the strength quantified by matrix element V_n. We note that potential $\hat{V}(x)$ does not commute with the TRS operation.

3.4.3 Solution of Dyson Equation. Conductance

In the presence of potential $\hat{V}(x)$ (3.101) the edge-state propagator obeys the equation

$$\hat{\mathcal{G}}(x, x') = \hat{G}(x, x') + \sum_{n=1\ldots N} \hat{G}(x, x_n) \begin{bmatrix} 0 & V_n \\ V_n & 0 \end{bmatrix} \hat{\mathcal{G}}(x_n, x'), \tag{3.102}$$

where

$$\hat{G}(x, x') = \begin{bmatrix} G_+(x, x') & 0 \\ 0 & G_-(x, x') \end{bmatrix} \tag{3.103}$$

is the Green function matrix in the absence of scattering [see Eq. (3.96)]. From Eqs. (3.102) and (3.103) we have two coupled equations for the diagonal, $\mathcal{G}_{++}(x, x')$, and off-diagonal, $\mathcal{G}_{-+}(x, x')$, elements:

$$\mathcal{G}_{++}(x, x') = G_+(x, x') + \sum_n G_+(x, x_n)V_n\mathcal{G}_{-+}(x_n, x'), \tag{3.104}$$

$$\mathcal{G}_{-+}(x, x') = \sum_n G_-(x, x_n)V_n\mathcal{G}_{++}(x_n, x'). \tag{3.105}$$

Eliminating $\mathcal{G}_{-+}(x, x')$ yields a closed Dyson equation for $\mathcal{G}_{++}(x, x')$:

$$\mathcal{G}_{++}(x, x') = G_{+}(x, x') + \sum_{n,m} G_{+}(x, x_n) V_n G_{-}(x_n, x_m) V_m \mathcal{G}_{++}(x_m, x'). \quad (3.106)$$

With known function $G_\tau(x, x')$ [see Eq. (3.96)] and for not large N, we can solve Eq. (3.106) analytically and calculate g. Leaving aside this algebra, we present the final results for the particular cases $N =$ 1, 2 and 3:

$$g = \frac{e^2}{h}\left(1 + \frac{V_1^2}{\hbar^2|v_+v_-|}\right)^{-2}, \quad (3.107)$$

$$g = \frac{e^2}{h}\left|1 + \frac{V_1^2 V_2^2}{\hbar^4 v_+^2 v_-^2} + \frac{V_1^2 + V_2^2 + V_1 V_2 e^{iQd_{12}}}{\hbar^2|v_+v_-|}\right|^{-2}, \quad (3.108)$$

$$g = \frac{e^2}{h}\left|1 + \frac{V_1^2 V_2^2 V_3^2}{\hbar^6|v_+v_-|^3} + \frac{\sum_{n=1}^{3} V_n^2 + V_1 V_2 e^{iQd_{12}} + V_1 V_3 e^{iQd_{13}} + V_2 V_3 e^{iQd_{23}}}{\hbar^2|v_+v_-|} + \frac{V_1^2 V_2^2 + V_1^2 V_3^2 + V_2^2 V_3^2 + V_1^2 V_2 V_3 e^{iQd_{23}}}{\hbar^4 v_+^2 v_-^2}\right|^{-2}, \quad (3.109)$$

where $Q = k_+ - k_-$ and $d_{nm} = x_m - x_n$. The wave vectors $k_\pm$ [see Eq. (3.96)] and velocities $v_\pm$ [see Eq. (3.95)] are obtained by solving numerically Eq. (3.93).

3.4.4 Transition from 2DTI to a Quantum Hall State

The interesting result is that all V-terms in Eqs. (3.107) - (3.109) diverge when dispersion $E_-(q)$ becomes flat at the gap energy with vanishing velocity, $v_- \to 0$ (see Fig. 3.7b). The most divergent terms are, clearly, those with the highest power of Vs. Generalizing this observation for arbitrary N, we can write the most divergent term as

$$\frac{V_1^2 \times \ldots \times V_N^2}{\hbar^{2N}|v_+v_-|^N} \quad (3.110)$$

It arises from simultaneous scattering from N potentials. The divergence of Eq. (3.110) means that the edge backscattering is greatly enhanced, leading to the suppressed conductance

$$g(E, B) \approx \frac{e^2}{h} \times \frac{\hbar^{4N}|v_+ v_-|^{2N}}{(V_1 \times \ldots \times V_N)^4} \ll \frac{e^2}{h}, \qquad v_- \to 0. \qquad (3.111)$$

Using Eq. (3.95) for the velocities $v_\pm(E, B)$ we find the qualitative energy and field dependence of the conductance near the band gap:

$$g(E, B) \propto \frac{(|\Delta| - |E|)^{2N}}{B^{2N}}, \qquad |E| \to |\Delta|. \qquad (3.112)$$

It shows a power-law magnetic field dependence B^{-2N}, determined by the number N of backscattering centers on the edge. This suggests a simple way to detect such defects in quasi-ballistic 2DTIs using standard magnetoresistance measurements.

In a wider range of energies and fields, the typical behavior of the conductance can be understood from Eq. (3.108) for two backscattering centers on the edge. Figure 3.8 shows the energy dependence of g at fixed magnetic field $B = 0.5$ T. Independently of disorder strength, the conductance drops to zero as the gap energy $|\Delta| = 1.5$ meV is approached. For relatively weak $V_{1,2}$ (upper curve) the transition is sharp because the initial conductance is close to

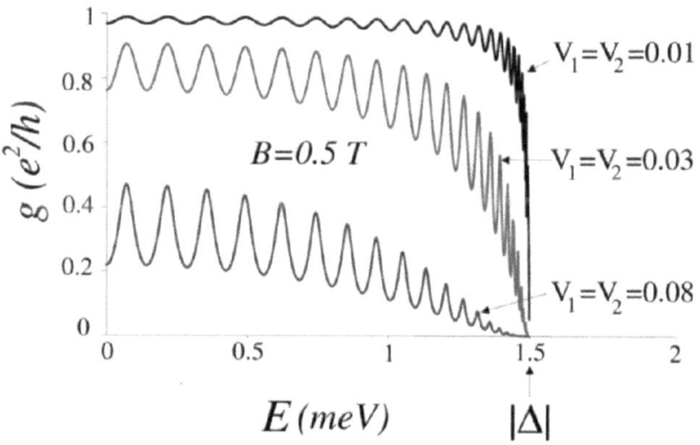

Figure 3.8 Longitudinal conductance (3.108) versus energy for $B = 0.5$ T, $|\Delta| = 1.5$ meV and $d_{12} = 3\mu$m. Disorder potential strengths $V_{1,2}$ are in units of meV$\cdot\mu$m. The exact vanishing of g indicates a topological transition from the 2DTI to a dissipationless quantum Hall state.

the ballistic value e^2/h. An increase in disorder strength $V_{1,2}$ (lower curve) smears the transition.

The exact vanishing of the longitudinal (dissipative) conductance indicates a *topological transition* from the 2DTI to a chiral quantum Hall state. The longitudinal conductance must vanish at $E = |\Delta|$ since higher energies $E > |\Delta|$ fall into a Landau gap in which the quantum Hall state is dissipationless and has only nonzero Hall conductivity. The transition is accompanied by Fabry–Perot-type oscillations due to interference of the counter-propagating channels which acquire the energy- and field-dependent phase difference $(k_+ - k_-)d_{12}$, in scattering between the defects. Summarizing our findings, we emphasize two chief factors needed for observing the topological transition studied above: edge backscattering and strong quantizing magnetic fields.

3.5 Problems

(1) Consider the Hamiltonian

$$\hat{H} = \begin{bmatrix} d_x\sigma_x + d_y\sigma_y + m\sigma_z & d_z\sigma_0 \\ d_z\sigma_0 & -d_x\sigma_x - d_y\sigma_y + m\sigma_z \end{bmatrix}, \tag{3.113}$$

where σ_x, σ_y and σ_z are Pauli matrices, while σ_0 is the 2×2 unit matrix. Apply subsequently the unitary transformations with

$$\hat{U}_1 = \frac{1}{\sqrt{2}} \begin{bmatrix} \sigma_0 & \sigma_z \\ -\sigma_z & \sigma_0 \end{bmatrix} \quad \text{and} \quad \hat{U}_2 = \begin{bmatrix} \sigma_0 & 0 \\ 0 & \sigma_z \end{bmatrix} \tag{3.114}$$

to obtain the block-diagonal Hamiltonian

$$\hat{H} = \begin{bmatrix} d_x\sigma_x + d_y\sigma_y + (m + d_z)\sigma_z & 0 \\ 0 & d_x\sigma_x + d_y\sigma_y + (m - d_z)\sigma_z \end{bmatrix}. \tag{3.115}$$

(2) Consider the Hamiltonian

$$\hat{H} = \begin{bmatrix} d_x\sigma_x - d_y\sigma_y + (d_z - m)\sigma_z & 0 \\ 0 & -d_x\sigma_x - d_y\sigma_y + (d_z + m)\sigma_z \end{bmatrix}. \tag{3.116}$$

Apply subsequently the unitary transformations with

$$\hat{U}_1 = \begin{bmatrix} 0 & \sigma_z \\ -i\sigma_y & 0 \end{bmatrix} \quad \text{and} \quad \hat{U}_2 = \begin{bmatrix} \sigma_0 & 0 \\ 0 & \sigma_z \end{bmatrix} \tag{3.117}$$

to cast the Hamiltonian to the form given by Eq. (3.115).

(3) Prove the commutation relation

$$P_x P_y - P_y P_x = \frac{ie\hbar}{c} B \tag{3.118}$$

for the components of the momentum operator $\boldsymbol{P}$ in the presence of a magnetic field [see Eq. (3.35)].

(4) Use Eq. (3.118) and the plane-wave expansion in Eq. (3.43) to derive the ordinary differential equations

$$\left[\partial_y^2 - \frac{\Delta^2 - E^2}{(\hbar v)^2} + \frac{eB}{c\hbar} - \left(k + \frac{eBy}{c\hbar}\right)^2\right] G_{\uparrow\uparrow,\tau} = \\ = \frac{E + \Delta\tau}{(\hbar v)^2}\delta(y - y'), \tag{3.119}$$

and

$$\left[\partial_y^2 - \frac{\Delta^2 - E^2}{(\hbar v)^2} - \frac{eB}{c\hbar} - \left(k + \frac{eBy}{c\hbar}\right)^2\right] G_{\downarrow\downarrow,\tau} = \\ = \frac{E - \Delta\tau}{(\hbar v)^2}\delta(y - y'). \tag{3.120}$$

Show that for the dimensionless coordinate $z = y/\ell$ and wave vector $q = -2k\ell \operatorname{sgn}(eB)$ [see Eqs. (3.48) and (3.50)] the equations above transform to Eqs. (3.44) and (3.46).

(5) Consider the second order differential equation

$$[\partial_x^2 + u(x)] f(x) = 0, \tag{3.121}$$

with two linearly independent solutions $f_1(x)$ and $f_2(x)$. Prove that for any $u(x)$ the Wronskian

$$W = \partial_x f_1(x) \cdot f_2(x) - f_1(x) \cdot \partial_x f_2(x) \tag{3.122}$$

is independent of x. Let us now introduce the function of two variables

$$G(x, x') = \frac{1}{W}\left[\Theta(x - x')f_1(x)f_2(x') + \Theta(x' - x)f_1(x')f_2(x)\right]. \quad (3.123)$$

Prove that $G(x, x')$ satisfies the equation

$$[\partial_x^2 + u(x)]G(x, x') = \delta(x - x'). \quad (3.124)$$

Equation (3.123) is, therefore, the Green function of a general second order differential equation.

Chapter 4

Three-Dimensional Topological Insulators and Weyl Semimetals

This part of the book builds upon the theory of quantum spin Hall insulators studied in Chapter 2. Quantum spin Hall insulators have higher-dimensional analogues—three-dimensional topological insulators (3DTIs). A 3DTI is a bulk band insulator with conducting gapless states at the surface of the material. To elucidate the origin of 3DTIs, in this chapter we adopt a microscopic model using Kane's $\boldsymbol{k} \cdot \boldsymbol{p}$ Hamiltonian. Its topological properties, notably Dirac-like surface states, deserve particular attention because $\boldsymbol{k} \cdot \boldsymbol{p}$ calculations are ubiquitous in solid-state physics. On par with 3DTIs, another class of 3D quantum materials—Weyl semimetals (WSs)—will be introduced. Starting with a tight-binding model on a cubic lattice, we shall explore rich topological properties of Weyl semimetals such as their band nodes, Chern numbers and surface Fermi arcs.

Topological Quantum Materials: Concepts, Models, and Phenomena
Grigory Tkachov

ISBN 978-981-4968-12-6 (Hardcover), 978-1-003-26641-9 (eBook)
www.jennystanford.com

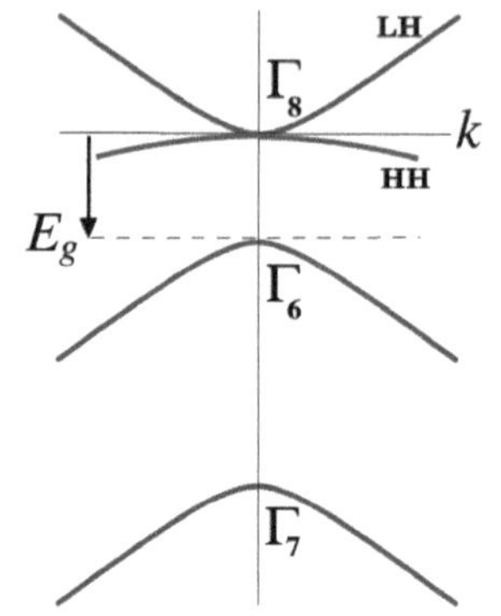

Figure 4.1 Schematic band structure of bulk unstrained HgTe.

4.1 Strained HgTe as a 3DTI. Helical Surface States

4.1.1 Kane's Hamiltonian and Boundary Conditions

To be concrete, we consider bulk HgTe. It is a text-book example of a narrow-gap semiconductor with a multiband electronic structure. As already mentioned in Chapter 2, the conduction and valence bands of HgTe are located close to the centre (the $k = 0$ point) of the Brillouin zone and have the s and p symmetries inherited from the atomic orbitals of Hg and Te, respectively. The $\boldsymbol{k} \cdot \boldsymbol{p}$ and spin–orbit splitting result in the three bands at $k = 0$: an s-type double degenerate Γ_6, a p-type four-fold degenerate Γ_8 with $j = 3/2$ and a p-type double degenerate Γ_7 with $j = 1/2$ (see also Ref. [86] and Fig. 4.1). Away from the zone centre the Γ_8 band splits into the light hole (LH) and heavy hole (HH) bands with the total angular momentum projections $\pm 1/2$ and $\pm 3/2$, respectively.

The degeneracy of the LH and HH at the Fermi point $k = 0$ is lifted by strain typically exerted by a substrate with a different lattice constant [25]. We intend to derive surface states of strained HgTe from Kane's $\boldsymbol{k} \cdot \boldsymbol{p}$ Hamiltonian (see e.g. Ref. [90]). For our purposes, the coupling to the remote Γ_7 band is not essential, so we can work with the six-component envelope wave function

$$f(\boldsymbol{r}) = \begin{bmatrix} f_1(\boldsymbol{r}) \\ f_2(\boldsymbol{r}) \\ f_3(\boldsymbol{r}) \\ f_4(\boldsymbol{r}) \\ f_5(\boldsymbol{r}) \\ f_6(\boldsymbol{r}) \end{bmatrix} \equiv \begin{bmatrix} |\Gamma_6, +\frac{1}{2}\rangle \\ |\Gamma_6, -\frac{1}{2}\rangle \\ |\Gamma_8, +\frac{3}{2}\rangle \\ |\Gamma_8, +\frac{1}{2}\rangle \\ |\Gamma_8, -\frac{1}{2}\rangle \\ |\Gamma_8, -\frac{3}{2}\rangle \end{bmatrix}. \tag{4.1}$$

Here f_1 and f_2 correspond to the s-type states of Γ_6 with the spin projections $\pm 1/2$, f_3 and f_6 represent the HH, while f_4 and f_5 the LH states of Γ_8. The corresponding 6×6 Kane Hamiltonian for the $\boldsymbol{k} \cdot \boldsymbol{p}$-coupled Γ_6 and Γ_8 bands can be written as

$$\hat{\mathcal{H}} = \begin{bmatrix} E_g & 0 & -P_+ & P_z & P_-/\sqrt{3} & 0 \\ 0 & E_g & 0 & -P_+/\sqrt{3} & P_z & P_- \\ -P_- & 0 & U & 0 & 0 & 0 \\ P_z & -P_-/\sqrt{3} & 0 & V & 0 & 0 \\ P_+/\sqrt{3} & P_z & 0 & 0 & V & 0 \\ 0 & P_+ & 0 & 0 & 0 & U \end{bmatrix}. \tag{4.2}$$

Here E_g is the direct gap between the Γ_6 and Γ_8 bands. It is negative, $E_g = -300$ meV, since HgTe has the inverted band structure [86]. The matrix elements $P_\pm$ and P_z are linear functions of the components of the electron momentum,

$$P_\pm = \sqrt{\frac{1}{2}} P\,(-i\partial_x \pm \partial_y), \qquad P_z = -\sqrt{\frac{2}{3}} P\, i\partial_z, \tag{4.3}$$

where P is the Kane parameter. The diagonal entries U and V are the energies of the HH and LH bands, respectively, measured from the Fermi level at $E = 0$. The case $U \neq V$ corresponds to the strained HgTe, with an energy gap $V - U$ between the LH and HH states. The terms of higher order in the electron momentum are omitted in Eq. (4.2) because near $k = 0$ they produce small corrections to the linear surface dispersion.

We assume that the material occupies the half-space $z < 0$ with the boundary to the vacuum at $z = 0$. The function $f(\boldsymbol{r})$ must satisfy a boundary condition at $z = 0$ that ensures the vanishing of the normal component of the particle current density

$$j_z(\boldsymbol{r}) = \tilde{f}(\boldsymbol{r})\hat{v}_z f(\boldsymbol{r}), \qquad \hat{v}_z = (z\hat{\mathcal{H}} - \hat{\mathcal{H}}z)/i\hbar, \tag{4.4}$$

where $\tilde{f}(\boldsymbol{r}) = [f_1^*(\boldsymbol{r}), f_2^*(\boldsymbol{r}), f_3^*(\boldsymbol{r}), f_4^*(\boldsymbol{r}), f_5^*(\boldsymbol{r}), f_6^*(\boldsymbol{r})]$ is the conjugated wave function, and the normal component of the velocity, $\hat{v}_z$, is given for Hamiltonian (4.2) by

$$\hat{v}_z = \sqrt{\frac{2}{3}}\frac{P}{\hbar}\begin{bmatrix} 0\,0\,0\,1\,0\,0 \\ 0\,0\,0\,0\,1\,0 \\ 0\,0\,0\,0\,0\,0 \\ 1\,0\,0\,0\,0\,0 \\ 0\,1\,0\,0\,0\,0 \\ 0\,0\,0\,0\,0\,0 \end{bmatrix}. \tag{4.5}$$

The equation above shows that the current j_z is carried by the coupled s and LH states. The HH drops out of Eq. (4.4). Therefore, the boundary problem can be reduced to the s-LH sector. In order to exclude the HH states, we write the eigenvalue equation $\hat{\mathcal{H}} f(\boldsymbol{r}) = E f(\boldsymbol{r})$ in components

$$E_g f_1 - P_+ f_3 + P_z f_4 + (P_-/\sqrt{3}) f_5 = E f_1, \tag{4.6}$$

$$E_g f_2 - (P_+/\sqrt{3}) f_4 + P_z f_5 + P_- f_6 = E f_2, \tag{4.7}$$

$$-P_- f_1 + U f_3 = E f_3, \tag{4.8}$$

$$P_z f_1 - (P_-/\sqrt{3}) f_2 + V f_4 = E f_4, \tag{4.9}$$

$$(P_+/\sqrt{3}) f_1 + P_z f_2 + V f_5 = E f_5, \tag{4.10}$$

$$P_+ f_2 + U f_6 = E f_6. \tag{4.11}$$

From Eqs. (4.8) and (4.11) we find the HH components as

$$f_3 = \frac{P_-}{U - E} f_1, \qquad f_6 = -\frac{P_+}{U - E} f_2. \tag{4.12}$$

Inserting these into the left-hand side of Eqs. (4.6), (4.7), (4.9) and (4.10), we obtain a closed system for the s and LH components, which in the compact notations reads

$$\hat{H}\psi(\boldsymbol{r}) = E\psi(\boldsymbol{r}), \tag{4.13}$$

where $\psi(\boldsymbol{r})$ is the four-component wave function

$$\psi(\boldsymbol{r}) = \begin{bmatrix} f_1(\boldsymbol{r}) \\ f_2(\boldsymbol{r}) \\ f_4(\boldsymbol{r}) \\ f_5(\boldsymbol{r}) \end{bmatrix}, \tag{4.14}$$

and $\hat{H}$ is the effective Hamiltonian for the coupled s and LH states

$$\hat{H} = \begin{bmatrix} E_g - \frac{P_+P_-}{U-E} & 0 & P_z & \frac{P_-}{\sqrt{3}} \\ 0 & E_g - \frac{P_-P_+}{U-E} & \frac{-P_+}{\sqrt{3}} & P_z \\ P_z & \frac{-P_-}{\sqrt{3}} & V & 0 \\ \frac{P_+}{\sqrt{3}} & P_z & 0 & V \end{bmatrix}. \tag{4.15}$$

We see that the coupling to the HH leads to the second order corrections in the electron momentum. These terms can be neglected in the vicinity of the $k = 0$ point.

In the following, we adopt the linearized version of Hamiltonian (4.15) in the Pauli matrix representation:

$$\hat{H} \approx \frac{E_g + V}{2}\sigma_0\tau_0 + \frac{E_g - V}{2}\sigma_0\tau_z - i\mathcal{B}\sigma_0\tau_x\partial_z + i\mathcal{A}(\sigma_y\partial_x - \sigma_x\partial_y)\tau_y, \tag{4.16}$$

where the matrices σ_j and τ_j act in spin and s-LH spaces, respectively, and the constants $\mathcal{A}$ and $\mathcal{B}$ are defined as

$$\mathcal{A} = \sqrt{\frac{1}{6}}P, \qquad \mathcal{B} = \sqrt{\frac{2}{3}}P. \tag{4.17}$$

The normal component of the current density [Eqs. (4.4) and (4.5)] in the basis of Eq. (4.14) is

$$j_z(\boldsymbol{r}) = \frac{\mathcal{B}}{\hbar}\tilde{\psi}(\boldsymbol{r})\sigma_0\tau_x\psi(\boldsymbol{r}). \tag{4.18}$$

The matrix structure of the equation above suggests the following boundary condition at the surface $z = 0$:

$$\psi(x, y, 0) = \sigma_0\tau_y\psi(x, y, 0). \tag{4.19}$$

It is easy to see that the current indeed vanishes at the boundary:

$$\begin{aligned} j_z(x, y, 0) &= \frac{\mathcal{B}}{\hbar}\tilde{\psi}(x, y, 0)\sigma_0\tau_x\psi(x, y, 0) & (4.20)\\ &= \frac{\mathcal{B}}{\hbar}\tilde{\psi}(x, y, 0)\sigma_0\tau_y\,\sigma_0\tau_x\,\sigma_0\tau_y\psi(x, y, 0) & (4.21)\\ &= -\frac{\mathcal{B}}{\hbar}\tilde{\psi}(x, y, 0)\sigma_0\tau_x\psi(x, y, 0) & \\ &= -j_z(x, y, 0), & (4.22) \end{aligned}$$

which necessarily leads to $j_z(x, y, 0) = 0$. Drawing a parallel to the edge states in Chapter 2, we can prove that boundary condition (4.19) is equivalent to a gap domain wall at $z = 0$.

4.1.2 Surface-State Dispersion, Helicity and Decay Length

The next step is to solve the boundary problem described by Eqs. (4.13), (4.16) and (4.19). We are looking for the solution propagating parallel to the surface $z = 0$ and decaying exponentially for $z \to -\infty$:

$$\psi(x, y, z) = \Phi_{\boldsymbol{k}} e^{i(k_x x + k_y y) + z/\lambda}, \tag{4.23}$$

$$\Phi_{\boldsymbol{k}} = \sigma_0 \tau_y \Phi_{\boldsymbol{k}}. \tag{4.24}$$

Again, the constraint (4.24) on the spinor $\Phi_{\boldsymbol{k}}$ follows from the boundary condition (4.19). Inserting Eq. (4.23) into Eq. (4.13) and using Eq. (4.24), we have

$$\left[\frac{E_g + V}{2} + \mathcal{A}(\sigma_x k_y - \sigma_y k_x)\right] \Phi_{\boldsymbol{k}} + \left[\frac{E_g - V}{2} + \frac{\mathcal{B}}{\lambda}\right] \tau_z \Phi_{\boldsymbol{k}} = E\Phi_{\boldsymbol{k}}. \tag{4.25}$$

The decay length is determined from the requirement for the second term to vanish identically,

$$\lambda = \frac{2\mathcal{B}}{V - E_g}. \tag{4.26}$$

Then, Eq. (4.25) reduces to

$$\hat{H}^s_{\boldsymbol{k}} \Phi_{\boldsymbol{k}} = E\Phi_{\boldsymbol{k}}, \qquad \hat{H}^s_{\boldsymbol{k}} = \frac{E_g + V}{2} + \mathcal{A}(\sigma_x k_y - \sigma_y k_x), \tag{4.27}$$

where $\hat{H}^s_{\boldsymbol{k}}$ can be identified as the surface Hamiltonian with wave vector $\boldsymbol{k} = [k_x, k_y, 0]$. Its eigenstates coincide with those of helicity matrix $(\boldsymbol{\sigma} \times \boldsymbol{n})_z$,

$$(\boldsymbol{\sigma} \times \boldsymbol{n})_z \Phi_{\boldsymbol{k}s} = s\Phi_{\boldsymbol{k}s}, \qquad \boldsymbol{n} = \frac{\boldsymbol{k}}{|\boldsymbol{k}|}, \qquad s = \pm 1, \tag{4.28}$$

while the energy eigenvalues are

$$E_{\boldsymbol{k}s} = \frac{E_g + V}{2} + s\mathcal{A}|\boldsymbol{k}|. \tag{4.29}$$

Let us summarize the results of this section:

(i) The surface state occurs in the middle of the gap between the bulk Γ_6 (s-type) and LH (p-type) bands and has a Dirac-like dispersion. The 3DTI phase is realized in the energy gap $V - U$ between the LH and HH bands, where only the surface state exists (see Fig. 4.2). Although our linear model is justified only in the vicinity of the $k = 0$ point, the conclusion regarding the 3DTI phase is still correct and agrees with numerical calculations based on extended models including higher order k-terms [91].

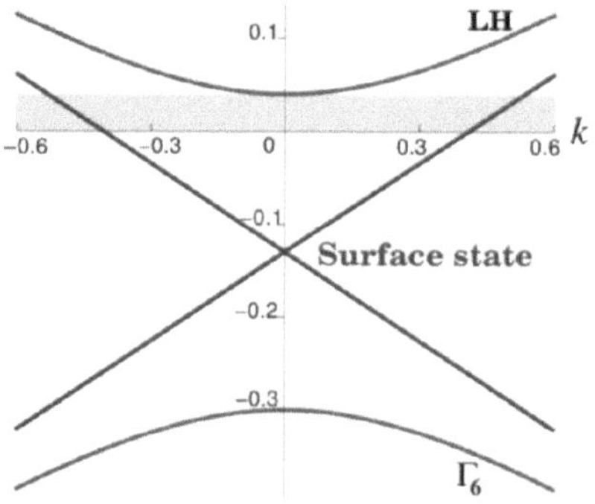

Figure 4.2 Surface and bulk bands of strained HgTe versus in-plane wave vector k (in nm^{-1}) from the reduced Hamiltonian (4.16) with $P = 780$ meV·nm, $E_g = -300$ meV, $V = 40$ meV and $U = 0$ meV. Shaded area indicates the strain-induced gap in which a 3DTI is realized.

(ii) In weakly strained HgTe with $V \ll |E_g|$, the surface-state decay length is determined by the gap energy E_g and the Kane matrix element P [see Eqs. (4.17) and (4.26)]. For typical values $E_g = -300$ meV and $P = 780$ meV·nm [86] the decay length is of order of a few nanometers:

$$\lambda \approx 4 \text{ nm}. \tag{4.30}$$

(iii) Because of topological constraint (4.24) imposed by the boundary condition, the surface state inherits only half of the bulk degrees of freedom, which is the electron spin in Eqs. (4.27) and (4.28). The spin helicity has major implications for electron transport in TI materials.

4.2 Bi-Based 3DTIs and Thin TI Materials

4.2.1 Surface States in Bi_2Se_3 and Bi_2Te_3

Another large group of materials which host topological surface states is Bi-based compounds, such as $Bi_{1-x}Sb_x$, Bi_2Se_3 and Bi_2Te_3 [31, 39, 40]. Both band structure calculations [31] and ARPES measurements [39, 40] reveal Dirac-like states at the surface of these materials. The low-energy dispersion and other properties of the surface carriers can be derived from the Hamiltonian of a bulk material [31] in the same way as done in Section 4.1 for HgTe. An illustrative example is the surface state in $Bi_2Se(Te)_3$ described by the Hamiltonian [46, 47]

$$\hat{H}^s_{\boldsymbol{k}} = \mathcal{A}(\sigma_x k_y - \sigma_y k_x) + \mathcal{M}_{\boldsymbol{k}}\sigma_z + \mathcal{D}\boldsymbol{k}^2 - E_F, \tag{4.31}$$

$$\mathcal{M}_{\boldsymbol{k}} = \frac{W}{2}(k_+^3 + k_-^3), \qquad k_\pm = k_x \pm i k_y, \tag{4.32}$$

where $\mathcal{A} > 0, \mathcal{D} > 0$ and $W > 0$ are band structure parameters, and E_F is the Fermi energy.

Equation (4.31) is an extended version of the Hamiltonian in Eq. (4.27). The quadratic term accounts for the broken particle-hole symmetry, whereas the cubic one $\mathcal{M}_{\boldsymbol{k}}$ reflects hexagonal warping of the surface bands in $Bi_2Se(Te)_3$ crystals [48, 49].

Upon the unitary transformation

$$\hat{H}^s_{\boldsymbol{k}} \to U\,\hat{H}^s_{\boldsymbol{k}} U^\dagger, \qquad U = \begin{bmatrix} i & 0 \\ 0 & 1 \end{bmatrix}, \tag{4.33}$$

Hamiltonian (4.31) takes the form

$$\hat{H}^s_{\boldsymbol{k}} = \mathcal{A}\boldsymbol{\sigma}\cdot\boldsymbol{k} + \mathcal{M}_{\boldsymbol{k}}\sigma_z + \mathcal{D}\boldsymbol{k}^2 - E_F. \tag{4.34}$$

The eigenvalues of Hamiltonian (4.34) yield the surface spectrum

$$E_\pm(k,\phi_{\boldsymbol{n}}) = \mathcal{D}k^2 - E_F \pm \sqrt{\mathcal{A}^2 k^2 + \frac{W^2 k^6}{2}(1 + \cos 6\phi_{\boldsymbol{n}})}. \tag{4.35}$$

It depends on the momentum direction specified by angle $\phi_{\boldsymbol{n}}$,

$$\boldsymbol{n} = \frac{\boldsymbol{k}}{|\boldsymbol{k}|} = (\cos\phi_{\boldsymbol{n}}, \sin\phi_{\boldsymbol{n}}, 0). \tag{4.36}$$

The surface dispersion relation (4.35) is plotted in Fig. 4.3.

4.2.2 TI Thin Films

Another representative example of quasi-2D helical systems is thin TI materials in the metallic regime where the Fermi level crosses the bulk conduction (or valence) band. Two typical examples are HgTe quantum wells and ultrathin films of $Bi_2Se(Te)_3$. Since the low-energy theories [13, 81, 82] for these materials share many similarities, the term "TI thin film" used below refers to both HgTe quantum wells and ultrathin films of $Bi_2Se(Te)_3$. The low-energy

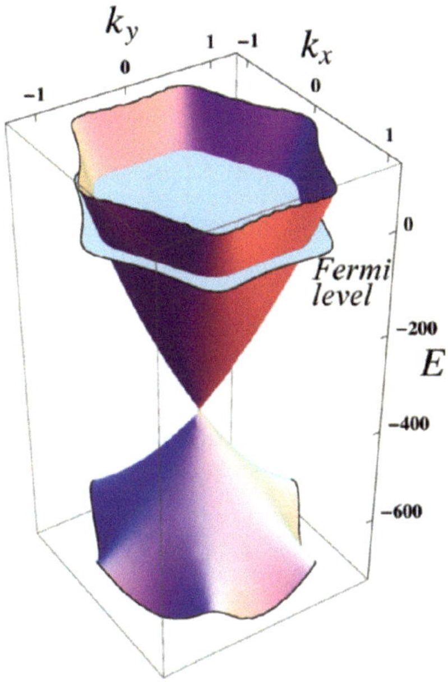

Figure 4.3 Energy bands (in meV) of a TI surface state [see Eq. (4.35)] versus in-plane wave-numbers k_x and k_y (in nm^{-1}). Cross section at $E = 0$ corresponds to the Fermi level; $\mathcal{A} = 300$ meV·nm, $\mathcal{D} = 75$ meV·nm^2, $E_F = 400$ meV and $W = 200$ meV·nm^3.

Hamiltonian for a TI thin film has been introduced in Chapter 2 and can be written as

$$\hat{H}^f_{\boldsymbol{k}} = \begin{bmatrix} \mathcal{A}\boldsymbol{\sigma}\cdot\boldsymbol{k} + \mathcal{D}\boldsymbol{k}^2 - E_F & \mathcal{M}_{\boldsymbol{k}}\sigma_0 \\ \mathcal{M}_{\boldsymbol{k}}\sigma_0 & -\mathcal{A}\boldsymbol{\sigma}\cdot\boldsymbol{k} + \mathcal{D}\boldsymbol{k}^2 - E_F \end{bmatrix}. \tag{4.37}$$

To distinguish it from the surface Hamiltonian, we add the superscript *f*. Recall that $\mathcal{M}_{\boldsymbol{k}}$ characterizes the hybridization of the top and bottom surfaces of the film. The TRS requires $\mathcal{M}_{\boldsymbol{k}}$ to be an even function of $\boldsymbol{k}$, e.g.

$$\mathcal{M}_{\boldsymbol{k}} = \Delta + \mathcal{B}\boldsymbol{k}^2, \tag{4.38}$$

where Δ and $\mathcal{B}$ are band structure parameters (hereafter, we choose $\mathcal{B} > 0$). After the unitary transformation

$$\hat{H}^f_{\boldsymbol{k}} \to U\,\hat{H}^f_{\boldsymbol{k}}U^{\dagger}, \qquad \hat{U} = \frac{1}{\sqrt{2}}\begin{bmatrix} \sigma_0 & \sigma_z \\ \sigma_z & -\sigma_0 \end{bmatrix}, \tag{4.39}$$

Hamiltonian (4.37) takes a block-diagonal form [13, 81, 82]

$$\hat{H}^f_{\boldsymbol{k}} = \begin{bmatrix} \hat{H}_{\boldsymbol{k}+} & 0 \\ 0 & \hat{H}_{\boldsymbol{k}-} \end{bmatrix}, \tag{4.40}$$

with $\hat{H}_{\boldsymbol{k}\pm}$ given by

$$\hat{H}_{\boldsymbol{k}\pm} = \pm\mathcal{A}\boldsymbol{\sigma}\cdot\boldsymbol{k} \pm \mathcal{M}_{\boldsymbol{k}}\sigma_z + \mathcal{D}\boldsymbol{k}^2 - E_F. \tag{4.41}$$

Unlike surface model (4.34), here the term $\mathcal{M}_{\boldsymbol{k}}$ plays the role of the Dirac mass. The eigenvalues of Hamiltonians (4.41) are

$$E^{(1,2)}(k) = \mathcal{D}k^2 - E_F \pm \sqrt{\mathcal{A}^2k^2 + (\Delta + \mathcal{B}k^2)^2}, \tag{4.42}$$

$$E^{(3,4)}(k) = \mathcal{D}k^2 - E_F \mp \sqrt{\mathcal{A}^2k^2 + (\Delta + \mathcal{B}k^2)^2}, \tag{4.43}$$

where $E^{(1,4)}$ and $E^{(2,3)}$ form double-degenerate conduction and valence bands, respectively. The bands are separated by an energy gap Δ at $\boldsymbol{k} = 0$, as illustrated in Fig. 4.4.

The energy spectra for the TI surface (Fig. 4.3) and thin film (Fig. 4.4) are topologically distinct at Dirac point $\boldsymbol{k} = 0$, which goes

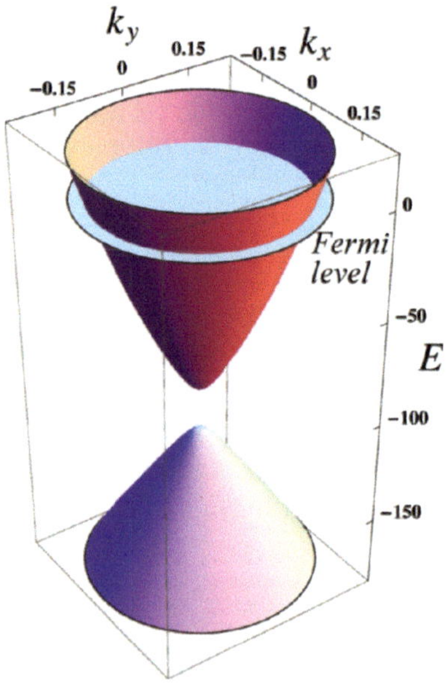

Figure 4.4 Energy bands (in meV) of a TI thin film [see Eqs. (4.42) and (4.43)] versus in-plane wave-numbers k_x and k_y (in nm^{-1}). Cross section at $E = 0$ corresponds to the Fermi level; $\mathcal{A} = 380$ meV·nm, $\mathcal{B} = 850$ meV·nm^2, $\mathcal{D} = 350$ meV·nm^2, $\mathcal{M} = -10$ meV and $E_F = 100$ meV.

back to the different symmetries of the "mass" terms in Eqs. (4.32) and (4.38). This distinction manifests itself in quantum transport properties of these materials, which will be the subject of the next chapter.

4.3 Weyl Semimetals

4.3.1 "Diabolical" Points in 3D *k*-Space

In the remainder of this chapter, we focus on another distinct class of topological quantum materials—Weyl semimetals. These are 3D crystalline solids in which electronic bands accidentally cross at isolated points in 3D **k**-space of the Brillouin zone. For example, the behavior of two such bands can be described by the Hamiltonian

$$\hat{h}_{\boldsymbol{k}} = \epsilon_{\boldsymbol{k}}\tau_0 + d^x_{\boldsymbol{k}}\tau_x + d^y_{\boldsymbol{k}}\tau_y + d^z_{\boldsymbol{k}}\tau_z,$$

where coefficients $d^x_{\boldsymbol{k}}$, $d^y_{\boldsymbol{k}}$ and $d^z_{\boldsymbol{k}}$ all vanish at a certain point $\boldsymbol{k} = \boldsymbol{k}_0$, so the two energy levels (eigenvalues of $\hat{h}_{\boldsymbol{k}}$) become degenerate and equal to $\epsilon_{\boldsymbol{k}_0}$. Above, τ_x, τ_y, τ_z and τ_0 are Pauli and unit matrices, respectively.

An accidental degeneracy point, called Weyl node, is stable in the sense that a perturbation of the form $\boldsymbol{m} \cdot \boldsymbol{\tau}$ only shifts the position of the node in 3D **k**-space of the Brillouin zone rather than lifting the band degeneracy (rendering the spectrum gapped). The band dispersion around such a point, dubbed "diabolical", has a conical shape. We can easily see this by linearizing $\hat{h}_{\boldsymbol{k}}$ around $\boldsymbol{k}_0$, e.g. for an isotropic case the Hamiltonian around the degeneracy point takes, up to an additive constant, the form $\hat{h}_{\boldsymbol{k}} = +\mathcal{A}\boldsymbol{\tau} \cdot (\boldsymbol{k} - \boldsymbol{k}_0)$ or $\hat{h}_{\boldsymbol{k}} = -\mathcal{A}\boldsymbol{\tau} \cdot (\boldsymbol{k} - \boldsymbol{k}_0)$. Save for the meaning of constant $\mathcal{A}$, these are the Hamiltonians for a pair of Weyl fermions that make up a Dirac fermion, such as a relativistic electron in the limit of zero rest mass.

The analogy to Dirac's theory deserves a separate comment. The solutions of the Dirac equation describe particles and antiparticles as positive- and negative-energy states with spin. The spin is not a conserved quantity, but its projection on the momentum vector—helicity—is. For the emergent Weyl fermions, the simplest form of the helicity is

$$\boldsymbol{\tau} \cdot \frac{\boldsymbol{k} - \boldsymbol{k}_0}{|\boldsymbol{k} - \boldsymbol{k}_0|}.$$

Another conserved quantity of massless Dirac's theory is the fermion chirality. The eigenstates of the chirality are referred to as "right-handed" and "left-handed" fermions, and are precisely the Weyl fermions. In the emergent theory, the "right-handed" and "left-handed" Weyl fermions are distinguished by prefactor $+1$ or -1 in the Hamiltonian near the node.

We should not take this analogy too literally because electronic states in crystalline solids do not have to obey the Lorentz invariance. In particular, the cone around the degeneracy point can be anisotropic and arbitrarily tilted (see also introductory chapter). Moreover, we can have a degeneracy at $\boldsymbol{k}_0^+$ with positive chirality and a negative-chirality partner at $\boldsymbol{k}_0^- \neq \boldsymbol{k}_0^+$. That is, unlike the relativistic massless Dirac theory, where the chiral partners occur at the same point, Weyl nodes are generally split in $\boldsymbol{k}$-space. Still, every positive-chirality node comes with its negative-chirality counterpart as long as the bands are defined on a periodic reciprocal lattice. This is again in accord with the Nielsen–Ninomiya theorem [80]. A rather nontrivial attribute of WSs is surface Fermi arcs connecting Weyl nodes of opposite chiralities (see e.g. review [57]). In the following, we illustrate these properties, beginning with a tight-binding lattice model for a WS.

4.3.2 Tight-Binding Model on a Cubic Lattice. Weyl Nodes

Recall the tight-binding model for a Chern insulator studied in Chapter 2. It invokes the nearest-neighbor-hopping of a spinless fermion on a square x, y lattice with an s-like and a p-like orbital at each site. We now consider a stack of such 2D lattices in the z-direction and allow hopping between them. The hopping in the z-direction is restricted to the nearest neighbors with the same-parity (s- or p-) orbitals. This 3D version of the Chern insulator is still described by a 2×2 Hamiltonian

$$\hat{h}_{\boldsymbol{k}} = \epsilon_{\boldsymbol{k}} \tau_0 + \boldsymbol{d}_{\boldsymbol{k}} \cdot \boldsymbol{\tau},$$

where $\boldsymbol{k}$ is a 3D wave vector, and $\boldsymbol{d_k}$ and $\epsilon_{\boldsymbol{k}}$ are defined similar to the 2D case:

$$\boldsymbol{d_k} = [\mathcal{A}\sin k_x, -\mathcal{A}\sin k_y, \mathcal{M}_{\boldsymbol{k}}], \tag{4.44}$$

$$\mathcal{M}_{\boldsymbol{k}} = \Delta + 2\mathcal{B}(3 - \cos k_x - \cos k_y - \cos k_z), \tag{4.45}$$

$$\epsilon_{\boldsymbol{k}} = \mathcal{C} + 2\mathcal{D}(3 - \cos k_x - \cos k_y - \cos k_z). \tag{4.46}$$

Constants Δ, $\mathcal{A}$, $\mathcal{B}$, $\mathcal{C}$ and $\mathcal{D}$ are related to the parameters of the tight-binding model as follows (cf. Chapter 2):

$$\Delta = \frac{V_s - V_p}{2} + 3(t_s - t_p), \quad \mathcal{A} = 2t_{sp}, \quad \mathcal{B} = -\frac{t_s - t_p}{2}, \tag{4.47}$$

$$\mathcal{C} = \frac{V_s + V_p}{2} + 3(t_s + t_p), \qquad \mathcal{D} = -\frac{t_s + t_p}{2}. \tag{4.48}$$

We are now in a position to examine the band degeneracies of the model. Just as in the 2D case, the system is an insulator at both Γ- and M-points with band gaps $\mathcal{M}_{[0,0,0]} = \Delta$ and $\mathcal{M}_{[\pi,\pi,0]} = \Delta + 8\mathcal{B}$, respectively. The band degeneracies can appear around the Γ-point on axis $[0, 0, k_z]$, where

$$\mathcal{M}_{[0,0,k_z]} = \Delta + 2\mathcal{B}(1 - \cos k_z). \tag{4.49}$$

Its zeros must satisfy the equation

$$\cos k_z = 1 + \frac{\Delta}{2\mathcal{B}},$$

with solutions

$$k_z = \pm k_0, \qquad k_0 = \arccos\left(1 + \frac{\Delta}{2\mathcal{B}}\right), \tag{4.50}$$

under conditions

$$\mathcal{B}\Delta < 0, \qquad |\Delta| < 2|\mathcal{B}|, \tag{4.51}$$

meaning a band inversion between Γ and M. For $|\Delta| \ll 2|\mathcal{B}|$, the nodes lie in the vicinity of the Γ-point at distance $k_0 \approx \sqrt{|\Delta/\mathcal{B}|}$. At the same time, on axis $[\pi, \pi, k_z]$ around the M-point the system remains gapped with

$$\mathcal{M}_{[\pi,\pi,k_z]} = \Delta + 2\mathcal{B}(5 - \cos k_z) = \mathcal{M}_{[0,0,k_z]} + 8\mathcal{B}.$$

Thus, in agreement with the Nielsen–Ninomiya theorem, our lattice model has two band degeneracy points

$$\boldsymbol{k}_0^{\pm} = [0, 0, \pm k_0]. \tag{4.52}$$

To look at the energy dispersion around them, we expand Eqs. (4.44)–(4.46) up to quadratic terms in $\boldsymbol{k}$. The resulting Hamiltonian can be cast into the form

$$\hat{h}_{\boldsymbol{k}} \approx (\mathcal{C} + \mathcal{D}\boldsymbol{k}^2)\tau_0 + \mathcal{A}k_x\tau_x - \mathcal{A}k_y\tau_y + \mathcal{B}(\boldsymbol{k}^2 - k_0^2)\tau_z. \tag{4.53}$$

Its eigenvalues are

$$E_{\boldsymbol{k}s} = \mathcal{C} + \mathcal{D}(k_x^2 + k_y^2 + k_z^2)$$
$$+ s\sqrt{\mathcal{A}^2(k_x^2 + k_y^2) + \mathcal{B}^2(k_x^2 + k_y^2 + k_z^2 - k_0^2)^2}, \quad s = \pm 1. \tag{4.54}$$

Here, s is the index of the 3D conduction and valence bands. In practice, it is convenient to pass a section either at $k_x = 0$ or at $k_y = 0$ and plot the reduced energy bands. A typical result is shown in Fig. 1.12 of the introductory chapter. We can see a conical band spectrum near each node. With increasing energy, the cones overlap, resulting in an anisotropic dispersion around the two nodes.

What about the chirality of the nodes? To answer this question, we linearize Hamiltonian (4.53) near $[0, 0, \pm k_0]$:

$$\hat{h}_{\boldsymbol{k}}^{+} = \mathcal{A}k_x\tau_x - \mathcal{A}k_y\tau_y - \mathcal{Q}(k_z + k_0)\tau_z,$$

$$\hat{h}_{\boldsymbol{k}}^{-} = \mathcal{A}k_x\tau_x - \mathcal{A}k_y\tau_y + \mathcal{Q}(k_z - k_0)\tau_z,$$

where we define $\mathcal{Q} = 2\mathcal{B}k_0$ and omit an unessential common constant $\mathcal{C} + \mathcal{D}k_0^2$. After unitary rotations

$$\hat{h}_{\boldsymbol{k}}^{+} \to \tau_x\hat{h}_{\boldsymbol{k}}^{+}\tau_x, \qquad \hat{h}_{\boldsymbol{k}}^{-} \to \tau_y\hat{h}_{\boldsymbol{k}}^{-}\tau_y,$$

the linearized Hamiltonians take the form

$$\hat{h}_{\boldsymbol{k}}^{\pm} = \pm[\mathcal{A}k_x\tau_x + \mathcal{A}k_y\tau_y + \mathcal{Q}(k_z \pm k_0)\tau_z]. \tag{4.55}$$

They acquire prefactors ± 1 akin to the Weyl Hamiltonians in Dirac's theory. We see therefore that the excitations emerging at $[0, 0, \pm k_0]$ mimic a "right-handed" and a "left-handed" Weyl fermion.

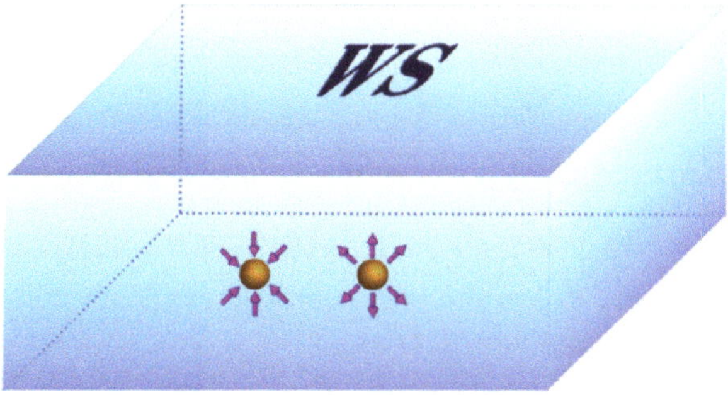

Figure 4.5 A pair of Weyl nodes as a source and a sink for Berry's curvature (indicated by arrows), realizing a monopole/antimonopole pair in reciprocal space.

4.3.3 Chern Numbers

Band degeneracies are point-like singularities with the topology analogous to monopole/antimonopole pairs (see also Fig. 4.5). To explore this analogy, we calculate the Chern invariant

$$n = \frac{1}{2\pi} \oiint (\nabla \times \boldsymbol{A}_{\boldsymbol{k}}) \cdot d\boldsymbol{S}_{\boldsymbol{k}}, \tag{4.56}$$

which, up to a prefactor, is the flux of Berry's curvature through a surface enclosing a Weyl node in 3D $\boldsymbol{k}$-space.

Consider, for example, point $[0, 0, -k_0]$ surrounded by a sphere

$$k_x^2 + k_y^2 + (k_z + k_0)^2 = const.$$

Berry's connection $\boldsymbol{A}_{\boldsymbol{k}} = -i\langle\psi_{\boldsymbol{k}}|\nabla|\psi_{\boldsymbol{k}}\rangle$ can be calculated by analogy with a 2D Chern insulator in Chapter 2. In particular, $\psi_{\boldsymbol{k}}$ is an eingenstate of $\hat{h}^+_{\boldsymbol{k}}$ and, consequently, of the helicity:

$$\boldsymbol{u}_{\boldsymbol{k}} \cdot \boldsymbol{\tau}\, \psi_{\boldsymbol{k}} = s\psi_{\boldsymbol{k}}.$$

Here, we use the spherical coordinates

$$k_x = k\cos\phi\sin\theta, \quad k_y = k\sin\phi\sin\theta, \quad k_z + k_0 = k\cos\theta,$$

and define helicity $\boldsymbol{u}_{\boldsymbol{k}} \cdot \boldsymbol{\tau}$ with unit vector

$$\boldsymbol{u}_{\boldsymbol{k}} = (u_\parallel \cos\phi,\, u_\parallel \sin\phi,\, u_z),$$

where

$$u_\parallel = \sqrt{1 - u_z^2}, \qquad u_z = \frac{\mathcal{Q}\cos\theta}{\sqrt{(\mathcal{A}\sin\theta)^2 + (\mathcal{Q}\cos\theta)^2}}. \tag{4.57}$$

Then, the normalized solution for $\psi_{\boldsymbol{k}}$ is given by[1]

$$\psi_{\boldsymbol{k}} = \frac{1}{\sqrt{2}} \begin{bmatrix} \sqrt{1+su_z} \\ se^{i\phi}\sqrt{1-su_z} \end{bmatrix}.$$

At the same time, Berry's connection in spherical coordinates is

$$\boldsymbol{A}_{\boldsymbol{k}} = A_{\boldsymbol{k}}^{(k)} \boldsymbol{e}_k + A_{\boldsymbol{k}}^{(\theta)} \boldsymbol{e}_\theta + A_{\boldsymbol{k}}^{(\phi)} \boldsymbol{e}_\phi,$$

where $\boldsymbol{e}_k$, $\boldsymbol{e}_\theta$ and $\boldsymbol{e}_\phi$ are the basis unit vectors. With known $\psi_{\boldsymbol{k}}$, we evaluate the components of $\boldsymbol{A}_{\boldsymbol{k}}$ as

$$A_{\boldsymbol{k}}^{(k)} = -i\,\langle \psi_{\boldsymbol{k}} | \frac{\partial}{\partial k} | \psi_{\boldsymbol{k}} \rangle = 0, \tag{4.58}$$

$$A_{\boldsymbol{k}}^{(\theta)} = -\frac{i}{k} \langle \psi_{\boldsymbol{k}} | \frac{\partial}{\partial \theta} | \psi_{\boldsymbol{k}} \rangle = 0, \tag{4.59}$$

and

$$A_{\boldsymbol{k}}^{(\phi)} = -\frac{i}{k\sin\theta} \langle \psi_{\boldsymbol{k}s} | \frac{\partial}{\partial \phi} | \psi_{\boldsymbol{k}s} \rangle = \frac{1-su_z}{2k\sin\theta}. \tag{4.60}$$

As only the ϕ-component is nonzero, Berry's curvature in spherical coordinates is given by

$$\nabla \times \boldsymbol{A}_{\boldsymbol{k}} = \frac{1}{k\sin\theta} \frac{\partial}{\partial\theta} \left(\sin\theta\, A_{\boldsymbol{k}}^{(\phi)} \right) \boldsymbol{e}_k - \frac{1}{k} \frac{\partial}{\partial k} \left(k A_{\boldsymbol{k}}^{(\phi)} \right) \boldsymbol{e}_\theta. \tag{4.61}$$

While it has both radial and angle components, only the former contributes to the integral over a sphere in Eq. (4.56):

$$n = -\frac{s}{2} \frac{1}{2\pi} \int_0^{2\pi} d\phi \int_0^{\pi} \frac{du_z}{d\theta} d\theta = s\,\mathrm{sgn}(\mathcal{Q}) = s\,\mathrm{sgn}(\mathcal{B}),$$

where, as usual, $s = -1$ for the valence band, and $\mathrm{sgn}(\mathcal{B}) = -\mathrm{sgn}(\Delta)$ for the inverted band structure. Hence, the Chern number

$$n = \mathrm{sgn}(\Delta) \qquad \text{for} \qquad [0,\, 0,\, -k_0]. \tag{4.62}$$

For the other node, we just replace $\mathcal{Q} \to -\mathcal{Q}$ and find

$$n = -\mathrm{sgn}(\Delta) \qquad \text{for} \qquad [0,\, 0,\, +k_0]. \tag{4.63}$$

These results can be interpreted as quantized topological charges of a monopole and an antimonopole in a 3D $\boldsymbol{k}$-space. As the total charge is zero, a pair of Weyl nodes act as a source and a sink for Berry's curvature. It is worth noting that the same topological

[1] To make notations simpler, we omit everywhere band index s.

numbers characterize two Chern species making up the BHZ model (cf. Chapter 2).

4.3.4 Fermi Arcs

As mentioned in the introductory chapter, the nontrivial bulk topology of a WS comes along with specific boundary modes. These are known as Fermi arcs. To understand what they are, we look at surface states of Hamiltonian (4.53), keeping all quadratic terms. The WS is assumed to occupy half-space $x \geq 0$, so $\boldsymbol{k}_{\parallel} = [0, k_y, k_z]$ remains a good quantum number, while k_x will be replaced by operator $-i\partial_x$. Hamiltonian (4.53) can be divided into two parts:

$$\hat{h}_{\boldsymbol{k}_{\parallel}}(x) = \hat{h}^{A}_{\boldsymbol{k}_{\parallel}}(x) + \hat{h}^{S}_{\boldsymbol{k}_{\parallel}}(x), \tag{4.64}$$

where $\hat{h}^{A}_{\boldsymbol{k}_{\parallel}}(x)$ and $\hat{h}^{S}_{\boldsymbol{k}_{\parallel}}(x)$ are the particle-hole asymmetric and symmetric Hamiltonians given by

$$\hat{h}^{A}_{\boldsymbol{k}_{\parallel}}(x) = (\mathcal{C} + \mathcal{D}\boldsymbol{k}_{\parallel}^2 - \mathcal{D}\partial_x^2)\tau_0 \tag{4.65}$$

and

$$\hat{h}^{S}_{\boldsymbol{k}_{\parallel}}(x) = -i\mathcal{A}\partial_x\tau_x - \mathcal{A}k_y\tau_y + \mathcal{B}(\boldsymbol{k}_{\parallel}^2 - \partial_x^2 - k_0^2)\tau_z. \tag{4.66}$$

This yields the eigenvalue equation

$$\hat{h}^{A}_{\boldsymbol{k}_{\parallel}}(x)\psi(x) + \hat{h}^{S}_{\boldsymbol{k}_{\parallel}}(x)\psi(x) = E\psi(x) \tag{4.67}$$

for energy E and two-component wave function $\psi(x)$. As boundary conditions we choose $\psi(x) = 0$ for $x \leq 0$ and $\psi(x) \to 0$ for $x \to +\infty$.

To determine E, we integrate Eq. (4.67) across the boundary:

$$\int_{-\infty}^{\infty} \hat{h}^{A}_{\boldsymbol{k}_{\parallel}}(x)\psi(x)dx + \int_{-\infty}^{\infty} \hat{h}^{S}_{\boldsymbol{k}_{\parallel}}(x)\psi(x)dx = E\int_{-\infty}^{\infty} \psi(x)dx. \tag{4.68}$$

In the first integral, the term $\partial_x\psi(x)$ vanishes at $\pm\infty$, which yields $(\mathcal{C} + \mathcal{D}\boldsymbol{k}_{\parallel}^2)\int_{-\infty}^{\infty}\psi(x)dx$. The second integral can be cast into a similar form by choosing $\psi(x)$ to be an eigenstate of the particle-hole symmetric Hamiltonian $\hat{h}^{S}_{\boldsymbol{k}_{\parallel}}(x)$:

$$\hat{h}^{S}_{\boldsymbol{k}_{\parallel}}(x)\psi(x) = E^{S}_{\boldsymbol{k}_{\parallel}}\psi(x), \tag{4.69}$$

with an eigenvalue $E^{S}_{\boldsymbol{k}_{\parallel}}$. Then, collecting the pre-integral factors in Eq. (4.68), we find the surface eigenenergy

$$E_{\boldsymbol{k}_{\parallel}} = \mathcal{C} + \mathcal{D}\boldsymbol{k}_{\parallel}^2 + E_{\boldsymbol{k}_{\parallel}}^{s}. \tag{4.70}$$

The problem now reduces to solving particle-hole symmetric Eq. (4.69). The substitution

$$\psi(x) = \Phi e^{-\varkappa x}$$

yields

$$[i\mathcal{A}\varkappa\tau_x - \mathcal{A}k_y\tau_y + \mathcal{B}(\boldsymbol{k}_{\parallel}^2 - \varkappa^2 - k_0^2)\tau_z]\Phi = E_{\boldsymbol{k}_{\parallel}}^{s}\Phi. \tag{4.71}$$

To determine $\varkappa$ and $E_{\boldsymbol{k}_{\parallel}}^{s}$ separately, we split Eq. (4.71) into two equations:

$$[\mathcal{A}\varkappa\tau_0 + \mathcal{B}(\varkappa^2 + k_0^2 - \boldsymbol{k}_{\parallel}^2)\tau_y]\Phi = 0, \tag{4.72}$$

and

$$-\mathcal{A}k_y\tau_y\Phi = E_{\boldsymbol{k}_{\parallel}}^{s}\Phi. \tag{4.73}$$

Note that in Eq. (4.72) we used $\tau_z = -i\tau_x\tau_y$. The solution for Φ is an eigenstate of τ_y, while those for $\varkappa$ and $E_{\boldsymbol{k}_{\parallel}}^{s}$ are given by

$$\varkappa_{1,2} = -\tau\frac{\mathcal{A}}{2\mathcal{B}} \pm \sqrt{\left(\frac{\mathcal{A}}{2\mathcal{B}}\right)^2 - (k_0^2 - \boldsymbol{k}_{\parallel}^2)}, \tag{4.74}$$

and

$$E_{\boldsymbol{k}_{\parallel}}^{s} = -\mathcal{A}k_y\tau, \tag{4.75}$$

with τ being an eigenvalue of τ_y.

Further, eigenvalue τ must be chosen such that the decay constants $\varkappa_1$ and $\varkappa_2$ have positive real parts. This is possible if the surface wave numbers lie inside the circle of radius k_0:

$$k_y^2 + k_z^2 \leq k_0^2. \tag{4.76}$$

Then, for

$$\tau = -\mathrm{sgn}(\mathcal{A}\mathcal{B}), \tag{4.77}$$

we obtain

$$\varkappa_{1,2} = \left|\frac{\mathcal{A}}{2\mathcal{B}}\right| \pm \sqrt{\left(\frac{\mathcal{A}}{2\mathcal{B}}\right)^2 - (k_0^2 - k_y^2 - k_z^2)}, \tag{4.78}$$

with $\mathrm{Re}(\varkappa_{1,2}) > 0$. In order for the wave function to vanish at the boundary, $x = 0$, the two fundamental solutions $e^{-\varkappa_1 x}$ and $e^{-\varkappa_2 x}$ must be taken with the opposite signs. Hence, the surface state

$$\psi(x) = \Phi(e^{-\varkappa_1 x} - e^{-\varkappa_2 x}).$$

In both terms, Φ is the eigenstate of τ_y with eigenvalue (4.77). Combining Eqs. (4.70), (4.75) and (4.77), we arrive at the energy dispersion of the surface mode

$$E_{[k_y,k_z]} = \mathcal{C} + \mathcal{D}(k_y^2 + k_z^2) + |\mathcal{A}|\mathrm{sgn}(\mathcal{B})k_y. \tag{4.79}$$

Two points of the surface spectrum, $[0, \pm k_0]$, deserve special attention. These are the projections of Weyl nodes (4.52) on the surface at $k_x = 0$. The surface mode connects these points, taking bulk value $E_{[0,\pm k_0]} = \mathcal{C} + \mathcal{D}k_0^2$ at each node. This is illustrated in Fig. 1.13 of the introductory chapter, where both bulk and surface bands are plotted[2] in the cross section $k_x = k_y = 0$. Because of the particle-hole asymmetry, the surface mode has the shape of a parabolic arc. If it is filled with electrons, the Fermi surface of the WS consists of bulk pockets and a surface arc between them, hence the name "Fermi arc".

4.3.5 Spin–Orbit Split Weyl Semimetal

We now extend our model by including the spin degree of freedom. This can be done solely on symmetry grounds if a WS possesses TRS, the typical case assumed hereafter. For spinless $\hat{h}_{\boldsymbol{k}}$, TR is realized by the complex conjugation combined with reflection $\boldsymbol{k} \to -\boldsymbol{k}$. That is, operator $\hat{h}^*_{-\boldsymbol{k}}$ is the Kramers partner to $\hat{h}_{\boldsymbol{k}}$. By analogy with QSHIs, the simplest Hamiltonian for WSs is a Kramers doublet

$$\begin{aligned}\hat{H}_{\boldsymbol{k}} &= \begin{bmatrix} \hat{h}_{\boldsymbol{k}} & 0 \\ 0 & \hat{h}^*_{-\boldsymbol{k}} \end{bmatrix} \\ &= \begin{bmatrix} \epsilon_{\boldsymbol{k}}\tau_0 + d^x_{\boldsymbol{k}}\tau_x + d^y_{\boldsymbol{k}}\tau_y + d^z_{\boldsymbol{k}}\tau_z & 0 \\ 0 & \epsilon_{\boldsymbol{k}}\tau_0 - d^x_{\boldsymbol{k}}\tau_x + d^y_{\boldsymbol{k}}\tau_y + d^z_{\boldsymbol{k}}\tau_z \end{bmatrix} \\ &= \epsilon_{\boldsymbol{k}}\tau_0\sigma_0 + d^x_{\boldsymbol{k}}\tau_x\sigma_z + d^y_{\boldsymbol{k}}\tau_y\sigma_0 + d^z_{\boldsymbol{k}}\tau_z\sigma_0,\end{aligned}$$

where σ_z and σ_0 are the Pauli and unit matrices in spin space. The spin degeneracy is generally lifted by spin–orbit interaction. We therefore add a spin–orbit term of the form [92]:

$$\hat{H}_{\mathrm{SO}} = \alpha\tau_y\sigma_y + \beta k_y\tau_x\sigma_y, \tag{4.80}$$

where α and β are constants.

---

[2] The parameters are $\Delta = -0.1$ eV, $\mathcal{A} = 1.5$ eVÅ, $\mathcal{B} = 12$ eVÅ^2, $\mathcal{C} = 0.1$ eV and $\mathcal{D} = 5$ eVÅ^2.

Our goal is to understand what happens to the Weyl nodes. To that end, we revisit solutions of the bulk eigenvalue equation

$$(\hat{H}_{\boldsymbol{k}} + \hat{H}_{\mathrm{SO}})\psi_{\boldsymbol{k}} = E\,\psi_{\boldsymbol{k}},$$

where $\psi_{\boldsymbol{k}}$ is a four-component wave function. The anticommuting terms can be grouped in

$$\hat{H}_0 = \mathcal{A}k_x\tau_x\sigma_z - \mathcal{A}k_y\tau_y\sigma_0 + \mathcal{B}(\boldsymbol{k}^2 - k_0^2)\tau_z\sigma_0 + \beta k_y\tau_x\sigma_y, \qquad (4.81)$$

so the equation can be written as

$$(\hat{H}_0 + \alpha\tau_y\sigma_y)\psi_{\boldsymbol{k}} = \tilde{E}\,\psi_{\boldsymbol{k}}, \qquad \tilde{E} = E - \mathcal{C} - \mathcal{D}\boldsymbol{k}^2. \qquad (4.82)$$

Squaring this equation yields

$$\begin{aligned}(\hat{H}_0 + \alpha\tau_y\sigma_y)^2\psi_{\boldsymbol{k}} &= (\hat{H}_0^2 + \alpha\hat{H}_0\tau_y\sigma_y + \alpha\tau_y\sigma_y\hat{H}_0 + \alpha^2\tau_0\sigma_0)\psi_{\boldsymbol{k}}\\ &= [\hat{H}_0^2 + 2\alpha\mathcal{A}(k_x\tau_z\sigma_x + k_y\tau_0\sigma_y) + \alpha^2\tau_0\sigma_0]\psi_{\boldsymbol{k}}\\ &= \tilde{E}^2\psi_{\boldsymbol{k}},\end{aligned}$$

where $\hat{H}_0^2$ is diagonal:

$$\hat{H}_0^2 = E_0^2\tau_0\sigma_0, \qquad E_0^2 = \mathcal{A}^2(k_x^2 + k_y^2) + \mathcal{B}^2(\boldsymbol{k}^2 - k_0^2)^2 + \beta^2k_y^2. \quad (4.83)$$

We then re-group the terms,

$$2\alpha\mathcal{A}(k_x\tau_z\sigma_x + k_y\tau_0\sigma_y)\psi_{\boldsymbol{k}} = (\tilde{E}^2 - E_0^2 - \alpha^2)\psi_{\boldsymbol{k}},$$

and square this equation, arriving at the equation for the energy levels

$$(\tilde{E}^2 - E_0^2 - \alpha^2)^2 = 4\alpha^2\mathcal{A}^2(k_x^2 + k_y^2).$$

Its solutions are

$$\tilde{E}_\pm^2 = \left(\mathcal{A}\sqrt{k_x^2 + k_y^2} \pm \alpha\right)^2 + \beta^2k_y^2 + \mathcal{B}^2(\boldsymbol{k}^2 - k_0^2)^2. \qquad (4.84)$$

Only the '$-$' branch has degeneracy points which lie at

$$k_x = \pm\frac{\alpha}{\mathcal{A}}, \quad k_y = 0, \quad k_z = \pm\sqrt{k_0^2 - \frac{\alpha^2}{\mathcal{A}^2}}, \qquad (4.85)$$

provided that $|\alpha| < |\mathcal{A}|k_0$. There are in total four Weyl points. For a TR-symmetric WS, this is the minimal number which results from spin splitting of the pair of nodes (4.52). Consequently, there are two Fermi arcs connecting the opposite-chirality nodes in Eq. (4.85).

To conclude, 3DTIs and WSs are both typical representatives of 3D topological quantum materials. The former are bulk band insulators, whereas the latter are gapless semimetals. This difference manifests itself in the behavior of topological boundary modes. While the surface states of 3DTIs have a conical band dispersion, the spectrum of Fermi arcs in WSs is highly anisotropic, being parabolic in one direction and a mixture of linear and quadratic terms in the other. Both types of boundary states have been the subjects of intense theoretical and experimental effort (see reviews [27, 57, 60]).

4.4 Problems

(1) Calculate commutator $z\hat{\mathcal{H}} - \hat{\mathcal{H}}z$ for the Kane Hamiltonian in Eq. (4.2) and prove Eq. (4.5) for the normal component of the carrier velocity in HgTe.

(2) Consider a film of strained HgTe occupying the region $-d \leq z \leq 0$. The film thickness, d, is assumed much larger than the penetration depth λ of the surface state [see Eq. (4.26)],

$$d \gg \lambda. \tag{4.86}$$

Under this condition we can neglect the overlap of the surface states originating from $z = 0$ and $z = -d$ and treat them independently of each other. By analogy with the solution obtained in Section 4.1.2 show that the ansatz

$$\psi(x, y, z) = \Phi_{\boldsymbol{k}} e^{i(k_x x + k_y y) - (z+d)/\lambda}, \qquad \Phi_{\boldsymbol{k}} = -\sigma_0 \tau_y \Phi_{\boldsymbol{k}} \tag{4.87}$$

describes the surface state localized at $z = -d$. Find its spectrum, decay length and helicity. Note that this solution satisfies the boundary condition $\psi(x, y, -d) = -\sigma_0 \tau_y \psi(x, y, -d)$ where the sign is opposite to that in Eq. (4.19) for $z = 0$. Prove that the "top" and "bottom" surface states are orthogonal to each other.

(3) Calculate Berry's connection and curvature for a WS with Hamiltonian (4.53) in the spherical coordinates

$$k_x = k \cos\phi \sin\theta, \quad k_y = k \sin\phi \sin\theta, \quad k_z = k \cos\theta.$$

Calculate Chern number (4.56) where the Berry curvature threads a sphere defined by

$$k_x^2 + k_y^2 + k_z^2 = const.$$

Explain the difference to the results of Section 4.3.3, using both mathematical and physical arguments.

Chapter 5

Surface Electron Transport and Magneto-Optics

Surface states play a key role in electron properties of 3D topological quantum materials. In this chapter, we examine representative transport and magneto-optical effects illustrating topological properties of the surface states. In particular, their spin-momentum locking prevents current carriers from being localized in a random potential of impurities, which is ubiquitously observed in experiments as weak antilocalization effect. We shall discuss both a simple qualitative picture and the detailed theory of this phenomenon. The remainder of the chapter is dedicated to the surface quantum Hall effect and related subjects such as emergent axion electrodynamics, the topological magnetoelectric effect and its manifestation in magneto-optical spectroscopy.

Topological Quantum Materials: Concepts, Models, and Phenomena
Grigory Tkachov

ISBN 978-981-4968-12-6 (Hardcover), 978-1-003-26641-9 (eBook)
www.jennystanford.com

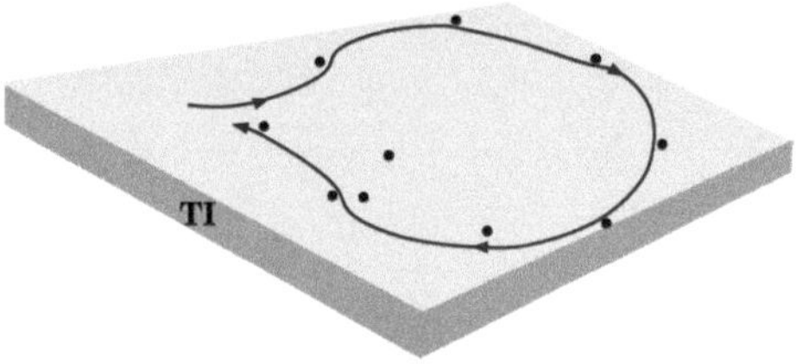

Figure 5.1 Illustration of carrier diffusion via multiple impurity scattering on a TI surface.

5.1 Electron Transport in Disordered TI Materials: Phenomenology

5.1.1 Diffusion and Spin-Momentum Locking

We now turn to transport properties of 2D helical states in disordered TI materials. The main source of disorder in these materials is randomly distributed static ionized impurities generating an inhomogeneous spin-independent potential, $V(\boldsymbol{r})$, for mobile charge carriers. As in the 1D case, the TRS-invariant potential $V(\boldsymbol{r})$ cannot cause 180° backscattering between surface states with opposite wave vectors, $\boldsymbol{k}$ and $-\boldsymbol{k}$. However, unlike the 1D case, scattering at any other angle is no longer forbidden and, therefore, through multiple collisions with impurities 2D helical carriers experience diffusion on a disordered surface (see also Fig. 5.1). The 2D diffusion of charged particles leads to a high electric conductivity, σ_c, compared with the conductance quantum,

$$\sigma_c \gg \frac{e^2}{h}. \tag{5.1}$$

The system of helical carriers in this regime is called a *helical metal*. In addition to the classical conductivity, σ_c, the carrier diffusion results in a quantum correction, $\delta\sigma$, known as the weak localization correction. It originates from coherent quantum interference at time scales Δt shorter than a phase-breaking time, τ_ϕ,

$$\Delta t \ll \tau_\phi. \tag{5.2}$$

The studies of such phase-coherent phenomena have emerged as an important tool of characterizing the underlying symmetries of disordered systems, from low-dimensional semiconductors [93] to graphene (e.g. [94–96]) and TI materials (e.g. [97–100]). In view of broad interest in the topic, we intend to present a systematic qualitative and quantitative analysis of the quantum-interference transport for different models of a 2D helical metal.

In a helical metal, the direction of the carrier spin $\boldsymbol{\sigma}$ is tied to the direction of its momentum $\boldsymbol{k}$. Such spin-momentum locking is described by the linear terms in the Hamiltonians in Eqs. (4.27), (4.31) and (4.41). These are reminiscent of the Bychkov–Rashba spin–orbit interaction in a conventional 2D electronic system confined in an asymmetric quantum well. The key distinction is the strength of the coupling constant $\mathcal{A}$. Its typical value can be estimated from Eq. (4.17) which gives $\mathcal{A}$ in terms of the Kane matrix element P for the HgTe surface states. For $P = 780$ meV·nm [86] we have $\mathcal{A} \approx 320$ meV·nm. For other materials the experimentally determined coupling constants are $\mathcal{A} = 355$ meV·nm for Bi_2Se_3 [49] and $\mathcal{A} = 255$ meV·nm for Bi_2Te_3 [40, 48]. These numbers are truly giant compared to a typical Bychkov–Rashba coupling $\alpha_{BR} \sim 1 \div 10$ meV·nm, e.g. in GaAs-based electronic systems. This large quantitative difference has two important physical consequences. The first is the formation of the conical band dispersion, as shown in Figs. 4.3 and 4.4. The second is the robustness of the spin-momentum locking against impurity scattering (and other spin-independent scattering) as long as the energy separation between the valence and conduction bands is larger than the spectrum broadening $\hbar/\tau$:

$$2\sqrt{\mathcal{A}^2 k_F^2 + \mathcal{M}_{k_F}^2} = 2(E_F - \mathcal{D}k_F^2) \gg \frac{\hbar}{\tau}, \tag{5.3}$$

where τ is the elastic life-time, and k_F is the wave number at which the conduction band crosses the Fermi level. Due to the large constant $\mathcal{A}$ condition (5.3) can be met already for the Fermi energy values, E_F, of several tens of meV. We note that for the Bychkov–Rashba spin–orbit interaction the condition opposite to Eq. (5.3) is usually assumed,

$$2\alpha_{BR} k_F \ll \frac{\hbar}{\tau}. \tag{5.4}$$

In the latter case, disorder induces interband mixing, enabling *spin relaxation*, i.e. the loss of the memory about an initial spin orientation after a large number of collisions with impurities. The typical processes of spin relaxation are the D'yakonov-Perel or Elliott–Yafet mechanisms.

The condition (5.3) simultaneously implies a metallic transport regime in which

$$2E_F\,\tau/\hbar \approx k_F v\tau_{tr} = k_F\ell_{tr} \gg 1, \tag{5.5}$$

where $\tau_{tr} \approx 2\tau$ and $\ell_{tr} = v\tau_{tr}$ are the transport momentum relaxation time and length, respectively, and v is the Fermi velocity. Thus, the large values of the coupling constant $\mathcal{A}$ enable metallic regime (5.5) without significant spin relaxation, i.e. the carrier spin continues to follow the momentum direction even after a large number of scattering events. Mathematically, this means that even in the presence of impurities the carrier state remains an eigenstate of the spin helicity, which for a 2D helical metal is

$$\boldsymbol{\sigma}\cdot\frac{\mathcal{A}\boldsymbol{k}+\mathcal{M}_k\mathbf{z}}{|\mathcal{A}\boldsymbol{k}+\mathcal{M}_k\mathbf{z}|}, \tag{5.6}$$

where $\mathbf{z}$ is the out-of-plane unit vector. This statement will be proved rigorously in the following sections.

Spin physics of 2D helical carriers is very attractive for potential applications. In the absence of spin relaxation, information encoded in the carrier spin can be transmitted over large distances. This may open new avenues for low-energy information processing and electronic devices.

5.1.2 Weak Localization and Antilocalization

At low temperatures a charge carrier can experience many elastic collisions before losing memory of its original phase. After a series of the scattering events the carrier trajectory can form a closed loop, as depicted in Fig. 5.2. Moreover, if the TRS is preserved, there is an equal probability for the particle to follow the same closed trajectory in the opposite direction. Such closed trajectories cause weak localization (WL) or antilocalization (WAL) depending on whether the spin–orbit coupling is zero or not. In the absence of the spin–orbit coupling, the two counter-propagating states are not protected against backscattering. Their interference yields a

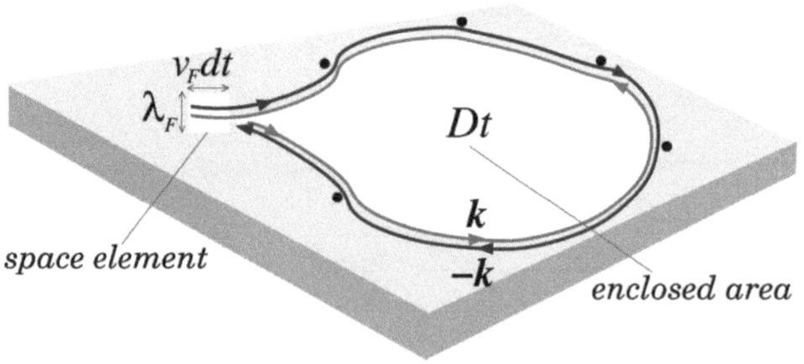

Figure 5.2 Illustration of closed electron trajectories giving rise to weak localization. Regions indicated in white are a space element of the trajectory, $\lambda_F vdt$, and the area enclosed by the trajectory, Dt, which we use to estimate the return probabilty $dP_{ret}(t)$ in Eq. (5.7).

negative correction, $\delta\sigma$, to the classical conductivity, σ_c. This is the WL effect, a precursor of the disorder-induced insulating phase. The relative change $\delta\sigma/\sigma_c$ is the measure of the classical return probability integrated from the shortest time-scale, τ, to the largest one, given by the dephasing time τ_φ:

$$\frac{\delta\sigma}{\sigma_c} \sim -\int_\tau^{\tau_\varphi} dP_{ret}(t), \qquad dP_{ret}(t) \sim \frac{\lambda_F vdt}{(\sqrt{Dt})^2}. \tag{5.7}$$

The negative sign here is the consequence of the fact that the interfering states have opposite momenta. The return probability $dP_{ret}(t)$ is proportional to the area of a space element of the trajectory, $\sim \lambda_F vdt$, divided by the typical area, $(\sqrt{Dt})^2$, swept by the particle trajectory after time t (λ_F is the Fermi wavelength and D is the diffusion constant). Conductivity correction (5.7) is logarithmic and has a universal prefactor [93]:

$$\delta\sigma = -\frac{2e^2}{2\pi h}\ln\frac{\tau_\varphi}{\tau}, \qquad \tau_\phi \gg \tau, \tag{5.8}$$

where the factor of 2 in the numerator is due to the spin degeneracy. The dephasing time τ_φ depends on the temperature T, increasing for $T \rightarrow 0$. The lower the temperature, the more insulating the system.

In contrast to Eq. (5.8), for 2D helical states the experiments and theoretical calculations show that the quantum conductivity

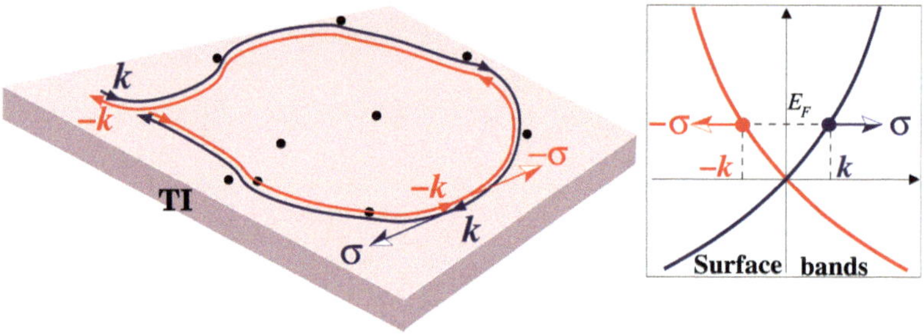

Figure 5.3 Example of closed electron trajectories on a 3DTI surface. Owing to spin-momentum locked band structure (shown in inset), the counter-propagating states are related by TRS and protected against backscattering, leading to WAL instead of WL.

correction is, as a rule, positive,

$$\delta\sigma > 0, \tag{5.9}$$

an effect termed *weak antilocalization* (WAL). It indicates that the system remains metallic in the low-temperature limit when $\tau_\varphi(T) \to \infty$. The WAL is intimately related to the absence of backscattering in a helical metal.

Consider, for example, the TI surface state described by Hamiltonian (4.34). Owing to the spin-momentum locking, the two states on the conjugated trajectories have not only the opposite wave vectors, $\boldsymbol{k}$ and $-\boldsymbol{k}$, but also the opposite spin polarizations (indicated by arrows $\rightarrow$ and $\leftarrow$ in Fig. 5.3). These states are related by TRS:

$$|-\boldsymbol{k}, \leftarrow\rangle = \mathbb{T}|\boldsymbol{k}, \rightarrow\rangle. \tag{5.10}$$

In this sense, there is a close analogy to the 1D helical edge states considered in Chapter 2. Since the 1D helical states are protected against backscattering, the same applies to the TR-conjugated closed trajectories on a 3DTI surface. The opposite orientation of the spins results in another sign "minus" in Eq. (5.7), yielding the WAL correction.

$$\delta\sigma = +\frac{e^2}{2\pi h}\ln\frac{\tau_\varphi}{\tau}. \tag{5.11}$$

This can also be viewed as the result of the π Berry phase accumulated along the closed loop [94]. The explicit calculation of $\delta\sigma$ with the account of the hexagonal warping under condition

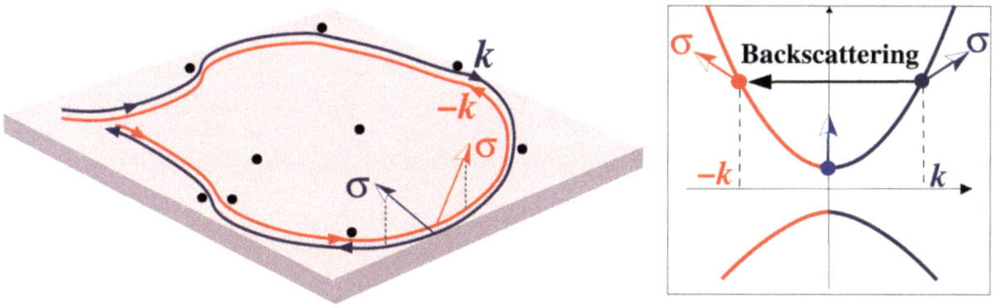

Figure 5.4 Example of closed electron trajectories in a TI thin film. Due to band spin polarization (see inset and Eq. (4.41) for H_+), counter-propagating states with $\boldsymbol{k}$, $\nearrow$ and $-\boldsymbol{k}$, $\nwarrow$ are no longer related by the TRS and experience backscattering, leading to suppressed WAL.

(5.3) was carried out in [98]. We note that conductivity correction $\delta\sigma$ (5.11) has the same form as that for a 2D electron system (2DES) with spin–orbit coupling [93]. The reason is that the surface states of 3DTIs with hexagonal warping and conventional 2D electron systems with spin–orbit interactions belong to the same—*symplectic*—universality class of disordered systems.

A different situation is encountered in TI thin films. The carrier trajectories described by Hamiltonian $H_{\boldsymbol{k}+}$ in Eq. (4.41) are sketched in Fig. 5.4. The mass term $\mathcal{M}_{\boldsymbol{k}}\sigma_z$ plays the role of a Zeeman field polarizing the spins out of plane, which is indicated by tilted arrows $\nearrow$ and $\nwarrow$ in Fig. 5.4. In other words, the spins of the counter-propagating states acquire equal out-of-plane components and are no longer antiparallel to each other. Clearly, these states are no longer related by TR:

$$|-\boldsymbol{k}, \nwarrow\rangle \neq \mathbb{T}|\boldsymbol{k}, \nearrow\rangle. \tag{5.12}$$

As a result, the counter-propagating states with $\boldsymbol{k}$ and $-\boldsymbol{k}$ experience backscattering which limits the carrier diffusion at low temperatures. The conductivity correction acquires a new time scale $\tau_{\mathcal{M}}$ akin to the dephasing time [97, 98]:

$$\delta\sigma_+ = \frac{e^2}{2\pi h} \ln \frac{\tau^{-1}}{\tau_{\mathcal{M}}^{-1} + \tau_{\varphi}^{-1}}, \qquad \tau_{\mathcal{M}},\ \tau_{\phi} \gg \tau, \tag{5.13}$$

$$\tau_{\mathcal{M}}^{-1} = \frac{2}{\tau} \frac{\mathcal{M}_{k_F}^2}{\mathcal{A}^2 k_F^2 + \mathcal{M}_{k_F}^2}, \quad k_F = \sqrt{2\pi n}. \tag{5.14}$$

where $\tau_{\mathcal{M}}^{-1}$ is the backscattering rate proportional to the total mass, $\mathcal{M}_{k_F} = \Delta + \mathcal{B}k_F^2$, at the Fermi level, and n is the 2D carrier density in the film.

$$\delta\sigma = \delta\sigma_+ + \delta\sigma_- = \frac{2e^2}{(2\pi)h} \ln \frac{\tau^{-1}}{\tau_{\mathcal{M}}^{-1} + \tau_\varphi^{-1}}. \tag{5.15}$$

Since the backscattering rate is temperature independent, the correction (5.13) does not diverge in the limit $T \to 0$ and is, generally, smaller than that in the symplectic case, Eq. (5.11). Disordered systems with such suppressed $\delta\sigma$ belong to the *unitary* universality class. That is, the two models of a 2D helical metal—a TI surface and a TI thin film—represent two distinct universality classes of disordered systems.

Usually, in the unitary class the TRS is broken. The TI thin films are quite unique as they allow one to probe the unitary behavior of the conductivity without an actual TRS breaking. This is because there is another carrier subsystem [see Eq. (4.41) for H_-] whose spin polarization is opposite to that of the '+' one. The net spin polarization of the '±' subsystems vanishes, whereas their net conductivity is twice that in Eq. (5.13):

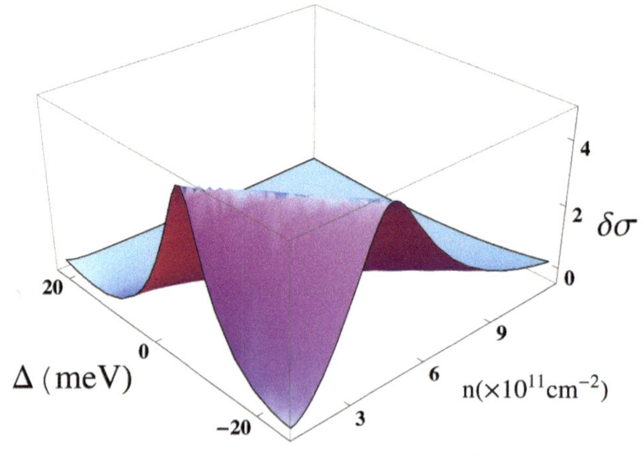

Figure 5.5 WAL conductivity correction $\delta\sigma$ [in units of $e^2/\pi h$, see Eq. (5.15)] versus carrier density n and band gap Δ.

Figure 5.5 shows the typical dependence of $\delta\sigma$ on carrier density n and band gap Δ. The next section contains the details of the calculation of the quantum conductivity correction.

5.2 Weak Antilocalization in TI Thin Films

5.2.1 Kubo Formula: Green's Functions

We begin with the TI thin-film model whose $\boldsymbol{k}$-space isotropy makes the calculations more transparent. The general formula for the electric conductivity in the linear approximation with respect to an external uniform electric field $\mathcal{E}e^{-i\omega t}$ is given by Kubo's theory (see e.g. Ref. [78]). We are interested in the longitudinal conductivity σ_{xx} that can be written as

$$\sigma_{xx} = \frac{e^2}{2\pi\omega a}\int dE[f(E) - f(E+\hbar\omega)] \times \sum_{\boldsymbol{k},\boldsymbol{k}'} \mathrm{Tr}[\hat{v}^x_{\boldsymbol{k}}\hat{G}^R_{\boldsymbol{k},\boldsymbol{k}'}(E+\hbar\omega)\hat{v}^x_{\boldsymbol{k}'}\hat{G}^A_{\boldsymbol{k}',\boldsymbol{k}}(E)], \tag{5.16}$$

where a is the area of the system, $f(E)$ is the Fermi-Dirac distribution function, Tr denotes the trace in spin space, $\hat{v}^x_{\boldsymbol{k}}$ is the x-component of the velocity operator

$$\hat{\boldsymbol{v}}_{\boldsymbol{k}} = \hbar^{-1}\partial_{\boldsymbol{k}}\hat{H}_{\boldsymbol{k}}, \tag{5.17}$$

and $\hat{G}^{R/A}_{\boldsymbol{k},\boldsymbol{k}'}(E)$ is the retarded/advanced Green function satisfying the equation

$$\sum_{\boldsymbol{k},\boldsymbol{k}_1}[(E_{R/A} - \hat{H}_{\boldsymbol{k}})\delta_{\boldsymbol{k},\boldsymbol{k}_1} - \hat{V}_{\boldsymbol{k},\boldsymbol{k}_1}]\hat{G}^{R/A}_{\boldsymbol{k}_1,\boldsymbol{k}'} = \delta_{\boldsymbol{k},\boldsymbol{k}'}\sigma_0, \tag{5.18}$$

where energy $E_{R/A} = E \pm i0$ includes an infinitesimal imaginary part, $i0$, with $\pm$ for the retarded and advanced functions, respectively. In Eqs. (5.17) and (5.18), $\hat{H}_{\boldsymbol{k}}$ is the Hamiltonian of the system in the absence of any inhomogeneities, while $\hat{V}_{\boldsymbol{k},\boldsymbol{k}'}$ is the matrix element of an inhomogeneous potential $\hat{V}(\boldsymbol{r})$, defined by

$$\hat{V}_{\boldsymbol{k},\boldsymbol{k}'} = a^{-1}\int d\mathrm{r}\,\hat{V}(\boldsymbol{r})e^{-i(\boldsymbol{k}-\boldsymbol{k}')\boldsymbol{r}}. \tag{5.19}$$

In the above equations and throughout the "hat" indicates 2×2 matrices in spin space.

For arbitrary potential $\hat{V}(\boldsymbol{r})$ the solution of Eq. (5.18) can be represented by an infinite series expansion in powers of matrix elements $\hat{V}_{\boldsymbol{k},\boldsymbol{k}'}$. For that purpose, we recast Eq. (5.18) in the integral form

$$\hat{G}^{R/A}_{\boldsymbol{k},\boldsymbol{k}'} = \hat{G}^{R/A}_{0\boldsymbol{k}} \delta_{\boldsymbol{k},\boldsymbol{k}'} + \sum_{\boldsymbol{k}_1} \hat{G}^{R/A}_{0\boldsymbol{k}} \hat{V}_{\boldsymbol{k},\boldsymbol{k}_1} \hat{G}^{R/A}_{\boldsymbol{k}_1,\boldsymbol{k}'}, \tag{5.20}$$

where $\hat{G}^{R/A}_{0\boldsymbol{k}}$ is the Green function of the homogeneous system, obeying the equation

$$(E_{R/A} - \hat{H}_{\boldsymbol{k}})\hat{G}^{R/A}_{0\boldsymbol{k}} = \sigma_0. \tag{5.21}$$

Iterating Eq. (5.20), we generate the series

$$\hat{G}^{R/A}_{\boldsymbol{k},\boldsymbol{k}'} = \hat{G}^{R/A}_{0\boldsymbol{k}} \delta_{\boldsymbol{k},\boldsymbol{k}'} + \hat{G}^{R/A}_{0\boldsymbol{k}} \hat{V}_{\boldsymbol{k},\boldsymbol{k}'} \hat{G}^{R/A}_{0\boldsymbol{k}'} \tag{5.22}$$

$$+ \sum_{\boldsymbol{k}_1} \hat{G}^{R/A}_{0\boldsymbol{k}} \hat{V}_{\boldsymbol{k},\boldsymbol{k}_1} \hat{G}^{R/A}_{0\boldsymbol{k}_1} \hat{V}_{\boldsymbol{k}_1,\boldsymbol{k}'} \hat{G}^{R/A}_{0\boldsymbol{k}'} \tag{5.23}$$

$$+ \sum_{\boldsymbol{k}_1,\boldsymbol{k}_2} \hat{G}^{R/A}_{0\boldsymbol{k}} \hat{V}_{\boldsymbol{k},\boldsymbol{k}_1} \hat{G}^{R/A}_{0\boldsymbol{k}_1} \hat{V}_{\boldsymbol{k}_1,\boldsymbol{k}_2} \hat{G}^{R/A}_{0\boldsymbol{k}_2} \hat{V}_{\boldsymbol{k}_2,\boldsymbol{k}'} \hat{G}^{R/A}_{0\boldsymbol{k}'} \tag{5.24}$$

$$+ \sum_{\boldsymbol{k}_1,\boldsymbol{k}_2,\boldsymbol{k}_3} \hat{G}^{R/A}_{0\boldsymbol{k}} \hat{V}_{\boldsymbol{k},\boldsymbol{k}_1} \hat{G}^{R/A}_{0\boldsymbol{k}_1} \hat{V}_{\boldsymbol{k}_1,\boldsymbol{k}_2} \hat{G}^{R/A}_{0\boldsymbol{k}_2} \hat{V}_{\boldsymbol{k}_2,\boldsymbol{k}_3} \hat{G}^{R/A}_{0\boldsymbol{k}_3} \hat{V}_{\boldsymbol{k}_3,\boldsymbol{k}'} \hat{G}^{R/A}_{0\boldsymbol{k}'} + \cdots \tag{5.25}$$

It is worth emphasizing that this is a matrix series in which different entries do not generally commute. Figure 5.6 gives a useful diagrammatic representation of the solution in Eqs. (5.22)–(5.25). It shows how to generate expansion terms of any required order.

In the equations above, the specifics of the system enter through Hamiltonian $\hat{H}_{\boldsymbol{k}}$ and Green's function $\hat{G}^{R/A}_{0\boldsymbol{k}}$. We now apply these general formulae to a TI thin film. Since its Hamiltonian (4.41) is block-diagonal, it suffices to consider only the '$+$' block, i.e. Hamiltonian $\hat{H}_{\boldsymbol{k}}$ in Eqs. (5.16)–(5.19) is identified as $\hat{H}_{\boldsymbol{k}+}$ from Eq. (4.41),

$$\hat{H}_{\boldsymbol{k}} = \hat{H}_{\boldsymbol{k}+}. \tag{5.26}$$

In this case, the velocity operator (5.17) is

$$\hat{\boldsymbol{v}}_{\boldsymbol{k}} = \hbar^{-1}(\mathcal{A}\boldsymbol{\sigma} + \partial_{\boldsymbol{k}}\mathcal{M}_{\boldsymbol{k}}\sigma_z + 2\mathcal{D}\boldsymbol{k}\sigma_0) \tag{5.27}$$

$$= \hbar^{-1}(\mathcal{A}\boldsymbol{\sigma} + 2\mathcal{B}\boldsymbol{k}\sigma_z + 2\mathcal{D}\boldsymbol{k}\sigma_0), \tag{5.28}$$

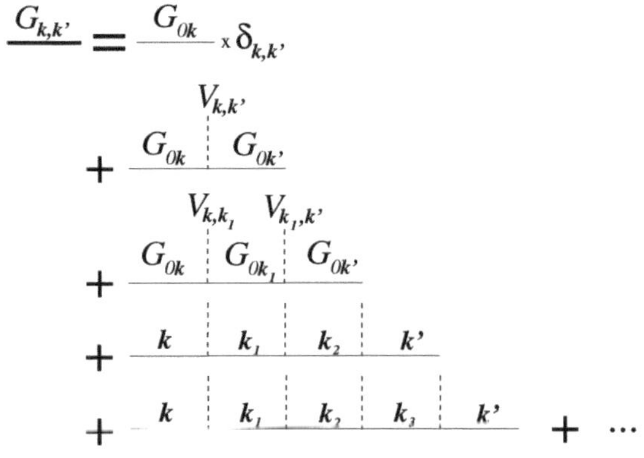

Figure 5.6 Diagrammatic representation of Green's function (5.22): the thick solid line corresponds to the exact Green function $\hat{G}^{R/A}_{k,k'}$, a thin solid line depicts the unperturbed Green function $\hat{G}^{R/A}_{0k}$ and a vertical dashed line represents the potential matrix element $\hat{V}_{k,k'}$. Summation is assumed over all the intermediate wave vectors $\boldsymbol{k}_1, \boldsymbol{k}_2, \boldsymbol{k}_3, \ldots$

and Eq. (5.21) for the unperturbed Green function $\hat{G}^{R/A}_{0k}$ reads

$$(E_{R/A} + E_F - \mathcal{D}\boldsymbol{k}^2 - \mathcal{A}\boldsymbol{\sigma}\cdot\boldsymbol{k} - \mathcal{M}_{\boldsymbol{k}}\sigma_z)\hat{G}^{R/A}_{0\boldsymbol{k}} = \sigma_0. \tag{5.29}$$

In order to invert this equation we apply matrix $E_{R/A} + E_F - \mathcal{D}\boldsymbol{k}^2 + \mathcal{A}\boldsymbol{\sigma}\cdot\boldsymbol{k} + \mathcal{M}_{\boldsymbol{k}}\sigma_z$ from the left:

$$\hat{G}^{R/A}_{0\boldsymbol{k}} = \frac{E_{R/A} + E_F - \mathcal{D}\boldsymbol{k}^2 + \mathcal{A}\boldsymbol{\sigma}\cdot\boldsymbol{k} + \mathcal{M}_{\boldsymbol{k}}\sigma_z}{[E_{R/A} - \xi_{\boldsymbol{k}}][E_{R/A} - \overline{\xi}_{\boldsymbol{k}}]}, \tag{5.30}$$

where

$$\xi_{\boldsymbol{k}} = \sqrt{\mathcal{A}^2\boldsymbol{k}^2 + \mathcal{M}^2_{\boldsymbol{k}}} + \mathcal{D}\boldsymbol{k}^2 - E_F,$$
$$\overline{\xi}_{\boldsymbol{k}} = -\sqrt{\mathcal{A}^2\boldsymbol{k}^2 + \mathcal{M}^2_{\boldsymbol{k}}} + \mathcal{D}\boldsymbol{k}^2 - E_F, \tag{5.31}$$

are the conduction- and valence-band dispersions, respectively.

We assume that the Fermi level lies in the conduction band, as depicted in Figs. 4.4 and 5.7. The corresponding Fermi surface is a closed line in momentum space, which is determined by

$$\xi_{\boldsymbol{k}_F} = \sqrt{\mathcal{A}^2\boldsymbol{k}_F^2 + \mathcal{M}^2_{\boldsymbol{k}_F}} + \mathcal{D}\boldsymbol{k}_F^2 - E_F = 0. \tag{5.32}$$

This is an equation for the Fermi wave vector, $\boldsymbol{k}_F$. At the same time, the valence band energy $\overline{\xi}_{\boldsymbol{k}}$ does not vanish for $\boldsymbol{k} = \boldsymbol{k}_F$, being always

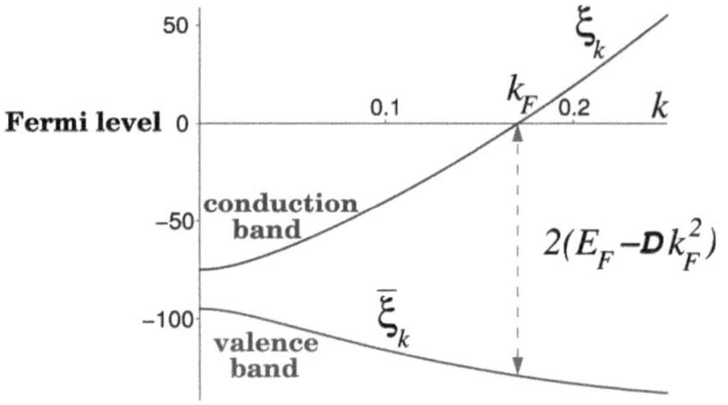

Figure 5.7 Conduction- and valence-band energies (in meV) versus wave number k (in nm^{-1}) for typical parameters $\mathcal{A} = 350$ meV nm, $\mathcal{B} = 500$ meV nm^2, $\mathcal{D} = 700$ meV nm^2, $\mathcal{M} = 10$ meV, and $E_F = 85$ meV; see Eq. (5.31).

negative (see also Fig. 5.7):

$$\bar{\xi}_{\boldsymbol{k}_F} = -\sqrt{\mathcal{A}^2\boldsymbol{k}_F^2 + \mathcal{M}_{\boldsymbol{k}_F}^2} + \mathcal{D}\boldsymbol{k}_F^2 - E_F \tag{5.33}$$

$$= -2\sqrt{\mathcal{A}^2\boldsymbol{k}_F^2 + \mathcal{M}_{\boldsymbol{k}_F}^2} \tag{5.34}$$

$$= -2(E_F - \mathcal{D}\boldsymbol{k}_F^2). \tag{5.35}$$

The difference $\xi_{\boldsymbol{k}_F} - \bar{\xi}_{\boldsymbol{k}_F}$ defines the energy separation between the conduction and valence bands for given $\boldsymbol{k}_F$,

$$\xi_{\boldsymbol{k}_F} - \bar{\xi}_{\boldsymbol{k}_F} = 2\sqrt{\mathcal{A}^2\boldsymbol{k}_F^2 + \mathcal{M}_{\boldsymbol{k}_F}^2} = 2(E_F - \mathcal{D}\boldsymbol{k}_F^2). \tag{5.36}$$

Since in many situations only the wave vectors close to $\boldsymbol{k}_F$ matter, we can simplify Eq. (5.30) as follows:

$$\hat{G}_{0\boldsymbol{k}}^{R/A} \approx \frac{E_{R/A} + \sqrt{\mathcal{A}^2\boldsymbol{k}_F^2 + \mathcal{M}_{\boldsymbol{k}_F}^2} + \mathcal{A}\boldsymbol{\sigma}\cdot\boldsymbol{k}_F + \mathcal{M}_{\boldsymbol{k}_F}\sigma_z}{[E_{R/A} - \xi_{\boldsymbol{k}}][E_{R/A} + 2\sqrt{\mathcal{A}^2\boldsymbol{k}_F^2 + \mathcal{M}_{\boldsymbol{k}_F}^2}]}, \quad \boldsymbol{k} \approx \boldsymbol{k}_F. \tag{5.37}$$

Also, for sufficiently large E_F the band separation (5.36) can be made larger than the characteristic energy scale of $E_{R/A}$, i.e.

$$2\sqrt{\mathcal{A}^2\boldsymbol{k}_F^2 + \mathcal{M}_{\boldsymbol{k}_F}^2} = 2(E_F - \mathcal{D}\boldsymbol{k}_F^2) \gg E_{R/A}. \tag{5.38}$$

In a disordered system, $E_{R/A} \sim \hbar/\tau$, where τ is the elastic lifetime [cf. Eq. (5.3)]. Under condition (5.38) Green's function (5.37) assumes a rather compact form

$$\hat{G}^{R/A}_{0\boldsymbol{k}} \approx \frac{\hat{P}_{\boldsymbol{n}}}{E_{R/A} - \xi_{\boldsymbol{k}}}, \qquad \xi_{\boldsymbol{k}} \approx \hbar v(|\boldsymbol{k}| - k_F), \tag{5.39}$$

$$\hat{P}_{\boldsymbol{n}} = \frac{\sigma_0 + \boldsymbol{\sigma} \cdot \mathbf{e}_{\boldsymbol{n}}}{2}, \quad \mathbf{e}_{\boldsymbol{n}} = \frac{\mathcal{A}k_F\boldsymbol{n} + \mathcal{M}_{k_F}\mathbf{z}}{\sqrt{\mathcal{A}^2k_F^2 + \mathcal{M}^2_{k_F}}}, \quad \boldsymbol{n} = \frac{\boldsymbol{k}}{|\boldsymbol{k}|}, \tag{5.40}$$

where the information about the carrier spin is contained in the matrix $\hat{P}_{\boldsymbol{n}}$. In particular, the unit vector $\mathbf{e}_{\boldsymbol{n}}$ indicates the direction of the carrier spin for given direction of the momentum, $\boldsymbol{n}$. That is, $\boldsymbol{\sigma} \cdot \mathbf{e}_{\boldsymbol{n}}$ is the spin helicity matrix, which is of primary importance throughout. It follows from Eq. (5.39) that

$$(\boldsymbol{\sigma} \cdot \mathbf{e}_{\boldsymbol{n}})\hat{G}^{R/A}_{0\boldsymbol{k}} = \hat{G}^{R/A}_{0\boldsymbol{k}}, \tag{5.41}$$

meaning that Green function (5.39) describes a state with the helicity eigenvalue equal to $+1$. It can be shown that the other eigenvalue, -1, corresponds to the valence band states. Since the states near the Fermi surface $|\boldsymbol{k}| \approx k_F$ matter most, the conduction band dispersion, $\xi_{\boldsymbol{k}}$, in Eq. (5.39) is linearized near k_F.

5.2.2 Potential Disorder

We further narrow our model by considering a spin-independent potential of the form

$$\hat{V}(\boldsymbol{r}) = V(\boldsymbol{r})\sigma_0, \tag{5.42}$$

where σ_0 denotes a 2×2 unit spin matrix, and $V(\boldsymbol{r})$ is a random function of position $\boldsymbol{r}$. We assume that the disorder originates from a large number of randomly distributed impurities with a homogeneous density in the thermodynamic limit. In this case, $V(\boldsymbol{r})$ has a Gaussian distribution which is fully characterized by a two-point correlation function. We assume short-ranged correlations modeled by

$$\zeta(\boldsymbol{r} - \boldsymbol{r}') = \langle\langle V(\boldsymbol{r})V(\boldsymbol{r}')\rangle\rangle = \frac{\hbar}{\pi N \tau_0}\delta(\boldsymbol{r} - \boldsymbol{r}'), \tag{5.43}$$

where $\langle\langle \ldots \rangle\rangle$ denotes averaging over the random positions of the impurities. For convenience of further calculations, the correlation function is parametrized in terms of the characteristic scattering time, τ_0, and the density of states (DOS) at the Fermi level,

$$N = \frac{k_F}{2\pi \hbar v}. \tag{5.44}$$

From Eqs. (5.19) and (5.42) we derive the correlation function for the matrix elements of the random potential,

$$\langle\langle \hat{V}_{\boldsymbol{k},\boldsymbol{k}_1} \otimes \hat{V}_{\boldsymbol{k}',\boldsymbol{k}_2} \rangle\rangle = \sigma_0 \otimes \sigma_0 \tag{5.45}$$

$$\times \int \frac{d\boldsymbol{r}_1 d\boldsymbol{r}_2}{a^2} \zeta(\boldsymbol{r}_1 - \boldsymbol{r}_2) e^{-i\mathrm{r}_1(\boldsymbol{k}-\boldsymbol{k}_1) - i\mathrm{r}_2(\boldsymbol{k}'-\boldsymbol{k}_2)} \tag{5.46}$$

$$= \sigma_0 \otimes \sigma_0 \int \frac{d\boldsymbol{r} d\boldsymbol{R}}{a^2} \zeta(\boldsymbol{r}) e^{-i\mathrm{r}(\boldsymbol{k}-\boldsymbol{k}_1) - i\boldsymbol{R}(\boldsymbol{k}-\boldsymbol{k}_1+\boldsymbol{k}'-\boldsymbol{k}_2)} \tag{5.47}$$

$$= \frac{\zeta_{\boldsymbol{k}-\boldsymbol{k}_1}}{a} \delta_{\boldsymbol{k}-\boldsymbol{k}_1, -\boldsymbol{k}'+\boldsymbol{k}_2} \sigma_0 \otimes \sigma_0, \tag{5.48}$$

where $\zeta_{\boldsymbol{k}}$ is the Fourier transform of the correlation function $\zeta(\boldsymbol{r})$,

$$\zeta_{\boldsymbol{k}} = \int \zeta(\mathbf{r}) \mathrm{e}^{-i\boldsymbol{k}\mathbf{r}} d\mathbf{r} \equiv \zeta = \frac{\hbar}{\pi N \tau_0}, \tag{5.49}$$

and $\sigma_0 \otimes \sigma_0$ indicates the direct product of the unit matrices. We will also use an alternative notation

$$\langle\langle V_{\alpha\alpha_1}(\boldsymbol{k}, \boldsymbol{k}_1) V_{\beta\beta_2}(\boldsymbol{k}', \boldsymbol{k}_2) \rangle\rangle = \frac{\zeta}{a} \delta_{\boldsymbol{k}-\boldsymbol{k}_1, -\boldsymbol{k}'+\boldsymbol{k}_2} \delta_{\alpha,\alpha_1} \delta_{\beta,\beta_2}, \tag{5.50}$$

where α and β are the spin indices. The wave vector selection rule is quite simple: the difference between the first two wave vectors plus that between the second two wave vectors equals to zero. The higher-even-order correlators are reduced to products of the second-order ones through pair contractions, e.g.

$$\begin{aligned} &\langle\langle \hat{V}_{\boldsymbol{k},\boldsymbol{k}_1} \otimes \hat{V}_{\boldsymbol{k}',\boldsymbol{k}_2} \otimes \hat{V}_{\boldsymbol{k}'',\boldsymbol{k}_3} \otimes \hat{V}_{\boldsymbol{k}''',\boldsymbol{k}_4} \rangle\rangle \\ &= \hat{V}_{\boldsymbol{k},\boldsymbol{k}_1} \otimes \hat{V}_{\boldsymbol{k}',\boldsymbol{k}_2} \otimes \hat{V}_{\boldsymbol{k}'',\boldsymbol{k}_3} \otimes \hat{V}_{\boldsymbol{k}''',\boldsymbol{k}_4} \\ &+ \hat{V}_{\boldsymbol{k},\boldsymbol{k}_1} \otimes \hat{V}_{\boldsymbol{k}',\boldsymbol{k}_2} \otimes \hat{V}_{\boldsymbol{k}'',\boldsymbol{k}_3} \otimes \hat{V}_{\boldsymbol{k}''',\boldsymbol{k}_4} \\ &+ \hat{V}_{\boldsymbol{k},\boldsymbol{k}_1} \otimes \hat{V}_{\boldsymbol{k}',\boldsymbol{k}_2} \otimes \hat{V}_{\boldsymbol{k}'',\boldsymbol{k}_3} \otimes \hat{V}_{\boldsymbol{k}''',\boldsymbol{k}_4}, \end{aligned} \tag{5.51}$$

where we indicate all possible pair correlators. The odd correlation functions for a Gaussian random process vanish

$$\langle\langle \hat{V}_{\boldsymbol{k},\boldsymbol{k}_1} \rangle\rangle = 0, \qquad \langle\langle \hat{V}_{\boldsymbol{k},\boldsymbol{k}_1} \otimes \hat{V}_{\boldsymbol{k}',\boldsymbol{k}_2} \otimes \hat{V}_{\boldsymbol{k}',\boldsymbol{k}_3} \rangle\rangle = 0, \qquad \ldots \tag{5.52}$$

5.2.3 Nonperturbative Treatment of Disorder: Diagrammatics

The strategy to calculate the conductivity is conceptually simple. We should insert the series expansion (5.22) - (5.25) for the Green functions into Kubo formula (5.16), and average the latter over disorder configurations. On this way we encounter two types of averaged Green functions: the retarded and advanced function averaged independently, i.e.

$$\langle\langle \hat{G}^R_{\mathbf{k},\mathbf{k}'}(E+\hbar\omega)\rangle\rangle, \qquad \langle\langle \hat{G}^A_{\mathbf{k},\mathbf{k}'}\rangle\rangle, \tag{5.53}$$

and the average direct product of the retarded and advanced functions

$$\langle\langle \hat{G}^R_{\mathbf{k},\mathbf{k}'}(E+\hbar\omega)\otimes \hat{G}^A_{\mathbf{k}',\mathbf{k}}\rangle\rangle. \tag{5.54}$$

We begin by calculating the Green functions (5.53). The first step is to average each term in the expansion (5.22) with the help of the Gaussian correlators from Eqs. (5.48) - (5.52). Diagrammatically, this amounts to connecting pairs of the dashes lines in Fig. 5.6 in all possible ways, as depicted in Fig. 5.8 (the superscripts R/A have been omitted for clarity). Only the terms with even powers of the disorder potential remain

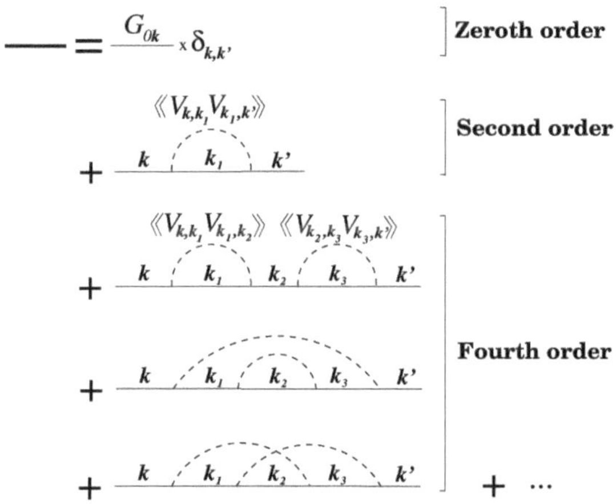

Figure 5.8 Diagrammatic representation of averaged Green function (5.22) up to fourth order in disorder potential.

$$\langle\langle \hat{G}_{\boldsymbol{k},\boldsymbol{k}'}\rangle\rangle = \hat{G}^{(0)}_{\boldsymbol{k},\boldsymbol{k}'} + \hat{G}^{(2)}_{\boldsymbol{k},\boldsymbol{k}'} + \hat{G}^{(4)}_{\boldsymbol{k},\boldsymbol{k}'} + \ldots \tag{5.55}$$

where $\hat{G}^{(0)}_{\boldsymbol{k},\boldsymbol{k}'}$, $\hat{G}^{(2)}_{\boldsymbol{k},\boldsymbol{k}'}$ and $\hat{G}^{(4)}_{\boldsymbol{k},\boldsymbol{k}'}$ are represented by the zeroth-, second- and fourth-order diagrams in Fig. 5.8. The corresponding analytical expressions are

$$\hat{G}^{(0)}_{\boldsymbol{k},\boldsymbol{k}'} = \hat{G}_{0\boldsymbol{k}}\delta_{\boldsymbol{k},\boldsymbol{k}'}, \tag{5.56}$$

$$\begin{aligned}
\hat{G}^{(2)}_{\boldsymbol{k},\boldsymbol{k}'} &= \hat{G}_{0\boldsymbol{k}} \sum_{\boldsymbol{k}_1} \hat{V}_{\boldsymbol{k},\boldsymbol{k}_1}\hat{G}_{0\boldsymbol{k}_1}\hat{V}_{\boldsymbol{k}_1,\boldsymbol{k}'}\hat{G}_{0\boldsymbol{k}'} \\
&= \hat{G}_{0\boldsymbol{k}}\delta_{\boldsymbol{k},\boldsymbol{k}'}\frac{\zeta}{a}\sum_{\boldsymbol{k}_1}\hat{G}_{0\boldsymbol{k}_1}\hat{G}_{0\boldsymbol{k}} \\
&= \hat{G}^{(0)}_{\boldsymbol{k},\boldsymbol{k}'}\zeta\int\frac{d\boldsymbol{k}_1}{(2\pi)^2}\hat{G}_{0\boldsymbol{k}_1}\hat{G}_{0\boldsymbol{k}},
\end{aligned} \tag{5.57}$$

$$\hat{G}^{(4)}_{\boldsymbol{k},\boldsymbol{k}'} = \hat{G}_{0\boldsymbol{k}} \sum_{\boldsymbol{k}_1,\boldsymbol{k}_2,\boldsymbol{k}_3} \hat{V}_{\boldsymbol{k},\boldsymbol{k}_1}\hat{G}_{0\boldsymbol{k}_1}\hat{V}_{\boldsymbol{k}_1,\boldsymbol{k}_2}\hat{G}_{0\boldsymbol{k}_2}\hat{V}_{\boldsymbol{k}_2,\boldsymbol{k}_3}\hat{G}_{0\boldsymbol{k}_3}\hat{V}_{\boldsymbol{k}_3,\boldsymbol{k}'}\hat{G}_{0\boldsymbol{k}'} \tag{5.58}$$

$$+\hat{G}_{0\boldsymbol{k}} \sum_{\boldsymbol{k}_1,\boldsymbol{k}_2,\boldsymbol{k}_3} \hat{V}_{\boldsymbol{k},\boldsymbol{k}_1}\hat{G}_{0\boldsymbol{k}_1}\hat{V}_{\boldsymbol{k}_1,\boldsymbol{k}_2}\hat{G}_{0\boldsymbol{k}_2}\hat{V}_{\boldsymbol{k}_2,\boldsymbol{k}_3}\hat{G}_{0\boldsymbol{k}_3}\hat{V}_{\boldsymbol{k}_3,\boldsymbol{k}'}\hat{G}_{0\boldsymbol{k}'} \tag{5.59}$$

$$+\hat{G}_{0\boldsymbol{k}} \sum_{\boldsymbol{k}_1,\boldsymbol{k}_2,\boldsymbol{k}_3} \hat{V}_{\boldsymbol{k},\boldsymbol{k}_1}\hat{G}_{0\boldsymbol{k}_1}\hat{V}_{\boldsymbol{k}_1,\boldsymbol{k}_2}\hat{G}_{0\boldsymbol{k}_2}\hat{V}_{\boldsymbol{k}_2,\boldsymbol{k}_3}\hat{G}_{0\boldsymbol{k}_3}\hat{V}_{\boldsymbol{k}_3,\boldsymbol{k}'}\hat{G}_{0\boldsymbol{k}'}. \tag{5.60}$$

Evaluating the indicated pair correlators with the help of Eq. (5.50), we have

$$\hat{G}^{(4)}_{\boldsymbol{k},\boldsymbol{k}'} = \hat{G}^{(0)}_{\boldsymbol{k},\boldsymbol{k}'}\left[\zeta\int\frac{d\boldsymbol{k}_1}{(2\pi)^2}\hat{G}_{0\boldsymbol{k}_1}\right]\hat{G}_{0\boldsymbol{k}}\left[\zeta\int\frac{d\boldsymbol{k}_3}{(2\pi)^2}\hat{G}_{0\boldsymbol{k}_3}\right]\hat{G}_{0\boldsymbol{k}} \tag{5.61}$$

$$+\hat{G}^{(0)}_{\boldsymbol{k},\boldsymbol{k}'}\left[\zeta\int\frac{d\boldsymbol{k}_1}{(2\pi)^2}\hat{G}_{0\boldsymbol{k}_1}\left[\zeta\int\frac{d\boldsymbol{k}_2}{(2\pi)^2}\hat{G}_{0\boldsymbol{k}_2}\right]\hat{G}_{0\boldsymbol{k}_1}\right]\hat{G}_{0\boldsymbol{k}} \tag{5.62}$$

$$+\hat{G}^{(0)}_{\boldsymbol{k},\boldsymbol{k}'}\left[\zeta\int\frac{d\boldsymbol{k}_1}{(2\pi)^2}\hat{G}_{0\boldsymbol{k}_1}\left[\zeta\int\frac{d\boldsymbol{k}_2}{(2\pi)^2}\hat{G}_{0\boldsymbol{k}_2}\hat{G}_{0\boldsymbol{k}_2-\boldsymbol{k}_1+\boldsymbol{k}}\right]\right]\hat{G}_{0\boldsymbol{k}}. \tag{5.63}$$

The obtained series (5.55) is not a perturbative expansion of powers of a certain small parameter. Generally, there is no small parameter because the Green function $\hat{G}_{0\boldsymbol{k}}$ (5.39) diverges at the Fermi surface $|\boldsymbol{k}| = k_F$ for $E \to 0$. It is quite obvious from Eq. (5.57) for the second-order term $\hat{G}^{(2)}_{\boldsymbol{k},\boldsymbol{k}'}$. As we cannot truncate the series at any finite order, the usual perturbative treatment fails.

At the same time, it is not possible to calculate the infinite sum exactly. The way out of this problem is to eliminate *irrelevant* terms and then try to sum up the rest of the infinite series by analogy with the Feynman diagrammatic technique. The irrelevant terms are those that *do* contain a small parameter compared with the previous order. As we just saw, this does not apply to the second-order term $\hat{G}^{(2)}_{\boldsymbol{k},\boldsymbol{k}'}$. However, an irrelevant term appears in the next—fourth—order. This is the last (crossed) diagram in Fig. 5.8. In its analytical expression (5.63) the $\boldsymbol{k}_2$ integration involves two Green functions, $\hat{G}_{0\boldsymbol{k}_2}$ and $\hat{G}_{0\boldsymbol{k}_2-\boldsymbol{k}_1+\boldsymbol{k}}$. Due to the wave vector difference, these functions cannot be simultaneously singular on the Fermi surface, leading to a suppressed contribution. To estimate it let us consider the integral

$$\zeta \int \frac{d\boldsymbol{k}_2}{(2\pi)^2} \hat{G}_{0\boldsymbol{k}_2} \hat{G}_{0\boldsymbol{k}_2-\boldsymbol{k}_1+\boldsymbol{k}} \approx \tag{5.64}$$

$$\frac{\zeta k_F}{2\pi} \int_{-\infty}^{\infty} \frac{d\kappa}{E - \hbar v \kappa} \overline{\frac{\hat{P}_{\boldsymbol{n}_2} \hat{P}_{\boldsymbol{n}_2-\boldsymbol{n}_1+\boldsymbol{n}}}{E - \hbar v(|(\kappa + k_F)\boldsymbol{n}_2 - \boldsymbol{k}_1 + \boldsymbol{k}| - k_F)}}, \tag{5.65}$$

where we switched to the integration over the wave vector difference $\kappa = k_2 - k_F$, extending the lower integration limit from $-k_F$ to $-\infty$, which is permitted under condition (5.38), and the bar denotes averaging over the directions of the unit vector $\boldsymbol{n}_2$,

$$\overline{(\ldots)} = \int_0^{2\pi} \frac{d\phi_{\boldsymbol{n}_2}}{2\pi} \ldots \tag{5.66}$$

Since the function in Eq. (5.65) is integrable for $\kappa \to \pm\infty$, the integration path can chosen around the pole of the first fraction, which yields

$$\pm 2\pi i \zeta N \overline{\frac{\hat{P}_{\boldsymbol{n}_2} \hat{P}_{\boldsymbol{n}_2-\boldsymbol{n}_1+\boldsymbol{n}}}{\hbar v(|(\frac{E}{\hbar v} + k_F)\boldsymbol{n}_2 - \boldsymbol{k}_1 + \boldsymbol{k}| - k_F) - E}}, \tag{5.67}$$

where $\pm$ are for the retarded and advanced functions, respectively. The wave vectors $\boldsymbol{k}$ and $\boldsymbol{k}_1$ can be taken on the Fermi surface,

$$\boldsymbol{k} = k_F \boldsymbol{n}, \qquad \boldsymbol{k}_1 = k_F \boldsymbol{n}_1, \tag{5.68}$$

because they are the arguments of two different Green functions, $\hat{G}_{0\boldsymbol{k}}$ and $\hat{G}_{0\boldsymbol{k}_1}$, each being singular on the Fermi surface [see Eq. (5.63)].

For $\hbar v k_F \gg E$ and ζ given by Eq. (5.49), the estimate of Eq. (5.67) is

$$\pm\frac{2i}{k_F v \tau_0}\overline{\frac{\hat{P}_{n_2}\hat{P}_{n_2-n_1+n}}{|\boldsymbol{n}_2-\boldsymbol{n}_1+\boldsymbol{n}|-1}}. \tag{5.69}$$

The prefactor is a small parameter $\sim 1/k_F v\tau_0 \sim 1/k_F\ell_{tr} \ll 1$ [cf., Eq. (5.5)]. Although the denominator $|\boldsymbol{n}_2-\boldsymbol{n}_1+\boldsymbol{n}|-1$ can vanish, this cannot compensate the smallness of the prefactor because this happens only for the special direction of $\boldsymbol{n}_2=\boldsymbol{n}_1-\boldsymbol{n}$, which has a small weight in the angle integral. Therefore, in the metallic regime with $k_F\ell_{tr} \gg 1$ the contribution of the crossed diagram [see Eq. (5.63)] is small compared to the second-order term (5.57),

$$\hat{G}^{(4)}_{\boldsymbol{k},\boldsymbol{k}'}|_{\text{crossed}} \sim \frac{1}{k_F\ell_{tr}}\hat{G}^{(2)}_{\boldsymbol{k},\boldsymbol{k}'} \ll \hat{G}^{(2)}_{\boldsymbol{k},\boldsymbol{k}'}. \tag{5.70}$$

As it only slightly renormalizes the second-order term, the crossed diagram can be neglected. Note that the other two fourth-order diagrams involve the integrals with a single Green function and are not small compared to $\hat{G}^{(2)}_{\boldsymbol{k},\boldsymbol{k}'}$.

To summarize, we see that all terms in the averaged series (5.55) are proportional to the Kronecker symbol $\delta_{\boldsymbol{k},\boldsymbol{k}'}$, allowing us to write the averaged Green function as

$$\langle\langle\hat{G}_{\boldsymbol{k},\boldsymbol{k}'}\rangle\rangle = \hat{G}_{\boldsymbol{k}}\delta_{\boldsymbol{k},\boldsymbol{k}'}. \tag{5.71}$$

The function $\hat{G}_{\boldsymbol{k}}$ is represented by an infinite series of *non-crossing* diagrams. This series sums up into a closed equation for $\hat{G}_{\boldsymbol{k}}$,

$$\hat{G}_{\boldsymbol{k}} = \hat{G}_{0\boldsymbol{k}} + \hat{G}_{0\boldsymbol{k}}\,\zeta\int\frac{d\boldsymbol{k}_1}{(2\pi)^2}\hat{G}_{\boldsymbol{k}_1}\hat{G}_{\boldsymbol{k}}, \tag{5.72}$$

By iteration we can check that Eq. (5.72) generates all possible non-crossing diagrams, e.g. Eqs. (5.57), (5.61) and (5.62), as depicted in Fig. 5.9. The non-perturbative approximate method employed

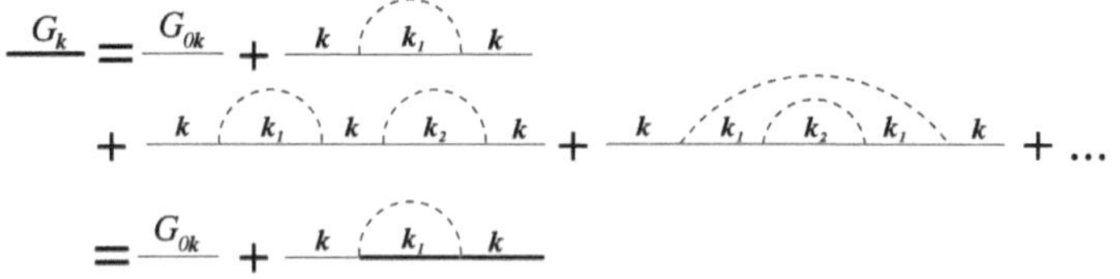

Figure 5.9 Diagrammatic representation of Eq. (5.72) for the averaged Green function $\hat{G}_{\boldsymbol{k}}$ (thick lines) in the self-consistent Born approximation.

above to derive Eq. (5.72) is known as the self-consistent Born approximation.

In the following, we outline the solution procedure for Eq. (5.72). We seek the solution in the form

$$\hat{G}_{\boldsymbol{k}} = \frac{\hat{P}_{\boldsymbol{n}}}{E - \xi_{\boldsymbol{k}} - \Sigma_{\boldsymbol{n}}}. \tag{5.73}$$

Here $\hat{P}_{\boldsymbol{n}}$ is the same projector matrix as in Eq. (5.39) for the unperturbed Green function. $\Sigma_{\boldsymbol{n}}$ is the unknown function called *self-energy*. It can depend only on the momentum direction $\boldsymbol{n}$ because $|\boldsymbol{k}|$ must be close to k_F. Inserting Eqs. (5.39) and (5.73) into Eq. (5.72) and using the identity

$$\hat{P}_{\boldsymbol{n}}\hat{P}_{\boldsymbol{n}_1}\hat{P}_{\boldsymbol{n}} = \hat{P}_{\boldsymbol{n}}\frac{1 + \mathbf{e}_{\boldsymbol{n}} \cdot \mathbf{e}_{\boldsymbol{n}_1}}{2}, \tag{5.74}$$

we obtain the equation for the self-energy:

$$\Sigma_{\boldsymbol{n}} = \zeta \int \frac{1}{E - \xi_{\boldsymbol{k}_1} - \Sigma_{\boldsymbol{n}_1}} \frac{1 + \mathbf{e}_{\boldsymbol{n}} \cdot \mathbf{e}_{\boldsymbol{n}_1}}{2} \frac{d\boldsymbol{k}_1}{(2\pi)^2}. \tag{5.75}$$

We again separate the angle averaging [see Eq. (5.66)] and transform the integral over $|\boldsymbol{k}_1|$ into integration over $\xi_{\boldsymbol{k}_1}$ with the limits from $-E_F$ to ∞. Under condition (5.38) the main contribution to the integral comes from the vicinity of the Fermi surface, $\xi_{\boldsymbol{k}_1} \approx 0$. We, therefore, can extend the lower limit from $-E_F$ to $-\infty$,

$$\Sigma_{\boldsymbol{n}} = \zeta \int_0^{2\pi} \frac{d\phi_{\boldsymbol{n}_1}}{2\pi} \frac{1 + \mathbf{e}_{\boldsymbol{n}} \cdot \mathbf{e}_{\boldsymbol{n}_1}}{2} N \int_{-\infty}^{\infty} d\xi \frac{1}{E - \xi - \Sigma_{\boldsymbol{n}_1}}. \tag{5.76}$$

Moreover, the real part of $\Sigma_{\boldsymbol{n}_1}$ can be absorbed into the integration over ξ, leading to a purely imaginary self-energy

$$\Sigma_{\boldsymbol{n}} = \mp i|\Im\Sigma_{\boldsymbol{n}}|, \tag{5.77}$$

$$|\Im\Sigma_{\boldsymbol{n}}| = \zeta N \int_0^{2\pi} \frac{d\phi_{\boldsymbol{n}_1}}{2\pi} \frac{1 + \mathbf{e}_{\boldsymbol{n}} \cdot \mathbf{e}_{\boldsymbol{n}_1}}{2} \int_{-\infty}^{\infty} d\xi \frac{|\Im\Sigma_{\boldsymbol{n}_1}|}{(E - \xi)^2 + |\Im\Sigma_{\boldsymbol{n}_1}|^2}, \tag{5.78}$$

where $\Im$ denotes the imaginary part and $\mp$ are for the retarded and advanced functions, respectively. The ξ integral equals to π, so that with Eq. (5.49) for ζ we have

$$|\Im\Sigma_{\boldsymbol{n}}| = \frac{\hbar}{\tau_0} \int_0^{2\pi} \frac{d\phi_{\boldsymbol{n}_1}}{2\pi} \frac{1 + \mathbf{e}_{\boldsymbol{n}} \cdot \mathbf{e}_{\boldsymbol{n}_1}}{2}. \tag{5.79}$$

Inserting Eqs. (5.77) and (5.79) back into Eq. (5.73) and reinstalling the superscripts R/A, we arrive at the final expressions for the disorder-averaged retarded and advanced Green functions,

$$\hat{G}_{\boldsymbol{k}}^{R/A} = \frac{\hat{P}_{\boldsymbol{n}}}{E_{R/A} - \xi_{\boldsymbol{k}}}, \qquad E_{R/A} = E \pm \frac{i\hbar}{2\tau}. \tag{5.80}$$

Thus, the disorder averaging introduces a finite single-particle lifetime, τ, which is given by

$$\frac{1}{\tau} = \frac{2}{\hbar}|\Im\Sigma_{\boldsymbol{n}}| = \frac{1}{\tau_0}\int_0^{2\pi}\frac{d\phi_{\mathbf{n}_1}}{2\pi}(1 + \mathbf{e}_{\boldsymbol{n}}\cdot\mathbf{e}_{\boldsymbol{n}_1}), \tag{5.81}$$

where unit vectors $\mathbf{n}$ and $\mathbf{n}_1$ specify the directions of the incident and scattered momentum states, respectively, while unit vector $\mathbf{e}_{\boldsymbol{n}}$ can be written as

$$\mathbf{e}_{\boldsymbol{n}} = \mathbf{e}_{\parallel} + \mathbf{e}_{\perp}, \quad \mathbf{e}_{\parallel} = \boldsymbol{n}\sqrt{1 - e_{\perp}^2}, \quad \mathbf{e}_{\perp} = \frac{\mathcal{M}_{k_F}\,\mathbf{z}}{\sqrt{\mathcal{A}^2 k_F^2 + \mathcal{M}_{k_F}^2}}, \tag{5.82}$$

with $e_{\parallel} \equiv |\mathbf{e}_{\parallel}|$ and $e_{\perp} \equiv |\mathbf{e}_{\perp}|$. The angle averaging in Eq. (5.81) yields

$$\tau = \frac{\tau_0}{1 + e_{\perp}^2} = \frac{\tau_0}{1 + \mathcal{M}_{k_F}^2/(\mathcal{A}^2 k_F^2 + \mathcal{M}_{k_F}^2)}. \tag{5.83}$$

This result is interesting because the time τ is shorter than the disorder-related time scale, τ_0 [see Eq. (5.49)], and involves the band structure parameters, the "mass" term $\mathcal{M}_{k_F}$ and constant $\mathcal{A}$. In order to understand this we recall that $\mathcal{M}_{k_F}$ effectively acts as a Zeeman field, polarizing the carrier spin out of plane. This enables backscattering into an opposite momentum state with $\mathbf{n}_1 = -\mathbf{n}$ [see Eq. (5.81) and Fig. 5.4], otherwise forbidden for massless Dirac fermions. Thus, the parameter $e_{\perp}^2$ in Eq. (5.83) quantifies the strength of the backscattering.

At this stage it is also convenient to introduce a related transport characteristic, the transport scattering time τ_{tr} [78]. It differs from τ (5.81) by an extra factor $1 - \boldsymbol{n}\cdot\boldsymbol{n}_1$ under the integral

$$\frac{1}{\tau_{tr}} = \frac{1}{\tau_0}\int_0^{2\pi}\frac{d\phi_{\mathbf{n}_1}}{2\pi}(1 - \boldsymbol{n}\cdot\boldsymbol{n}_1)(1 + \mathbf{e}_{\boldsymbol{n}}\cdot\mathbf{e}_{\boldsymbol{n}_1}) = \frac{1 + 3e_{\perp}^2}{2\tau_0}. \tag{5.84}$$

This time determines the diffusion coefficient $D = v^2\tau_{tr}/2$ and the classical conductivity via the Einstein relation,

$$\sigma_c = e^2 N D = \frac{e^2}{h}k_F v \tau_{tr} = \frac{e^2}{h}k_F \ell_{tr}. \tag{5.85}$$

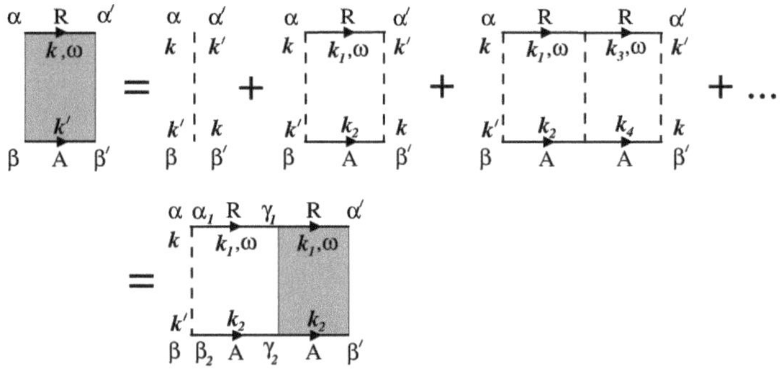

Figure 5.10 Bethe–Salpeter equation for Cooperon (shaded area). Vertical dashed lines represent the disorder correlator, e.g. $\langle\langle V_{\alpha\alpha_1}(\mathbf{k}, \mathbf{k}_1) V_{\beta\beta_2}(\mathbf{k}', \mathbf{k}_2)\rangle\rangle$. Horizontal lines R and A denote dressed retarded and advanced Green functions, e.g. $G^{R}_{\alpha_1,\gamma_1}(\mathbf{k}_1, E + \hbar\omega)$ and $G^{A}_{\beta_2,\gamma_2}(\mathbf{k}_2, E)$

The reason why τ_{tr}, rather than τ, enters the conductivity is that the factor $1 - \mathbf{n} \cdot \mathbf{n}_1$ accounts for the non-equilibrium distribution of the carriers around the direction of the applied electric field. This effect is not captured in life-time τ (5.81). In Section 5.2.7 the calculation of the classical conductivity is explained in more detail.

5.2.4 Cooperon

We now turn to averaging the direct product of the retarded and advanced Green functions [see Eq. (5.54)]. This reduces to calculating a specific two-particle Green function $C_{\alpha\beta\alpha'\beta'}(\mathbf{k}, \mathbf{k}', \omega)$, known as Cooperon. Similar to the single-particle Green function, the Cooperon can be expressed by an infinite series of diagrams,

$$C_{\alpha\beta\alpha'\beta'}(\mathbf{k}, \mathbf{k}', \omega) = C^{(2)}_{\alpha\beta\alpha'\beta'}(\mathbf{k}, \mathbf{k}', \omega) + C^{(4)}_{\alpha\beta\alpha'\beta'}(\mathbf{k}, \mathbf{k}', \omega) + \cdots, \quad (5.86)$$

as depicted in Fig. 5.10. As the crossed diagrams can again be omitted, the disorder averaging amounts to connecting the retarded and advanced functions by vertical dashed lines. This generates a series of "ladder" diagrams in which the retarded and advanced functions are simultaneously "dressed" according to Fig. 5.9. Since the zeroth order term $\langle\langle \hat{G}^{R}_{\mathbf{k},\mathbf{k}'}(E + \hbar\omega)\rangle\rangle \otimes \langle\langle \hat{G}^{A}_{\mathbf{k}',\mathbf{k}}\rangle\rangle$ involves the independently averaged functions, we have extracted it, which makes the subsequent calculation of the conductivity correction easier. The ladder diagrams sum up into a closed equation for $C_{\alpha\beta\alpha'\beta'}(\mathbf{k}, \mathbf{k}', \omega)$, known as the Bethe–Salpeter equation:

$$\begin{aligned} C_{\alpha\beta\alpha'\beta'}(\mathbf{k}, \mathbf{k}', \omega) &= \langle\langle V_{\alpha\alpha'}(\mathbf{k}, \mathbf{k}')V_{\beta\beta'}(\mathbf{k}', \mathbf{k})\rangle\rangle \\ &\quad + \sum_{\alpha_1,\beta_2,\mathbf{k}_1,\mathbf{k}_2} \langle\langle V_{\alpha\alpha_1}(\mathbf{k}, \mathbf{k}_1)V_{\beta\beta_2}(\mathbf{k}', \mathbf{k}_2)\rangle\rangle \\ &\quad \times G^{R}_{\alpha_1,\gamma_1}(\mathbf{k}_1, E + \hbar\omega)G^{A}_{\beta_2,\gamma_2}(\mathbf{k}_2, E)C_{\gamma_1\gamma_2,\alpha',\beta'}(\mathbf{k}_1, \mathbf{k}_2, \omega). \end{aligned} \tag{5.87}$$

With correlation function (5.50) the above equation reads

$$\begin{aligned} C_{\alpha\beta\alpha'\beta'}(\mathbf{k}, \mathbf{k}', \omega) &= \frac{\zeta}{a}\delta_{\alpha\alpha'}\delta_{\beta\beta'} \\ &\quad + \zeta\int \frac{d\mathbf{k}_1}{(2\pi)^2} G^{R}_{\alpha,\gamma_1}(\mathbf{k}_1, E + \hbar\omega)G^{A}_{\beta,\gamma_2}(\mathbf{k}+\mathbf{k}' - \mathbf{k}_1, E) \\ &\quad \times C_{\gamma_1\gamma_2,\alpha',\beta'}(\mathbf{k}_1, \mathbf{k}+\mathbf{k}' - \mathbf{k}_1, \omega). \end{aligned} \tag{5.88}$$

We note that the right-hand side is a function of the wave vector sum $\mathbf{k}+\mathbf{k}'$. Therefore, the Cooperon depends on the *sum* of its wave vector arguments:

$$C_{\alpha\beta\alpha'\beta'}(\mathbf{k}, \mathbf{k}', \omega) = C_{\alpha\beta\alpha'\beta'}(\mathbf{q}, \omega), \qquad \mathbf{q} = \mathbf{k}+\mathbf{k}'. \tag{5.89}$$

For $\mathbf{q} \to 0$ the wave vectors $\mathbf{k}$ and $\mathbf{k}'$ are antiparallel to each other. This is reminiscent of a Cooper pair, hence the name "Cooperon". In the weak localization theory the Cooperon (5.89) with $\mathbf{q} \to 0$ describes quantum interference of two states with the antiparallel momenta in a disordered medium.

It is convenient to change the normalization of the Cooperon,

$$C_{\alpha\beta\alpha'\beta'}(\mathbf{q}, \omega) \to \frac{\zeta}{a}\frac{\tau_0}{\tau^2}C_{\alpha\beta\alpha'\beta'}(\mathbf{q}, \omega), \tag{5.90}$$

and write Eq. (5.88) as

$$\begin{aligned} C_{\alpha\beta\alpha'\beta'}(\mathbf{q}, \omega) &= \frac{\tau^2}{\tau_0}\delta_{\alpha\alpha'}\delta_{\beta\beta'} + \zeta\int \frac{d\mathbf{k}}{(2\pi)^2} \\ &\quad \times \sum_{\gamma'\delta'} G^{R}_{\alpha\gamma'}(\mathbf{k}, E + \hbar\omega)G^{A}_{\beta\delta'}(\mathbf{q} - \mathbf{k}, E)C_{\gamma'\delta'\alpha'\beta'}(\mathbf{q}, \omega). \end{aligned} \tag{5.91}$$

One way to solve this equation is to expand the Cooperon in the orthonormal basis of the two-electron spin states,

$$C_{\alpha\beta\alpha'\beta'}(\mathbf{q}, \omega) = \sum_{ij} C^{ij}(\mathbf{q}, \omega)\, \Psi^{i}_{\alpha\beta}\Psi^{j*}_{\alpha'\beta'}, \tag{5.92}$$

where the basis functions, $\Psi^i_{\alpha\beta}$, can be chosen as follows:

$$\Psi^j = \frac{\sigma_j \sigma_y}{\sqrt{2}}, \qquad j = 0, x, y, z, \qquad \sum_{\alpha\beta} \Psi^j_{\alpha\beta} \Psi^{i*}_{\alpha\beta} = \delta_{ji}. \quad (5.93)$$

The index $j = 0$ labels the spin-singlet state, while $j = x, y$ and z correspond to the three spin-triplet states. Consequently, there is one singlet Cooperon with the amplitude $C^{00}(\boldsymbol{q}, \omega)$ and 15 triplet Cooperons. The next step is to derive equations for the Cooperon amplitudes $C^{ij}(\boldsymbol{q}, \omega)$. This part of calculations is relegated to Appendix A.1. Here we quote the result:

$$C^{ij}(\boldsymbol{q}, \omega) = \frac{\tau^2}{\tau_0}\delta_{ij} + \frac{\tau}{\tau_0} \sum_{s=0,x,y,z} \mathrm{Tr}\langle \hat{P}_{-\boldsymbol{n}} \sigma_i \hat{P}_{\boldsymbol{n}} \sigma_s \rangle C^{sj}(\boldsymbol{q}, \omega), \quad (5.94)$$

where $\hat{P}_{\boldsymbol{n}}$ is the spin matrix defined in Eq. (5.40), and brackets $\langle \ldots \rangle$ stand for the integral

$$\langle \ldots \rangle = \int_0^{2\pi} \frac{d\phi_{\boldsymbol{n}}}{2\pi} \ldots \times (1 - i\tau\omega + i\tau v \boldsymbol{n} \cdot \boldsymbol{q})^{-1}, \quad (5.95)$$

taken over the momentum directions on the Fermi surface.

5.2.5 Cooperon for a Conventional Metal

Since we have not used any concrete matrix $\hat{P}_{\boldsymbol{n}}$, Eq. (5.94) is valid also for a conventional metal with a spin-degenerate quadratic spectrum. In this case, the carrier Green function is given by Eq. (5.39) where $\hat{P}_{\boldsymbol{n}}$ should be replaced by the unit spin matrix, σ_0. Consequently, for the elastic scattering time τ one finds $\tau = \tau_0/2$. Inserting

$$\hat{P}_{\boldsymbol{n}} = \sigma_0 \qquad \text{and} \qquad \tau = \frac{\tau_0}{2} \quad (5.96)$$

into Eq. (5.94) and using $\mathrm{Tr}(\sigma_i \sigma_s) = 2\delta_{is}$, we find the Cooperon amplitudes for a conventional metal:

$$C^{ij}(\boldsymbol{q}, \omega) = \frac{\tau/2}{1 - \langle 1 \rangle}\delta_{ij}, \quad (5.97)$$

where $\langle 1 \rangle$ is the angle integral

$$\begin{aligned}\langle 1 \rangle &= \int_0^{2\pi} \frac{d\phi_{\boldsymbol{n}}}{2\pi}[1 - i\tau\omega + i\tau v \boldsymbol{n} \cdot \boldsymbol{q}]^{-1} \\ &= \int_0^{2\pi} \frac{d\phi_{\boldsymbol{n}}}{2\pi}[1 + (i\tau\omega - i\tau v \boldsymbol{n} \cdot \boldsymbol{q}) + (i\tau\omega - i\tau v \boldsymbol{n} \cdot \boldsymbol{q})^2 + \cdots].\end{aligned} \quad (5.98)$$

In the diffusive case defined by the conditions

$$\tau v|\boldsymbol{n}\cdot\boldsymbol{q}| \ll 1, \qquad \tau\omega \ll 1, \tag{5.99}$$

it is sufficient to expand Eq. (5.98) up to the second order, which after angle averaging yields

$$\langle 1\rangle \approx 1 + i\tau\omega - \frac{\tau^2 v^2 \boldsymbol{q}^2}{2}. \tag{5.100}$$

Inserting this into Eq. (5.97), we have

$$C^{ij}(\boldsymbol{q},\omega) = \frac{1/2}{D_0\boldsymbol{q}^2 - i\omega}\delta_{ij}, \qquad D_0 = \frac{v^2\tau}{2}, \tag{5.101}$$

where D_0 is the diffusion coefficient for a conventional metal, and the factor of 1/2 reflects the chosen normalization [see Eq. (5.90)].

A few comments about Eq. (5.101) are in order. First, the matrix of Cooperon amplitudes C^{ij} is diagonal. Second, the Cooperons are divergent for $\boldsymbol{q}, \omega \to 0$. This eventually leads to the logarithmic divergence of the WL correction in Eq. (5.8). Third, the ac electric field produces dephasing $\hbar\omega$ in energies of the counter-propagating states. The dephasing is accounted for by the imaginary term $-i\omega$ in the Cooperon denominator. If the dephasing is dominated by other sources, typically by electron-electron or electron-phonon interactions, then $-i\omega$ is replaced by dephasing rate $\tau_\phi^{-1}(T)$. It is a function of temperature depending on the type of the interaction.

5.2.6 Cooperon for a Helical Metal

For a helical metal $\hat{P}_{\boldsymbol{n}}$ is a projector matrix defined in Eq. (5.40). Inserting $\hat{P}_{\boldsymbol{n}}$ from Eq. (5.40) into Eq. (5.94) and evaluating the traces of the Pauli matrices we find

$$\begin{aligned} C^{0j} &= \frac{\tau^2}{\tau_0}\delta_{0j} + \frac{\tau}{2\tau_0}\langle 1 - \mathbf{e}_+\cdot\mathbf{e}_-\rangle C^{0j} \\ &\quad + \frac{\tau}{2\tau_0}\sum_{b=x,y,z}\langle(\mathbf{e}_+ - \mathbf{e}_- + i\mathbf{e}_+\times\mathbf{e}_-)\cdot\mathbf{b}\rangle C^{bj}, \end{aligned} \tag{5.102}$$

$$\begin{aligned} C^{aj} &= \frac{\tau^2}{\tau_0}\delta_{aj} + \frac{\tau}{2\tau_0}\langle(\mathbf{e}_+ - \mathbf{e}_- - i\mathbf{e}_+\times\mathbf{e}_-)\cdot\mathbf{a}\rangle C^{0j} \\ &\quad + \frac{\tau}{2\tau_0}\sum_{b=x,y,z}\langle(1+\mathbf{e}_+\cdot\mathbf{e}_-)\mathbf{a}\cdot\mathbf{b} - i(\mathbf{e}_+ + \mathbf{e}_-)\cdot\mathbf{a}\times\mathbf{b} \\ &\quad - (\mathbf{e}_+\cdot\mathbf{a})(\mathbf{e}_-\cdot\mathbf{b}) - (\mathbf{e}_-\cdot\mathbf{a})(\mathbf{e}_+\cdot\mathbf{b})\rangle C^{bj}. \end{aligned} \tag{5.103}$$

We separated the amplitudes C^{0j} and $C^{(a,b)j}$, where a and b both run over the triplet indices x, y, and z only. Respectively, vectors **a** and **b** run over the unit vector basis of the Cartesian system. We also introduced a shorthand notation for the unit vectors

$$\mathbf{e}_{\pm} = \mathbf{e}_{\pm n}. \tag{5.104}$$

Equations (5.102) and (5.103) are the central results of this section. Unlike in ordinary metals [see Eq. (5.97)], the Cooperon matrix C^{ij} is nondiagonal, and the singlet and triplet Cooperons are coupled to each other in Eqs. (5.102) and (5.103). This is the consequence of the spin-momentum locking of helical states. Although our starting point was the TI thin film model, the applicability of Eqs. (5.102) and (5.103) is in fact much broader. For instance, they also describe a TI surface with the hexagonal warping. The warping anisotropy is encoded in the dependence of $\mathbf{e}_n$ on the momentum direction $\boldsymbol{n}$. This case is discussed in the next section.

In the following, we solve Eqs. (5.102) and (5.103) for the particular case of TI thin films. Our goal is to understand how the carrier backscattering affects the Cooperons. As discussed in the preceding subsection, the backscattering strength is characterized by out-of-plane component $\mathbf{e}_{\perp}$ of unit vectors (5.104). It is therefore convenient to separate the in- and out-of-plane components,

$$\mathbf{e}_{\pm} = \pm\mathbf{e}_{\parallel} + \mathbf{e}_{\perp}, \tag{5.105}$$

where $\mathbf{e}_{\parallel}$ and $\mathbf{e}_{\perp}$ are defined by Eq. (5.82). In view of the identities

$$\mathbf{e}_{+} + \mathbf{e}_{-} = 2\mathbf{e}_{\perp}, \quad \mathbf{e}_{+} - \mathbf{e}_{-} = 2\mathbf{e}_{\parallel}, \tag{5.106}$$

$$\mathbf{e}_{+} \cdot \mathbf{e}_{-} = 1 - 2\mathbf{e}_{\parallel}^2, \quad \mathbf{e}_{+} \times \mathbf{e}_{-} = 2\mathbf{e}_{\parallel} \times \mathbf{e}_{\perp}, \tag{5.107}$$

Eqs. (5.102) and (5.103) reduce to

$$C^{0j} = \frac{\tau^2}{\tau_0}\delta_{0j} + \frac{\tau}{\tau_0}\langle \mathbf{e}_{\parallel}^2\rangle C^{0j} + \frac{\tau}{\tau_0}\sum_{b=x,y,z}\langle(\mathbf{e}_{\parallel} + i\mathbf{e}_{\parallel}\times\mathbf{e}_{\perp})\cdot\mathbf{b}\rangle C^{bj}, \tag{5.108}$$

$$\begin{aligned} C^{aj} &= \frac{\tau^2}{\tau_0}\delta_{aj} + \frac{\tau}{\tau_0}\langle(\mathbf{e}_{\parallel} - i\mathbf{e}_{\parallel}\times\mathbf{e}_{\perp})\cdot\mathbf{a}\rangle C^{0j} \\ &+ \frac{\tau}{\tau_0}\sum_{b=x,y,z}\langle \mathbf{e}_{\perp}^2(\mathbf{a}\cdot\mathbf{b}) \\ &- i\mathbf{e}_{\perp}\cdot(\mathbf{a}\times\mathbf{b}) + (\mathbf{e}_{\parallel}\cdot\mathbf{a})(\mathbf{e}_{\parallel}\cdot\mathbf{b}) - (\mathbf{e}_{\perp}\cdot\mathbf{a})(\mathbf{e}_{\perp}\cdot\mathbf{b})\rangle C^{bj}. \end{aligned} \tag{5.109}$$

Explicitly, these equations are

$$\left[\frac{\tau_0}{\tau} - \langle e_{\parallel}^2\rangle\right] C^{0j} - \langle e_x + ie_y e_{\perp}\rangle C^{xj} - \langle e_y - ie_x e_{\perp}\rangle C^{yj} = \tau\delta_{0j}, \tag{5.110}$$

$$-\langle e_x - ie_y e_{\perp}\rangle C^{0j} + \left[\frac{\tau_0}{\tau} - \langle e_{\perp}^2 + e_x^2\rangle\right] C^{xj} - \langle e_x e_y - ie_{\perp}\rangle C^{yj} = \tau\delta_{xj}, \tag{5.111}$$

$$-\langle e_y + ie_x e_{\perp}\rangle C^{0j} - \langle e_y e_x + ie_{\perp}\rangle C^{xj} + \left[\frac{\tau_0}{\tau} - \langle e_{\perp}^2 + e_y^2\rangle\right] C^{yj} = \tau\delta_{yj}, \tag{5.112}$$

$$C^{zj} = \frac{\tau^2}{\tau_0}\delta_{zj}. \tag{5.113}$$

The decoupled Cooperons C^{zj} are momentum- and frequency-independent and can be safely neglected. To calculate the conductivity correction we need the diagonal Cooperons $C^{00}(\boldsymbol{q},\omega)$, $C^{xx}(\boldsymbol{q},\omega)$ and $C^{yy}(\boldsymbol{q},\omega)$. These are obtained from the coupled Eqs. (5.110) - (5.112) in which index j should be subsequently set to 0, x and y. The coefficients in Eqs. (5.110) - (5.112) are evaluated by expanding the denominator in Eq. (5.95) in powers of the Cooperon wave vector, $\boldsymbol{q}$, and frequency, ω, under condition (5.99). In doing so we keep the lowest order terms that yield the nonzero angle average $\langle\ldots\rangle$ in Eqs. (5.110) - (5.112). Under these conditions one obtain the following expressions for the diagonal Cooperons:

$$C^{00}(\boldsymbol{q},\omega) = \frac{1}{D\boldsymbol{q}^2 + \tau_{\mathcal{M}}^{-1} - i\omega}, \qquad \tau_{\mathcal{M}}^{-1} = \frac{2e_{\perp}^2}{\tau}, \tag{5.114}$$

$$C^{xx}(\boldsymbol{q},\omega) = \frac{2\tau}{e_{\parallel}^2}\frac{2 - e_{\parallel}^4\cos^2\phi_{\boldsymbol{q}}}{2 + 5e_{\perp}^2 + e_{\perp}^4} + \frac{2\tau(\tau_{\mathcal{M}}^{-1} - i\omega)}{e_{\parallel}^4}\frac{(1+3e_{\perp}^2)^2 - e_{\parallel}^6\sin^2\phi_{\boldsymbol{q}}}{2 + 5e_{\perp}^2 + e_{\perp}^4}C^{00}(\boldsymbol{q},\omega), \tag{5.115}$$

$$C^{yy}(\boldsymbol{q},\omega) = \frac{2\tau}{e_{\parallel}^2}\frac{2 - e_{\parallel}^4\sin^2\phi_{\boldsymbol{q}}}{2 + 5e_{\perp}^2 + e_{\perp}^4} + \frac{2\tau(\tau_{\mathcal{M}}^{-1} - i\omega)}{e_{\parallel}^4}\frac{(1+3e_{\perp}^2)^2 - e_{\parallel}^6\cos^2\phi_{\boldsymbol{q}}}{2 + 5e_{\perp}^2 + e_{\perp}^4}C^{00}(\boldsymbol{q},\omega). \tag{5.116}$$

In Eq. (5.114) $D = D_0(2 + 5e_\perp^2 + e_\perp^4)/e_\parallel^2$ is the diffusion coefficient with the prefactor defined in Eq. (5.101). In Eqs. (5.115) and (5.116) $\phi_{\boldsymbol{q}}$ is the angle indicating the direction of the Cooperon momentum: $\boldsymbol{q} = |\boldsymbol{q}| \cdot (\cos\phi_{\boldsymbol{q}}, \sin\phi_{\boldsymbol{q}}, 0)$.

Now, we can see how the backscattering influences the Cooperons. There are two main effects. Firstly, the backscattering introduces a new time scale, $\tau_{\mathcal{M}}$, that determines the rate of backscattering $\tau_{\mathcal{M}}^{-1}$. The latter enters the singlet Cooperon C^{00} in the same fashion as the dephasing rate $-i\omega$ does. The backscattering eliminates the divergence of C^{00} at $\boldsymbol{q}, \omega \to 0$, meaning that at time scales larger than $\tau_{\mathcal{M}}$ the quantum interference is suppressed. In order to be observable the time $\tau_{\mathcal{M}}$ itself must be larger than the elastic life time:

$$\frac{\tau}{\tau_{\mathcal{M}}} = 2e_\perp^2 \ll 1, \tag{5.117}$$

which requires small values of parameter $e_\perp^2$. Under this condition the diffusion coefficient is

$$D \approx 2D_0. \tag{5.118}$$

The factor of 2 is specific to the Dirac model and closely related to the vertex corrections. Secondly, the backscattering modifies the triplet Cooperons C^{xx} and C^{yy} in Eqs. (5.115) and (5.116). Despite their complicated structure, Eqs. (5.115) and (5.116) are easy to interpret. The first terms never become divergent and can be safely neglected. The second terms are proportional to $C^{00}(\boldsymbol{q}, \omega)$ with the prefactors depending on the backscattering rate $\tau_{\mathcal{M}}^{-1}$ and $-i\omega$. For $\omega \to 0$ they scale with the small parameter $e_\perp^2$. It can be shown that the backscattering also generates other triplet Cooperons, C^{xy} and C^{yx}, which are proportional to $e_\perp^3$.

5.2.7 Quantum Conductivity Correction

With the averaged Green functions $\hat{G}_{\boldsymbol{k}}^{R,A}$ and the Cooperon $C_{\alpha\beta\alpha'\beta'}(\boldsymbol{q}, \omega)$ at hand, we proceed to calculate the conductivity from Kubo formula (5.16). Our goal is to express the disorder-averaged conductivity in terms of the known $\hat{G}_{\boldsymbol{k}}^{R,A}$ and $C_{\alpha\beta\alpha'\beta'}(\boldsymbol{q}, \omega)$. Inserting series expansion (5.22) - (5.25) for the exact Green functions into

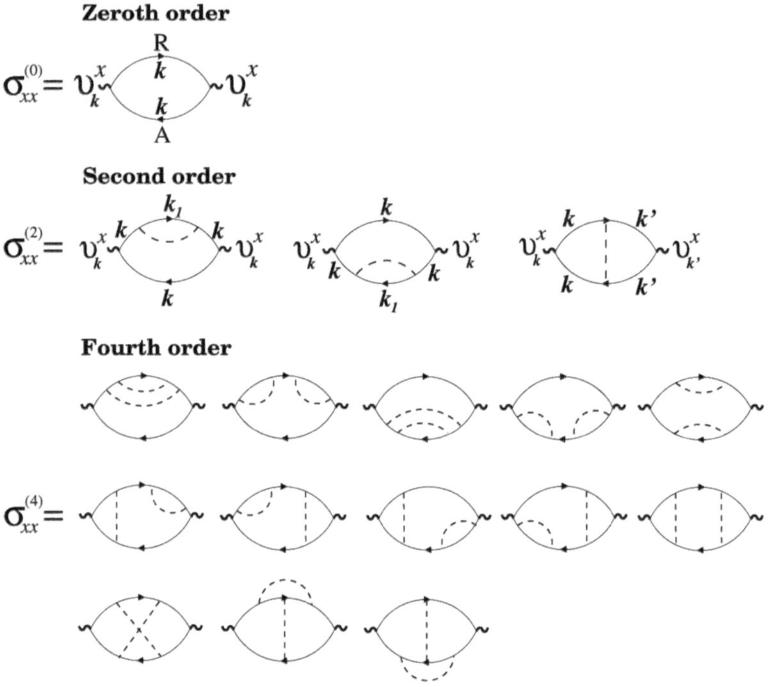

Figure 5.11 Diagrammatic representation of Kubo conductivity up to fourth order in disorder potential.

Kubo formula (5.16) and averaging it over disorder configurations, we obtain an infinite series:

$$\langle\langle \sigma_{xx} \rangle\rangle = \sigma_{xx}^{(0)} + \sigma_{xx}^{(2)} + \sigma_{xx}^{(4)} + \ldots, \tag{5.119}$$

where the superscript indicates the order in the scattering potential. The terms up to the fourth order are shown diagrammatically in Fig. 5.11. Since Kubo formula (5.16) involves the traces over the momentum and spin variables, each term can be pictured as a "bubble" formed by the retarded and advanced Green functions connecting the velocity operators $\hat{v}_k^x$ in the direction around the loop (which is indicated by the arrows in Fig. 5.11).

In order to illustrate the summation procedure we begin with the second order diagrams. The first two of them are corrections to the bare retarded and advanced Green functions (cf. Fig. 5.9).

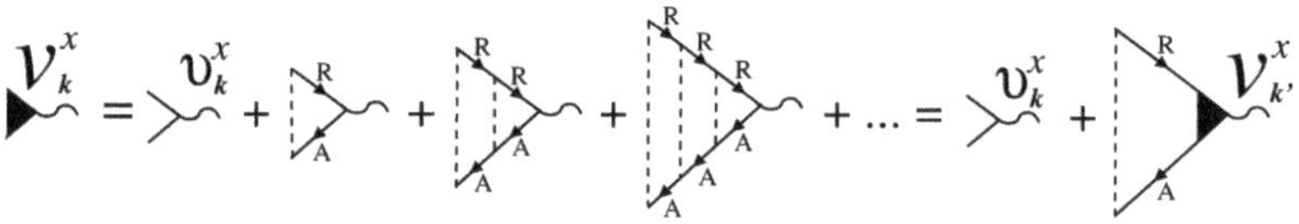

Figure 5.12 Ladder diagrams for vertex corrections to bare velocity operator $\hat{\upsilon}^x_{\mathbf{k}}$. Black triangle denotes the renormalized operator, $\hat{\mathcal{V}}^x_{\mathbf{k}}$.

These diagrams can be absorbed into the zeroth order term with the correspondingly modified retarded and advanced functions. The third diagram leads to the renormalization of one of the velocity operators. This is known as a vertex correction. Such corrections are conventionally represented by ladder diagrams in Fig. 5.12. Proceeding to the fourth order terms in Fig. 5.11, we see that the first two rows are corrections to the bare Green functions and/or the vertex corrections renormalizing $\hat{\upsilon}^x_{\mathbf{k}}$ according to Fig. 5.12. Again, these can be absorbed into the zeroth-order term by simultaneously renormalizing the Green functions and one of the velocity operators. Similar corrections appear in all orders and can be summed up according to the afore-described recipe. This summation scheme leads to the classical conductivity of a disordered metal, σ_c, which is depicted in Fig. 5.13. The corresponding analytical expression is

$$\sigma_C = \frac{e^2}{2\pi\omega}\int dE[f(E)-f(E+\hbar\omega)]\int\frac{d\mathbf{k}}{(2\pi)^2}\mathrm{Tr}[\hat{\upsilon}^x_{\mathbf{k}}\hat{G}^R_{\mathbf{k},E+\hbar\omega}\hat{\mathcal{V}}^x_{\mathbf{k}}\hat{G}^A_{\mathbf{k},E}], \tag{5.120}$$

where $\hat{G}^{R,A}_{\mathbf{k}}$ are the averaged Green functions given by Eq. (5.80), and $\hat{\mathcal{V}}^x_{\mathbf{k}}$ denotes the velocity operator renormalized by the ladder diagrams in Fig. 5.12. These diagrams sum up into the equation for $\hat{\mathcal{V}}^x_{\mathbf{k}}$:

$$\hat{\mathcal{V}}^x_{\mathbf{k}} = \hat{\upsilon}^x_{\mathbf{k}} + \zeta\int\frac{d\mathbf{k}}{(2\pi)^2}\hat{G}^R_{\mathbf{k},E+\hbar\omega}\hat{\mathcal{V}}^x_{\mathbf{k}}\hat{G}^A_{\mathbf{k},E}. \tag{5.121}$$

It is possible to show that the solution for $\hat{\mathcal{V}}^x_{\mathbf{k}}$ differs from the bare velocity just by a prefactor depending on the ratio of the transport and elastic scattering times (see e.g. Ref. [78] for more details),

$$\hat{\mathcal{V}}^x_{\mathbf{k}} = \left(\frac{\tau_{tr}}{\tau}\right)\hat{\upsilon}^x_{\mathbf{k}}, \qquad \frac{\tau_{tr}}{\tau} = \frac{2(1+e_\perp^2)}{1+3e_\perp^2}, \tag{5.122}$$

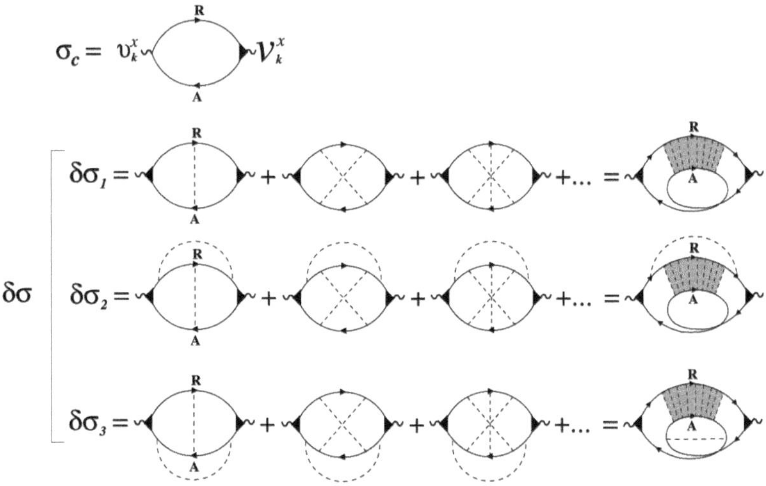

Figure 5.13 Diagrammatic representation of the classical conductivity, σ_c, and the quantum correction, $\delta\sigma$. Shaded area denotes the Cooperon.

where τ and τ_{tr} were introduced earlier in Eqs. (5.81) and (5.84). After this, the classical conductivity in Eq. (5.85) can be readily calculated, yielding for $\omega\tau \ll 1$ Einstein relation (5.85) introduced in Section 5.2.3.

We now turn to the last three fourth-order diagrams in Fig. 5.11. Clearly these are neither Green function nor vertex corrections. They cannot be incorporated into σ_c by renormalizing $G_{\boldsymbol{k}}^{R/A}$ or $\hat{v}_{\boldsymbol{k}}^{x}$. Therefore, the last three diagrams in Fig. 5.11 belong to the *quantum correction* to the classical conductivity. More precisely, the sum of such diagrams from all orders represents the quantum correction, $\delta\sigma$, as depicted in Fig. 5.13. The relevant diagrams are those in which all the dashed lines connecting the retarded and advanced functions cross each other within the bubble. Such diagrams are related to the Cooperon. Indeed, by twisting the A line as shown in Fig. 5.13 we see that all the crossed lines sum up into the ladder diagram for the Cooperon (shaded area, cf. Fig. 5.10). Thus, the conductivity correction $\delta\sigma$ consists of three contributions,

$$\delta\sigma = \delta\sigma_1 + \delta\sigma_2 + \delta\sigma_3, \tag{5.123}$$

The diagrams for $\delta\sigma_1$, $\delta\sigma_2$ and $\delta\sigma_3$, which all incorporate the Cooperon, are also called the *Hikami boxes*. Since the diagrams for the last two corrections in Fig. 5.13 are similar, $\delta\sigma_3$ appears to be equal to $\delta\sigma_2$. This reduces the problem to calculating $\delta\sigma_1$ and $\delta\sigma_2$. The corresponding analytical expressions are

$$\delta\sigma_1 = \frac{e^2\hbar}{\pi N\tau^2}\int\frac{dE}{2\pi\omega}[f(E)-f(E+\hbar\omega)]\int\frac{d\boldsymbol{q}}{(2\pi)^2}C_{\beta\beta'\gamma\gamma'}(\boldsymbol{q},\omega)$$
$$\times\int\frac{d\boldsymbol{k}}{(2\pi)^2}(\hat{G}^A_{\boldsymbol{k},E}\hat{V}^x_{\boldsymbol{k}}\hat{G}^R_{\boldsymbol{k},E+\hbar\omega})_{\gamma'\beta}(\hat{G}^R_{\boldsymbol{q}-\boldsymbol{k},E+\hbar\omega}\hat{V}^x_{\boldsymbol{q}-\boldsymbol{k}}\hat{G}^A_{\boldsymbol{q}-\boldsymbol{k},E})_{\gamma\beta'}, \tag{5.124}$$

$$\delta\sigma_2 = \frac{e^2\hbar^2}{\pi^2N^2\tau_0\tau^2}\int\frac{dE}{2\pi\omega}[f(E)-f(E+\hbar\omega)]\int\frac{d\boldsymbol{q}}{(2\pi)^2}C_{\beta\beta'\gamma\gamma'}(\boldsymbol{q},\omega)$$
$$\times\int\frac{d\boldsymbol{k}}{(2\pi)^2}\int\frac{d\boldsymbol{k}'}{(2\pi)^2}(\hat{G}^A_{\boldsymbol{k},E}\hat{V}^x_{\boldsymbol{k}}\hat{G}^R_{\boldsymbol{k},E+\hbar\omega}\hat{G}^R_{\boldsymbol{q}-\boldsymbol{k}',E+\hbar\omega})_{\gamma'\beta}$$
$$\times(\hat{G}^R_{\boldsymbol{q}-\boldsymbol{k},E+\hbar\omega}\hat{G}^R_{\boldsymbol{k}',E+\hbar\omega}\hat{V}^x_{\boldsymbol{k}'}\hat{G}^A_{\boldsymbol{k}',E})_{\gamma\beta'}, \tag{5.125}$$

where summation over the spin indices is assumed. Since in the $\boldsymbol{k}$ integrals only the vicinity of the Fermi surface matters, we change the integration variables: $\frac{d\boldsymbol{k}}{(2\pi)^2}\to\frac{d\phi_n}{2\pi}Nd\xi$, and integrate over ξ. The calculation procedure follows the same steps as in the case of the Cooperon equation (see Appendix A.2). The results are

$$\delta\sigma_1 = \frac{8e^2D_0}{hv^2}\overline{[\hat{P}_{\boldsymbol{n}}\hat{V}^x_{\boldsymbol{n}}\hat{P}_{\boldsymbol{n}}]_{\gamma'\beta}[\hat{P}_{-\boldsymbol{n}}\hat{V}^x_{-\boldsymbol{n}}\hat{P}_{-\boldsymbol{n}}]_{\gamma\beta'}}\int\frac{d\boldsymbol{q}}{(2\pi)^2}C_{\beta\beta'\gamma\gamma'}(\boldsymbol{q},\omega), \tag{5.126}$$

$$\delta\sigma_2 = -\frac{8e^2D_0\tau}{hv^2\tau_0}\overline{[\hat{P}_{\boldsymbol{n}}\hat{V}^x_{\boldsymbol{n}}\hat{P}_{\boldsymbol{n}}]_{\gamma'\beta_1}[\hat{P}_{-\boldsymbol{n}}]_{\gamma\gamma_1}}$$
$$\times\overline{[\hat{P}_{-\boldsymbol{n}'}]_{\beta_1\beta}[\hat{P}_{\boldsymbol{n}'}\hat{V}^x_{\boldsymbol{n}'}\hat{P}_{\boldsymbol{n}'}]_{\gamma_1\beta'}}\int\frac{d\boldsymbol{q}}{(2\pi)^2}C_{\beta\beta'\gamma\gamma'}(\boldsymbol{q},\omega), \tag{5.127}$$

where the bar, as usual, denotes the angle averaging.

5.2.8 Conductivity Correction for a Conventional Metal

Equations (5.126) and (5.127) recover the well known result, Eq. (5.8), for the WL conductivity correction for a conventional metal with the spin-degenerate quadratic dispersion. In this case, we use $\hat{P}_{\boldsymbol{n}} = \sigma_0$, $\hat{V}^x_{\boldsymbol{n}} = \hat{v}^x_{\boldsymbol{n}} = v\,n_x\sigma_0$ and Eqs. (5.92) and (5.101) for the

Cooperon. Inserting these into Eqs. (5.126) and (5.127) and doing the angle averaging, we have

$$\delta\sigma_1 = -\frac{4e^2 D_0}{h}\int \frac{d\boldsymbol{q}}{(2\pi)^2}[C^{xx} + C^{yy} + C^{zz} - C^{00}]$$

$$= -\frac{2e^2}{(2\pi)h}\int_0^{D_0 q_m^2} d(D_0 q^2)\frac{1}{D_0 q^2 - i\omega}, \tag{5.128}$$

$$\delta\sigma_2 = 0. \tag{5.129}$$

It is worth noting that the negative sign in front of Eq. (5.128) reflects the opposite directions of the velocities of the interfering states. The correction $\delta\sigma_2$ vanishes because the angle averaging in Eq. (5.127) is reduced to the product of the average velocities, each being zero. When evaluating the integral, we should keep in mind that the expression for the Cooperon is valid in the diffusive regime for small values of $\tau D_0 q^2$ [cf., Eq. (5.99)]. Therefore, the integration must be cut when $\tau D_0 q^2$ becomes of order of 1, which yields the upper integration limit $D_0 q_m^2 = \tau^{-1}$. Replacing further $-i\omega \to \tau_\phi^{-1}$ for generality, we obtain Eq. (5.8) from Eq. (5.128).

5.2.9 Conductivity Correction for a Helical Metal

In the case of a helical metal, $\hat{P}_{\boldsymbol{n}}$ in Eqs. (5.126) and (5.127) is a projector matrix defined in Eq. (5.40). With the help of the identity

$$\hat{P}_{\boldsymbol{n}}\hat{\mathcal{V}}_{\boldsymbol{n}}^x \hat{P}_{\boldsymbol{n}} = \frac{\tau_{tr}}{\tau} v\, n_x \hat{P}_{\boldsymbol{n}}, \tag{5.130}$$

Eqs. (5.126) and (5.127) can be simplified to

$$\delta\sigma_1 = -\frac{8e^2 D_0}{h}\left(\frac{\tau_{tr}}{\tau}\right)^2 \overline{n_x^2[\hat{P}_{\boldsymbol{n}}]_{\gamma'\beta}[\hat{P}_{-\boldsymbol{n}}]_{\gamma\beta'}}\int \frac{d\boldsymbol{q}}{(2\pi)^2} C_{\beta\beta'\gamma\gamma'}(\boldsymbol{q},\omega), \tag{5.131}$$

$$\delta\sigma_2 = -\frac{8e^2 D_0}{h}\frac{\tau_{tr}^2}{\tau\tau_0}\overline{n_x[\hat{P}_{\boldsymbol{n}}]_{\gamma'\beta_1}[\hat{P}_{-\boldsymbol{n}}]_{\gamma\gamma_1}}$$
$$\times\, \overline{n_x[\hat{P}_{-\boldsymbol{n}'}]_{\beta_1\beta}[\hat{P}_{\boldsymbol{n}'}]_{\gamma_1\beta'}}\int \frac{d\boldsymbol{q}}{(2\pi)^2} C_{\beta\beta'\gamma\gamma'}(\boldsymbol{q},\omega). \tag{5.132}$$

Unlike the case of a conventional metal, $\hat{P}_{\boldsymbol{n}}$ contains the spin helicity $\boldsymbol{\sigma}\cdot\mathbf{e}_{\boldsymbol{n}}$, which contributes to the angle averaging, leading to a nonzero correction $\delta\sigma_2$. We proceed by expanding the Cooperon in the basis of the singlet and triplet states [see Eq. (5.92)] and performing the summations over the spin indices. After some algebra the conductivity corrections in Eqs. (5.131) and (5.132) can be expressed in terms of the singlet and triplet Cooperon amplitudes:

$$\delta\sigma_1 = -\frac{4e^2 D_0}{h}\left(\frac{\tau_{tr}}{\tau}\right)^2 \int \frac{d\boldsymbol{q}}{(2\pi)^2}\big[-\overline{n_x^2}(1-e_\perp^2)C^{00} \\ +\overline{n_x^2(1-e_y^2)}C^{xx} + \overline{n_x^2(1-e_x^2)}C^{yy} + \overline{n_x^2}ie_\perp(C^{xy}-C^{yx})\big], \quad (5.133)$$

$$\delta\sigma_2 = \frac{2e^2 D_0}{h}\frac{\tau_{tr}^2}{\tau\tau_0}e_\parallel^2(\overline{n_x^2})^2 \int \frac{d\boldsymbol{q}}{(2\pi)^2}\big[-(1+e_\perp^2)C^{00} \\ +C^{xx} + e_\perp^2 C^{yy} + ie_\perp(C^{xy}-C^{yx})\big]. \quad (5.134)$$

Noticing further that on average over the directions of $\boldsymbol{q}$ the triplet Cooperons (5.115) and (5.116) coincide, we can write the net conductivity correction as

$$\begin{aligned}\delta\sigma &= \delta\sigma_1 + 2\delta\sigma_2 \\ &= -\frac{2e^2 D_0}{h}\left(\frac{\tau_{tr}}{\tau}\right)^2 \int \frac{qdq}{2\pi}\Big[-\frac{1-e_\perp^2}{2}\overline{C^{00}} \\ &\quad +\frac{1+3e_\perp^2}{2}\overline{C^{xx}} + \frac{ie_\perp(1+3e_\perp^2)}{2(1+e_\perp^2)}(\overline{C^{xy}}-\overline{C^{xy}})\Big], \quad (5.135)\end{aligned}$$

where the bar denotes the angle averaging $\overline{(\ldots)} = \int_0^{2\pi}\ldots d\phi_{\boldsymbol{q}}/2\pi$ with $\phi_{\boldsymbol{q}}$ defined in the text after Eq. (5.116). Note that Eq. (5.135) has been obtained without the assumption that $e_\perp^2$ is small.

In Eq. (5.135) the singlet Cooperon C^{00} enters with the negative sign because in the singlet channel the spin projections of the interfering states are antiparallel. This can also be viewed as the manifestation of an intrinsic π Berry phase [94]. Given the other minus sign from the opposite velocities, the singlet contribution to $\delta\sigma$ is positive, which corresponds to the WAL effect. In contrast, in the triplet channels the interference occurs between the states with the parallel spin projections. The triplet contribution to $\delta\sigma$ is negative and proportional to $e_\perp^2$ characterizing the out-of-plane spin polarization. For sufficiently small $e_\perp^2 \ll 1$ it suffices to keep only the singlet Cooperon (5.114). Then, the q integral is calculated in the same manner as described in Section 5.2.8:

$$\delta\sigma \approx \frac{e^2}{2\pi h} 2 \int_0^{Dq_m^2} d(Dq^2) \frac{1/2}{Dq^2 + \tau_{\mathcal{M}}^{-1} - i\omega} \tag{5.136}$$

$$= \frac{e^2}{2\pi h} \ln \frac{\tau^{-1}}{\tau_{\mathcal{M}}^{-1} + \tau_{\phi}^{-1}}. \tag{5.137}$$

Here we took $D \approx 2D_0$ and $\tau_{tr}/\tau \approx 2$ in the leading order in $e_{\perp}^2 \ll 1$ [see Eqs. (5.118) and (5.122), respectively]. The factor of 2 from the scattering times cancels with the factor of 1/2 that originates from the correction $\delta\sigma_2$. As a result, in the leading order in the small parameter $e_{\perp}^2$ the net correction in Eq. (5.137) has the standard universal prefactor. If we now add the contribution of the '−' subsystem in Eq. (4.41), we arrive at Eq. (5.15) for the WAL correction for a TI thin film.

5.3 Weak Antilocalization on TI Surfaces

As already mentioned, the surface states in Bi_2Se_3 or Bi_2Te_3 have a single-cone anisotropic spectrum [see also Eq. (4.35) and Fig. 4.3] described by Hamiltonian (4.34) with the hexagonal warping given by Eq. (4.32). Since the Hamiltonian is TR-invariant, it is quite natural to expect the WAL conductivity correction to be typical of the symplectic symmetry class of disordered systems. Here we demonstrate this by straightforward calculation, assuming again that the Fermi level lies in the conduction band (see Figs. 4.3 and 5.7) and the band separation is large enough to enable the metallic transport regime [see Eqs. (5.3) and (5.5)]. The warping will be treated as a weak perturbation of the surface bands under condition

$$\frac{W^2 k^4}{2\mathcal{A}^2} \ll 1, \tag{5.138}$$

which is justified for not very large surface Fermi energies. In this case the main effect of the warping on the spectrum is to increase the carrier energy $E(k, \phi_{\mathbf{n}})$ on average over all the angles $\phi_{\mathbf{n}}$ [see Eq. (4.35)]. This amounts to replacing $\mathcal{M}_{\mathbf{k}}^2$ by its angle average,

$$\mathcal{M}_{\mathbf{k}}^2 \Longrightarrow \overline{\mathcal{M}_{\mathbf{k}}^2} = \frac{W^2 k^6}{2}, \tag{5.139}$$

in Eq. (4.35). In fact, the same replacement can be done in all *even* functions of $\mathcal{M}_{\mathbf{k}}$, e.g. in the Fermi momentum, Fermi velocity, DOS etc. This allows us to use Eq. (5.80) for the retarded and advanced Green functions, Cooperon equations (5.102) and (5.103) as well as the Hikami boxes (5.126) and (5.127) derived in the preceding section.

The angular dependence of the warping is accounted for in the carrier helicity $\boldsymbol{\sigma} \cdot \mathbf{e}_{\boldsymbol{n}}$, where the vector $\mathbf{e}_{\boldsymbol{n}}$ can be written as

$$\mathbf{e}_{\boldsymbol{n}} = \mathbf{e}_{\parallel} + \mathbf{e}_{\perp}, \quad \mathbf{e}_{\parallel} \approx \boldsymbol{n}\sqrt{1 - e_{\perp}^2}, \quad \mathbf{e}_{\perp} \approx \frac{\mathcal{M}_{k_F \boldsymbol{n}}}{\mathcal{A}k_F}\mathbf{z}. \tag{5.140}$$

The warping generates the vector component $\mathbf{e}_{\perp}$ that describes the out-of-plane angular modulation of the spin texture on the Fermi surface. This is the main nontrivial effect of the warping which needs to be taken into account in the calculations of the transport properties of the TI surface state.

5.3.1 Scattering Times

The elastic and transport scattering times for the surface carriers can be calculated from Eqs. (5.81) and (5.84), respectively. Since $\mathbf{e}_{\boldsymbol{n}}$ is an odd function of the unit wave vector $\boldsymbol{n}$, there is no backscattering. The elastic life-time is independent of the warping and coincides with the disorder-related time scale, τ_0,

$$\tau = \tau_0. \tag{5.141}$$

However, the warping enters the transport scattering time through the angle average of the two anisotropic factors:

$$\begin{aligned} \frac{1}{\tau_{tr}} &= \frac{1 - \overline{(\boldsymbol{n} \cdot \boldsymbol{n}_1)\,(\mathbf{e}_{\boldsymbol{n}} \cdot \mathbf{e}_{\boldsymbol{n}_1})}}{\tau_0} \\ &= \frac{1 - e_{\parallel}^2/2}{\tau_0} = \frac{1}{2\tau_0}\left[1 + \frac{\overline{\mathcal{M}_{k_F \cdot \boldsymbol{n}}^2}}{\mathcal{A}^2 k_F^2}\right]. \end{aligned} \tag{5.142}$$

This means that the conductivity corrections $\delta\sigma_1$ (5.131) and $\delta\sigma_2$ (5.132) also depend on the warping through the prefactor $(\tau_{tr}/\tau)^2$. We shall see, however, that in the net conductivity $\delta\sigma = \delta\sigma_1 + 2\delta\sigma_2$ the different warping terms cancel each other.

5.3.2 Cooperon

Our next step is to specify the Cooperon equations (5.102) and (5.103) for the surface state. Introducing the notations $\mathbf{e}_{\pm \boldsymbol{n}} = \mathbf{e}_{\pm} = \pm\mathbf{e}$ and using the identities

$$\mathbf{e}_{-} = -\mathbf{e}_{+}, \quad \mathbf{e}_{+} - \mathbf{e}_{-} = 2\mathbf{e}, \quad \mathbf{e}_{+} \cdot \mathbf{e}_{-} = -1, \quad \mathbf{e}_{+} \times \mathbf{e}_{-} = 0, \tag{5.143}$$

we rewrite Eqs. (5.102) and (5.103) as

$$C^{0j} = \tau_0 \delta_{0j} + \langle 1 \rangle C^{0j} + \sum_{b=x,y,z} \langle \mathbf{e} \cdot \mathbf{b} \rangle C^{bj}, \tag{5.144}$$

$$C^{aj} = \tau_0 \delta_{aj} + \langle \mathbf{e} \cdot \mathbf{a} \rangle C^{0j} + \sum_{b=x,y,z} \langle (\mathbf{e} \cdot \mathbf{a})(\mathbf{e} \cdot \mathbf{b}) \rangle C^{bj}, \tag{5.145}$$

or, explicitly,

$$[1 - \langle 1 \rangle]\, C^{0j} - \langle e_x \rangle C^{xj} - \langle e_y \rangle C^{yj} = \tau_0 \delta_{0j}, \tag{5.146}$$

$$-\langle e_x \rangle C^{0j} + \left[1 - \langle e_x^2 \rangle\right] C^{xj} = \tau_0 \delta_{xj}, \tag{5.147}$$

$$-\langle e_y \rangle C^{0j} + \left[1 - \langle e_y^2 \rangle\right] C^{yj} = \tau_0 \delta_{yj}, \tag{5.148}$$

$$\left[1 - \langle e_{\perp}^2 \rangle\right] C^{zj} = \tau_0 \delta_{zj}. \tag{5.149}$$

Again, the coefficients in these equations can be evaluated by expanding the denominator in Eq. (5.95) and averaging over the directions of the unit wave vector $\boldsymbol{n}$. Then, for the diagonal Cooperons we find

$$C^{00}(\boldsymbol{q}, \omega) = \frac{1}{D\boldsymbol{q}^2 - i\omega}, \qquad D = \frac{v_F^2 \tau_{tr}}{2}, \tag{5.150}$$

$$\overline{C^{xx}} = \overline{C^{yy}} = \tau_{tr} \left(1 - \frac{e_{\parallel}^2}{4}\right) - \frac{e_{\parallel}^2}{4} i \tau_{tr} \omega C^{00}(\boldsymbol{q}, \omega), \tag{5.151}$$

$$C^{zz} = \frac{\tau_0}{1 - \langle e_{\perp}^2 \rangle}. \tag{5.152}$$

It is instructive to compare these results to those obtained for the TI thin film [see Eqs. (5.114)–(5.116)]. The most essential difference is the absence of the relaxation gap in the singlet Cooperon C^{00}, Eq. (5.150). This reflects the absence of the backscattering for the gapless TI surface states, irrespective of the warping. Although the latter has been treated perturbatively, it is possible to prove the gapless character of the singlet Cooperon for the warping of arbitrary strength, which is the consequence of the TRS. The warping only modifies the diffusion constant D through the transport scattering time (5.142). We also note that the relevant contribution of the triplet Cooperons C^{xx} and C^{yy} [see Eq. (5.151)] is proportional to the small parameter $-i\omega\tau_{tr} \sim \tau_{tr}/\tau_{\phi} \ll 1$ and, therefore, can be neglected. The same is true for the other triplet Cooperon, C^{zz}.

5.3.3 Quantum Conductivity Correction

We now turn to the conductivity correction. In order to calculate it we need to do the angle averaging in Eqs. (5.131) and (5.132) for the surface-state spin helicity defined by the vector $\mathbf{e}_n$ in Eq. (5.140). Repeating the calculations outlined in Section 5.2.9, we express the conductivity corrections $\delta\sigma_1$ and $\delta\sigma_2$ in terms of the diagonal Cooperons:

$$\delta\sigma_1 = \frac{e^2 D_0}{\pi\hbar}\left(\frac{\tau_{tr}}{\tau}\right)^2 \int \frac{d\boldsymbol{q}}{(2\pi)^2}\Big[2\langle n_x^2\rangle C^{00} \\ -2\langle n_x^2 e_x^2\rangle C^{xx} - 2\langle n_x^2 e_y^2\rangle C^{yy} - 2\langle n_x^2 e_\perp^2\rangle C^{zz}\Big], \tag{5.153}$$

$$\delta\sigma_2 = -\frac{e^2 D_0}{\pi\hbar}\left(\frac{\tau_{tr}}{\tau}\right)^2 \left(\frac{\mathbf{e}_\parallel}{2}\right)^2 \int \frac{d\boldsymbol{q}}{(2\pi)^2}\Big[C^{00} - C^{xx}\Big], \tag{5.154}$$

where $D_0 = v^2\tau/2$ involves the elastic life-time rather than τ_{tr} [cf. Eq. (5.150) for D]. Then, keeping only the singlet Cooperon C^{00}, we obtain the following expression for the net correction:

$$\delta\sigma_{xx} = \delta\sigma_{xx}^{(1)} + 2\delta\sigma_{xx}^{(2)} = \frac{e^2 D_0}{\pi\hbar}\left(\frac{\tau_{tr}}{\tau}\right)^2\left(1 - \frac{e_\parallel^2}{2}\right)\int \frac{d\boldsymbol{q}}{(2\pi)^2} C^{00}(\boldsymbol{q}) \\ = \frac{e^2 D}{\pi\hbar}\int \frac{d\boldsymbol{q}}{(2\pi)^2} C^{00}(\boldsymbol{q}) = \frac{e^2}{2\pi h}\ln\frac{\tau_\phi}{\tau}. \tag{5.155}$$

The important observation here is that the warping-dependent parameters cancel each other [see also Eq. (5.142)]:

$$D_0\left(\frac{\tau_{tr}}{\tau}\right)^2\left(1 - \frac{e_\parallel^2}{2}\right) = D\,\frac{\tau_{tr}}{\tau}\,\frac{\tau}{\tau_{tr}} = D, \tag{5.156}$$

so that $\delta\sigma$ retains the universal form typical of the symplectic symmetry class of disordered systems. This conclusion is in fact independent of the warping strength.

5.3.4 Magnetoconductivity

So far we have treated the surface states in 3DTIs as purely two-dimensional. There is, however, a finite length λ of order of a few nm over which they penetrate into the bulk of the material. Since λ is rather small, the surface WAL conductivity is sensitive to the orientation of an external magnetic field with respect to the surface of the material. Such anisotropic B-field dependence can be used to detect the surface states in 3DTIs. Defining the magnetoconductivity as $\Delta\sigma(B) = \delta\sigma(B) - \delta\sigma(0)$, one can obtain the following B-field dependences for perpendicular ($\perp$) and parallel ($\parallel$) field orientations:

$$\Delta\sigma_{\perp}(B) = \frac{e^2}{2\pi h}\left[\ln\frac{B_{\perp}}{B} - \psi\left(\frac{1}{2} + \frac{B_{\perp}}{B}\right)\right], \quad B_{\perp} = \frac{\hbar}{4|e|\,\ell_{\varphi}^2}, \tag{5.157}$$

$$\Delta\sigma_{\parallel}(B) = -\frac{e^2}{2\pi h}\ln\left(1 + \frac{B^2}{B_{\parallel}^2}\right), \quad B_{\parallel} = \frac{\hbar}{\sqrt{2}|e|\lambda\ell_{\varphi}^2}, \tag{5.158}$$

where $\psi(x)$ is the digamma function. The equations above contain the same phase-coherence length $\ell_{\varphi} = \sqrt{D\tau_{\varphi}}$. However, the magnetic-field scales, $B_{\perp}$ and $B_{\parallel}$, on which the magnetoconductivity decreases, are distinctly different. The field $B_{\perp}$ corresponds to the Aharonov–Bohm magnetic flux of order of the quantum h/e through a typical area $\sim \ell_{\varphi}^2$ swept by the interfering trajectories, while $B_{\parallel}$ corresponds to the same flux h/e, but through a significantly smaller area $\sim \lambda\ell_{\varphi}$ which is proportional to the surface-state penetration length into the bulk.

In order to identify the surface state one should extract its penetration length λ from Eqs. (5.157) and (5.158). Excluding ℓ_{φ} from Eqs. (5.157) and (5.158) we can express λ in terms of two parameters, $B_{\perp}$ and $B_{\parallel}$, which can be obtained independently from fitting the corresponding experimental data:

$$\lambda = \sqrt{\frac{2\hbar}{|e|}\frac{B_{\perp}}{B_{\parallel}^2}}. \tag{5.159}$$

The knowledge of the penetration length is of significant practical importance. It helps to estimate the critical thickness of the sample at which the two surface states start to overlap and the gapless TI state disappears.

To conclude this section, we should mention that the WAL in 3DTIs is suppressed if an energy gap opens in the surface spectrum as a result of TRS breaking [97]. This should apply to 3DTIs with magnetically doped surfaces. The suppressed WAL is also expected in ultrathin TIs films where it can be attributed to the overlap of the states from the opposite surfaces. Besides, quantum transport in electrically gated TI films may depend sensitively on the coupling between surface and bulk states, which can lead to both WAL and WL regimes.

5.4 Quantum Hall Effect on TI Surfaces

We have just discussed the quantum magnetoresistance of a 3DTI surface subjected to a perpendicular magnetic field. It originates from disorder-induced quantum interference and manifests itself in arbitrary weak magnetic fields. If the magnitude of the magnetic field increases, another quantum transport phenomenon—the quantum Hall effect (QHE)—can be observed. Essentially, the problem goes back to finding the Landau levels of a conventional 2D electron system with Rashba spin–orbit coupling, hereafter referred to as Rashba 2DES.

5.4.1 Recalling Landau Levels of a Rashba 2DES

The Hamiltonian of a Rashba 2DES in a magnetic field is

$$\hat{H} = \frac{1}{2m}\sigma_0\left(\hat{\boldsymbol{p}} - \frac{e}{c}\boldsymbol{A}\right)^2 + \alpha_{so}\left[\sigma_x\left(\hat{p}_y - \frac{e}{c}A_y\right) - \sigma_y\left(\hat{p}_x - \frac{e}{c}A_x\right)\right], \tag{5.160}$$

where α_{so} is the spin–orbit coupling constant, and $\boldsymbol{A}$ is the vector potential of magnetic field $\boldsymbol{B} = \nabla \times \boldsymbol{A}$. An alternative form of the Hamiltonian above is

$$\hat{H} = \frac{1}{2m}\sigma_0\left(\hat{\boldsymbol{p}} - \frac{e}{c}\boldsymbol{A}\right)^2 + \alpha_{so}\boldsymbol{\sigma}\cdot\left(\hat{\boldsymbol{p}} - \frac{e}{c}\boldsymbol{A}\right). \tag{5.161}$$

It is obtained by the unitary transformation of Eq. (5.160):

$$\hat{H} \to U\hat{H}U^{\dagger}, \qquad U = \begin{pmatrix} 1 & 0 \\ 0 & i \end{pmatrix}. \tag{5.162}$$

The Rashba coupling takes the form of the scalar product of vector $\boldsymbol{\sigma}$ and momentum $\hat{\boldsymbol{p}} - \frac{e}{c}\boldsymbol{A}$. In the following, we use this, more symmetric form.

The system is placed in a uniform magnetic field directed perpendicular to plane (x, y):

$$\boldsymbol{B} = [0, 0, B] = B\mathbf{z}. \tag{5.163}$$

This can be represented by the vector potential in the gauge

$$\boldsymbol{A} = [0, Bx, 0] = Bx\mathbf{y}, \tag{5.164}$$

where $\mathbf{y}$ is the unit vector in y-direction. In this gauge, the Hamiltonian is

$$\hat{H} = \frac{\hbar^2}{2m}\sigma_0\left(-i\nabla + \frac{x}{\ell^2}\mathbf{y}\right)^2 + \alpha_{so}\hbar\boldsymbol{\sigma}\cdot\left(-i\nabla + \frac{x}{\ell^2}\mathbf{y}\right) \tag{5.165}$$

$$= \frac{\hbar^2}{2m}\sigma_0\left[-\partial_x^2 + \left(-i\partial_y + \frac{x}{\ell^2}\right)^2\right] + \alpha_{so}\hbar\sigma_x\left(-i\partial_x\right) + \alpha_{so}\hbar\sigma_y\left(-i\partial_y + \frac{x}{\ell^2}\right), \tag{5.166}$$

where we introduce the magnetic length

$$\ell = \sqrt{\frac{c\hbar}{|e|B}}. \tag{5.167}$$

For example, for $B = 10$ T the magnetic length $\ell \approx 8$ nm.

We then write the Hamiltonian as the 2×2 matrix

$$\hat{H} = \frac{\hbar^2}{2m}\begin{pmatrix} -\partial_x^2 + \left(-i\partial_y + \frac{x}{\ell^2}\right)^2 & 0 \\ 0 & -\partial_x^2 + \left(-i\partial_y + \frac{x}{\ell^2}\right)^2 \end{pmatrix} \tag{5.168}$$

$$+\alpha_{so}\hbar\begin{pmatrix} 0 & -i\partial_x \\ -i\partial_x & 0 \end{pmatrix} + \alpha_{so}\hbar\begin{bmatrix} 0 & -i\left(-i\partial_y + \frac{x}{\ell^2}\right) \\ i\left(-i\partial_y + \frac{x}{\ell^2}\right) & 0 \end{bmatrix} \tag{5.169}$$

$$= \begin{pmatrix} \frac{\hbar^2}{2m}\left[-\partial_x^2 + \left(-i\partial_y + \frac{x}{\ell^2}\right)^2\right] & \alpha_{so}\hbar\left[-i\partial_x - i\left(-i\partial_y + \frac{x}{\ell^2}\right)\right] \\ \alpha_{so}\hbar\left[-i\partial_x + i\left(-i\partial_y + \frac{x}{\ell^2}\right)\right] & \frac{\hbar^2}{2m}\left[-\partial_x^2 + \left(-i\partial_y + \frac{x}{\ell^2}\right)^2\right] \end{pmatrix}. \tag{5.170}$$

The wave function can be written as

$$\psi(x, y) = \chi(x)e^{iky}, \tag{5.171}$$

so the Hamiltonian becomes

$$\hat{H} = \begin{pmatrix} \frac{\hbar^2}{2m}\left[-\partial_x^2 + \left(k + \frac{x}{\ell^2}\right)^2\right] & \alpha_{so}\hbar\left[-i\partial_x - i\left(k + \frac{x}{\ell^2}\right)\right] \\ \alpha_{so}\hbar\left[-i\partial_x + i\left(k + \frac{x}{\ell^2}\right)\right] & \frac{\hbar^2}{2m}\left[-\partial_x^2 + \left(k + \frac{x}{\ell^2}\right)^2\right] \end{pmatrix}. \tag{5.172}$$

The next step is to introduce the dimensionless coordinate and wave number

$$X = \frac{x}{\ell}, \qquad X_0 = k\ell, \tag{5.173}$$

and express the Hamiltonian in terms of X and X_0:

$$\hat{H} = \begin{pmatrix} \frac{\hbar^2}{2m\ell^2}\left[-\partial_X^2 + (X + X_0)^2\right] & -i\frac{\alpha_{so}\hbar}{\ell}\left[\partial_X + X + X_0\right] \\ i\frac{\alpha_{so}\hbar}{\ell}\left[-\partial_X + X + X_0\right] & \frac{\hbar^2}{2m\ell^2}\left[-\partial_X^2 + (X + X_0)^2\right] \end{pmatrix}. \tag{5.174}$$

The eigenvalues of this Hamiltonian yield the Landau levels for the Rashba 2DES. To find the eigenvalues, we introduce the ladder operators

$$a = \frac{1}{\sqrt{2}}(\partial_X + X + X_0), \qquad a^\dagger = \frac{1}{\sqrt{2}}(-\partial_X + X + X_0). \tag{5.175}$$

It is easy to prove the following relations:

$$\begin{aligned} a^\dagger a &= \frac{1}{2}(-\partial_X + X + X_0)(\partial_X + X + X_0) \\ &= \frac{1}{2}[-\partial_X^2 - \partial_X(X + X_0) + (X + X_0)\partial_X + (X + X_0)^2] \\ &= \frac{1}{2}[-\partial_X^2 - 1 + (X + X_0)^2], \end{aligned} \tag{5.176}$$

$$\begin{aligned} aa^\dagger &= \frac{1}{2}(\partial_X + X + X_0)(-\partial_X + X + X_0) \\ &= \frac{1}{2}[-\partial_X^2 + \partial_X(X + X_0) - (X + X_0)\partial_X + (X + X_0)^2] \\ &= \frac{1}{2}[-\partial_X^2 + 1 + (X + X_0)^2], \end{aligned} \tag{5.177}$$

and, finally,

$$aa^\dagger - a^\dagger a = 1. \tag{5.178}$$

As expected, we have obtained the commutator of the ladder operators of the harmonic oscillator. Inserting (5.175) und (5.178) into (5.174), we arrive at the following represenattion:

$$\hat{H} = \begin{pmatrix} \epsilon_2 \left[a^\dagger a + \frac{1}{2}\right] & -i\epsilon_1 a \\ i\epsilon_1 a^\dagger & \epsilon_2 \left[a^\dagger a + \frac{1}{2}\right] \end{pmatrix}. \tag{5.179}$$

The energies

$$\epsilon_1 = \frac{\sqrt{2}\alpha_{so}\hbar}{\ell} = \alpha_{so}\sqrt{\frac{2\hbar|e|B}{c}}, \qquad \epsilon_2 = \frac{\hbar^2}{m\ell^2} = \frac{\hbar|e|B}{mc} \tag{5.180}$$

are the scales of the linear and quadratic terms in the Hamiltonian.

Next, we need to solve the eigenvalue equation, $\hat{H}\chi = E\chi$, which consists of two equations for spinor components $\chi_\uparrow$ and $\chi_\downarrow$:

$$\begin{pmatrix} \epsilon_2 \left[a^\dagger a + \frac{1}{2}\right] & -i\epsilon_1 a \\ i\epsilon_1 a^\dagger & \epsilon_2 \left[a^\dagger a + \frac{1}{2}\right] \end{pmatrix} \begin{pmatrix} \chi_\uparrow \\ \chi_\downarrow \end{pmatrix} = E \begin{pmatrix} \chi_\uparrow \\ \chi_\downarrow \end{pmatrix}. \tag{5.181}$$

We seek the solution in the form

$$\chi_n = \begin{pmatrix} g_\uparrow|n-1\rangle \\ g_\downarrow|n\rangle \end{pmatrix} \quad \Longleftrightarrow \quad \chi_\uparrow = g_\uparrow|n-1\rangle,\ \chi_\downarrow = g_\downarrow|n\rangle, \tag{5.182}$$

where $|n\rangle$ and $|n-1\rangle$ are the eigenstates of the harmonic oscillator, $g_\uparrow$ and $g_\downarrow$ are constants, and n is a natural number, i.e. $1, 2, \ldots$. Eigenstates $|n\rangle$ and $|n-1\rangle$ are related to each other by simple operations involving a and $a^\dagger$:

$$a|n\rangle = \sqrt{n}|n-1\rangle, \quad a^\dagger|n-1\rangle = \sqrt{n}|n\rangle, \quad a^\dagger a|n\rangle = n|n\rangle. \tag{5.183}$$

Since the harmonic oscillator does not have the state with $n = -1$, Eq. (5.182) does not hold for $n = 0$. This case requires a special form of the solution:

$$\chi_0 = \begin{pmatrix} 0 \\ g_\downarrow|0\rangle \end{pmatrix} \quad \Longleftrightarrow \quad \chi_\uparrow = 0, \qquad \chi_\downarrow = g_\downarrow|0\rangle. \tag{5.184}$$

It describes a spin-polarized state with spin projection $s_z = -\frac{\hbar}{2}$.

Writing Eq. (5.181) in components, we have

$$g_\uparrow\left(\epsilon_2\left[a^\dagger a+\frac{1}{2}\right]-E\right)|n-1\rangle+g_\downarrow(-i\epsilon_1)a|n\rangle=0,$$
$$g_\uparrow(i\epsilon_1)a^\dagger|n-1\rangle+g_\downarrow\left(\epsilon_2\left[a^\dagger a+\frac{1}{2}\right]-E\right)|n\rangle=0.$$

The ladder operators act according to rules (5.183), with the result

$$g_\uparrow\left(\epsilon_2\left[n-1+\frac{1}{2}\right]-E\right)|n-1\rangle+g_\downarrow(-i\epsilon_1)\sqrt{n}|n-1\rangle=0,$$
$$g_\uparrow(i\epsilon_1)\sqrt{n}|n\rangle+g_\downarrow\left(\epsilon_2\left[n+\frac{1}{2}\right]-E\right)|n\rangle=0.$$

Eventually, we have a system of two algebraic equations

$$g_\uparrow\left(\epsilon_2\left[n-\frac{1}{2}\right]-E\right)+g_\downarrow(-i\epsilon_1\sqrt{n})=0, \qquad (5.185)$$
$$g_\uparrow(i\epsilon_1\sqrt{n})+g_\downarrow\left(\epsilon_2\left[n+\frac{1}{2}\right]-E\right)=0. \qquad (5.186)$$

Consider first $n\neq 0$. A nontrivial solution requires the vanishing of the determinant of the system,

$$\left(\epsilon_2\left[n-\frac{1}{2}\right]-E\right)\left(\epsilon_2\left[n+\frac{1}{2}\right]-E\right)-\epsilon_1^2 n=0. \qquad (5.187)$$

Solving for E yields the energy levels

$$E_n=\epsilon_2 n\pm\sqrt{\left(\frac{\epsilon_2}{2}\right)^2+\epsilon_1^2 n}, \qquad n=1,2,3,\ldots \qquad (5.188)$$

Using Eq. (5.180), we also find the magnetic-field dependence

$$E_n(B)=\frac{\hbar|e|B}{mc}n\pm\sqrt{\left(\frac{\hbar|e|B}{2mc}\right)^2+\frac{2\alpha_{so}^2\hbar|e|B}{c}n}, \qquad n=1,2,3,\ldots \qquad (5.189)$$

In the special case where $n=0$, spinor (5.184) has no $\uparrow$-component. Therefore, constant $g_\uparrow$ in (5.185) und (5.186) must be set to zero

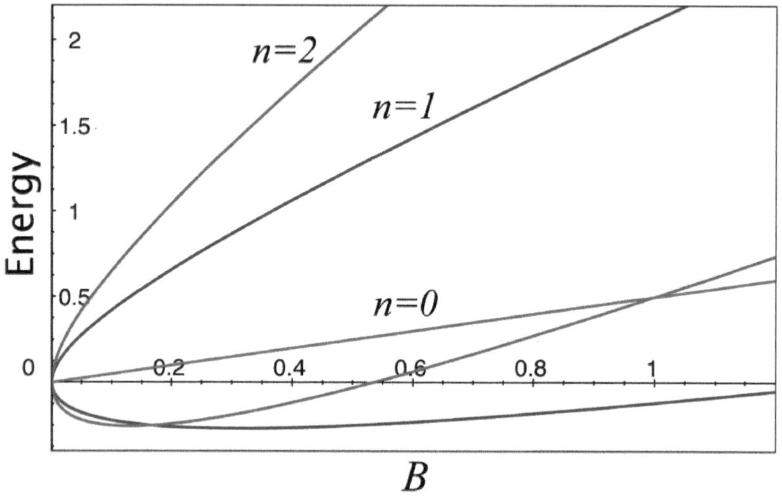

Figure 5.14 Landau levels of a Rashba 2DES as function of magnetic field B [see (5.189) und (5.191)].

simultaneously with n. This leads to the following equation:

$$g_{\downarrow}\left(\frac{\epsilon_2}{2} - E\right) = 0. \tag{5.190}$$

Its nontrivial solution (i.e. the one with $g_{\downarrow} \neq 0$) is

$$E_0(B) = \frac{\epsilon_2}{2} = \frac{\hbar|e|B}{2mc}, \qquad n = 0. \tag{5.191}$$

Equations (5.189) and (5.191) are the main results of this subsection. We see that the states with different wave number k have the same energy depending only on integer n. These discrete energies form a sequence of the Landau levels (plotted in Fig. 5.14). It is essential that the spin–orbit interaction has a particle-hole symmetry. For this reason, the Landau levels with $n \neq 0$ come in pairs, one from the '+' branch and one from the '−' branch in Eq. (5.189), whereas zeroth level (5.191) has no partner. In a Rashba 2DES, the quadratic part of the Hamiltonian overwhelms the linear one. This explains the non-monotonic behavior of the Landau levels of the '−' branch in Fig. 5.14.

5.4.2 Landau Quantization of Surface States: Half-Integer-Quantized Hall Conductivity

Equations (5.189) and (5.191) can, on an equal footing, describe the surface states of a 3DTI. In this case, however, the linear part of the Hamiltonian is dominant. In the main approximation, we set $m \to \infty$ resp. $\epsilon_2 \to 0$ and obtain the Landau levels (LLs)

$$E_n = \pm\sqrt{\frac{2\alpha_{so}^2 \hbar |e| B}{c} n}, \qquad n = 1, 2, 3, \ldots, \tag{5.192}$$

$$E_0 = 0, \qquad n = 0. \tag{5.193}$$

The pairs of LLs (5.192) are symmetric with respect to zeroth level $E_0 = 0$. To take advantage of this emerging particle-hole symmetry, we hereafter label the negative LLs by negative integers, while the positive LLs by positive ones, such that

$$n = 0, \pm 1, \pm 2, \ldots \tag{5.194}$$

counts all LLs, including the zeroth one. Then, the solutions for the LLs and eigenspinors can be re-written in a more compact form

$$E_n = \mathrm{sgn}(n)\hbar\Omega\sqrt{2|n|}, \quad \overline{|n\rangle} = \begin{bmatrix} -iC_{\uparrow n}\,||n| - 1\rangle \\ C_{\downarrow n}\,||n|\rangle \end{bmatrix}, \tag{5.195}$$

$$C_{\uparrow n} = \mathrm{sgn}(n)/\sqrt{2}, \qquad C_{\downarrow n} = 1/\sqrt{2}, \qquad n \neq 0, \tag{5.196}$$

and

$$E_0 = 0, \quad \overline{|0\rangle} = \begin{bmatrix} 0 \\ |0\rangle \end{bmatrix}, \quad C_{\uparrow 0} = 0,\ C_{\downarrow 0} = 1, \quad n = 0. \tag{5.197}$$

Above, $\Omega = \frac{\alpha_{so}}{\ell}$ is a frequency associated with magnetic length ℓ.

Having the LLs and eigenspinors, we now calculate the Hall component of the conductivity tensor in Eq. (2.33). In the LL basis, the Hall conductivity can be written as

$$\sigma_{xy}(\omega) = \frac{i\hbar}{2\pi\ell^2} \sum_{n,n'} \frac{f(E_n) - f(E_{n'})}{E_n - E_{n'}} \frac{\langle\overline{n}|j_x\overline{|n'\rangle}\langle\overline{n'}|j_y\overline{|n\rangle}}{E_n - E_{n'} + i\hbar/\tau + \hbar\omega}, \tag{5.198}$$

where $j_{x,y} = e\alpha_{so}\sigma_{x,y}$ are the components of the surface current operator, and τ accounts for the level broadening, e.g. due to disorder. Using matrices $\sigma_\pm = (\sigma_x \pm i\sigma_y)/2$, we write the current matrix elements as

$$\langle\overline{n'}|j_x|\overline{n}\rangle = ev[\langle\overline{n'}|\sigma_-|\overline{n}\rangle + \langle\overline{n'}|\sigma_+|\overline{n}\rangle],$$
$$\langle\overline{n'}|j_y|\overline{n}\rangle = iev[\langle\overline{n'}|\sigma_-|\overline{n}\rangle - \langle\overline{n'}|\sigma_+|\overline{n}\rangle],$$

and find the scalar products

$$\langle\overline{n'}|\sigma_\pm|\overline{n}\rangle = \pm i C_{\downarrow(\uparrow)n} C_{\uparrow(\downarrow)n'} \delta_{|n'|,|n|\pm 1}. \tag{5.199}$$

These selection rules allow the transitions between LLs with $|n'| = |n| \pm 1$. Accordingly, Hall conductivity (5.198) takes the form

$$\sigma_{xy}(\omega) = \frac{e^2\Omega^2}{h} \sum_{n,n'} [f(E_n) - f(E_{n'})] \frac{|\langle\overline{n'}|\sigma_+|\overline{n}\rangle|^2 - |\langle\overline{n'}|\sigma_-|\overline{n}\rangle|^2}{(E_{n'} - E_n)^2/\hbar^2 + (\tau^{-1} - i\omega)^2}. \tag{5.200}$$

The QHE occurs in strong magnetic fields under condition

$$\sqrt{2}\Omega\tau = 4R_0\sigma_c\sqrt{\frac{|B|}{\rho\Phi_0}} \gg 1. \tag{5.201}$$

The realization of this condition depends on the classical surface conductivity $\sigma_c = (e^2/h)(k_F\ell_{tr}/2)$ and surface carrier density ρ; $R_0 = h/2e^2$ is the resistance quantum. Further, we focus on the zero temperature case. At zero temperature LLs below the Fermi level E_F are occupied,

$$-N_h \le n \le N, \qquad f(E_n) = 1, \tag{5.202}$$

while those above E_F are empty (see also Fig. 5.15):

$$N + 1 \le n \le N_p, \qquad f(E_n) = 0, \tag{5.203}$$

where N is the index of the highest occupied LL given by

$$N = \mathrm{Int}\left(\frac{E_F^2}{2\hbar^2\Omega^2}\right) = \mathrm{Int}\left(2\pi\rho\ell^2\right) = \mathrm{Int}\left(\frac{\rho\Phi_0}{|B|}\right), \tag{5.204}$$

where Int(...) denotes the integer part. For the purpose of further calculation, we formally introduce index $-N_h$ for the lowest hole LL and index N_p for the highest particle LL. Equation (5.200) can then be written as

$$\sigma_{xy}(\omega) = \frac{e^2}{h} 2\Omega^2 \left[\sum_{n'=N+1}^{N_p} \sum_{n=-N}^{N} \frac{|\langle \overline{n'}|\sigma_+|\overline{n}\rangle|^2 - |\langle \overline{n'}|\sigma_-|\overline{n}\rangle|^2}{(E_{n'} - E_n)^2/\hbar^2 + (\tau^{-1} - i\omega)^2} \right. \tag{5.205}$$

$$\left. + \sum_{n'=N+1}^{N_p} \sum_{n=-N_h}^{-N-1} \frac{|\langle \overline{n'}|\sigma_+|\overline{n}\rangle|^2 - |\langle \overline{n'}|\sigma_-|\overline{n}\rangle|^2}{(E_{n'} - E_n)^2/\hbar^2 + (\tau^{-1} - i\omega)^2} \right], \tag{5.206}$$

where indices n and n' run over occupied and empty LLs, respectively. In particular, the sum in Eq. (5.205) contains both *intra*band (within the conduction band) and *inter*band (from the valence to conduction band) transitions from the occupied LLs in the symmetric interval $-N \leq n \leq N$. The non-zero contributions come from the intraband transition $N \to N+1$ with matrix element $\langle \overline{N+1}|\sigma_+|\overline{N}\rangle$ and from the interband transition $-N \to N+1$ with matrix element $\langle \overline{N+1}|\sigma_+|\overline{-N}\rangle$, as also shown in Fig. 5.15. The sum in Eq. (5.206) only involves the interband transitions from the rest of the occupied LLs $-N_h \leq n \leq -N-1$ to empty LLs

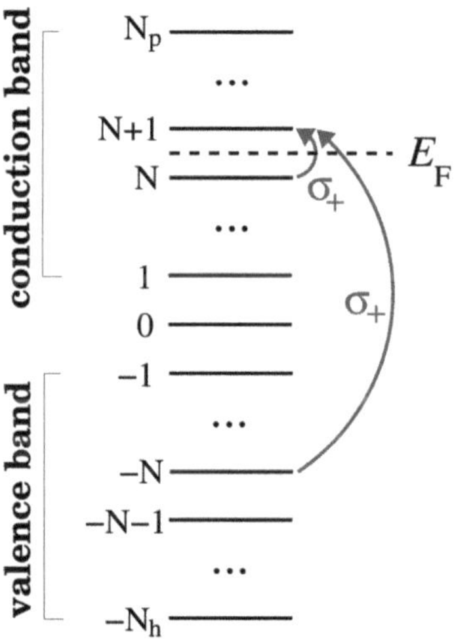

Figure 5.15 Schematic of surface Landau levels in Eqs. (5.195) and (5.197). Arrows indicate transitions $N \to N+1$ (intraband) and $-N \to N+1$ (interband) which yield the half-integer-quantized conductivity $\sigma_{xy} = (e^2/h)(N+1/2)$ (5.208); N is the index of the highest occupied Landau level (5.204). Adapted from [102].

$N+1 \leq n' \leq N_p$. This sum is nonzero only if the numbers of the particle and hole LLs are different, i.e. if $N_p \neq N_h$. To see this we rename the summation index $n \to -n$ and then use $|\langle\overline{n'}|\sigma_-|\overline{-n}\rangle| = |\langle\overline{n'}|\sigma_-|\overline{n}\rangle| = |\langle\overline{n}|\sigma_+|\overline{n'}\rangle|$, which yields

$$\sigma_{xy}(\omega) = \frac{e^2}{h}2\Omega^2\left[\frac{1/4}{(E_{N+1}-E_N)^2/\hbar^2+(\tau^{-1}-i\omega)^2}\right.$$
$$+\frac{1/4}{(E_{N+1}+E_N)^2/\hbar^2+(\tau^{-1}-i\omega)^2}$$
$$\left.+\sum_{n'=N+1}^{N_p}\sum_{n=N+1}^{N_h}\frac{|\langle\overline{n'}|\sigma_+|\overline{n}\rangle|^2-|\langle\overline{n}|\sigma_+|\overline{n'}\rangle|^2}{(E_{n'}+E_n)^2/\hbar^2+(\tau^{-1}-i\omega)^2}\right]. \tag{5.207}$$

The sum in Eq. (5.207) involves an anti-symmetric function under $n \leftrightarrow n'$ and, therefore, vanishes for symmetric limits $N_p = N_h$, which is the actual case. In other words, the interband transitions induced by the spin-raising and -lowering operators $\sigma_\pm$ between the particle-hole symmetric LLs $n \leq -N-1$ and $n' \geq N+1$ exactly compensate each other. In QHE regime (5.201) and for vanishing frequency ω, the Hall conductivity becomes

$$\sigma_{xy} = \frac{e^2}{h}\left[\frac{1/4}{(\sqrt{N+1}-\sqrt{N})^2}+\frac{1/4}{(\sqrt{N+1}+\sqrt{N})^2}\right]$$
$$= \frac{e^2}{h}\left[N+\frac{1}{2}\right]. \tag{5.208}$$

It is quantized in half-integer units of fundamental constant e^2/h [101]. In Fig. 5.16 we plot the dc Hall conductivity σ_{xy} in the entire range of magnetic fields and for finite temperature T, using Eq. (5.200) with $\omega = 0$.

It is not difficult to reconcile the above result with the TKNN theory allowing only an integer-quantized σ_{xy}. Indeed, a 3DTI film has two surface layers, one at the top and one at the bottom of the sample, which generally contribute on an equal footing to the total Hall conductivity

$$\sigma_{xy} = \sigma_{xy}^t + \sigma_{xy}^b, \tag{5.209}$$

where the conductivities of the top and bottom layers, $\sigma_{xy}^{t,b}$, are both given by Eq. (5.208) with, generally, different classical conductivities

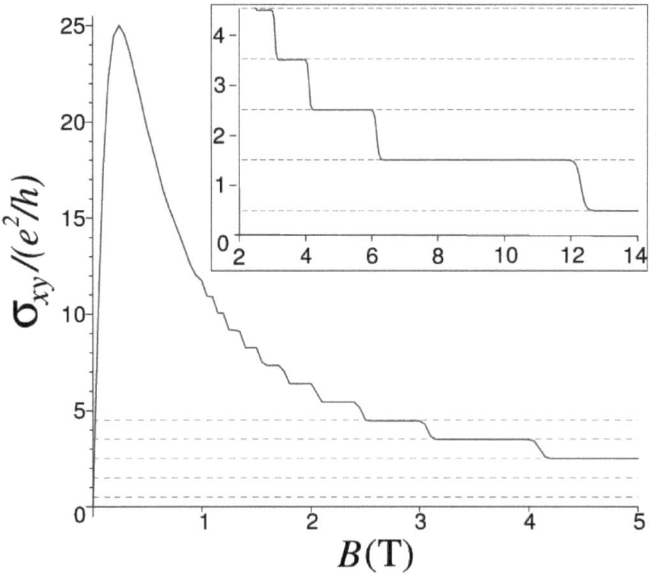

Figure 5.16 DC Hall conductivity σ_{xy} [given by Eq. (5.200) for $\omega = 0$] versus magnetic field B in Tesla. Inset: same dependence in high-field region; surface state parameters are $R_0\sigma_c = 25$, $\rho = 3 \cdot 10^{11}\text{cm}^{-2}$, $\hbar\alpha_{so} = 330$ meV·nm and temperature $T = 2$K. Adapted from [102].

$\sigma_c^{t,b}$ and carrier densities $\rho^{t,b}$. Although, separately, each surface layer exhibits a half-integer Hall quantization, the plateaus of the total conductivity are given by integer numbers, e.g. Fig. 5.17 shows a plateau sequence e^2/h, $2e^2/h$, $4e^2/h$, $5e^2/h\ldots$ The QHE in a 3DTI can thus have a rather unusual sequence of odd and even plateaus depending on the parameters of the top and bottom surface layers. The bulk energy gap is a necessary prerequisite for the observation of the surface QHE in transport which is, otherwise, dominated by bulk carriers.

5.4.3 AC Conductivities: Cyclotron Resonance

For the sake of completeness, here we briefly discuss the ac Hall $\sigma_{xy}(\omega)$ and diagonal $\sigma_{xx}(\omega)$ conductivities. At zero temperature $T = 0$ for the particle-hole symmetric LL spectrum, they can be obtained

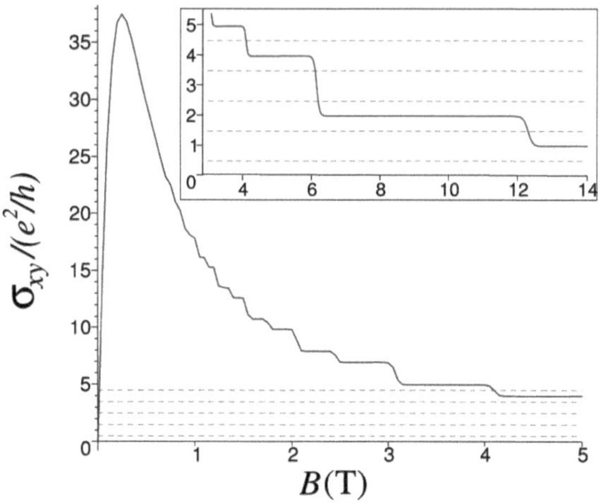

Figure 5.17 Two-surface Hall conductivity σ_{xy} [given by Eqs. (5.200) and (5.209) for $\omega = 0$] versus magnetic field B in Tesla. Inset: same dependence in high-field region; top (t) and bottom (b) surface state parameters are $R_0\sigma_c^t = 25$, $\rho^t = 3 \cdot 10^{11}\text{cm}^{-2}$ and $R_0\sigma_c^b = 12.5$, $\rho^b = 1.5 \cdot 10^{11}\text{cm}^{-2}$. For both surfaces $\hbar\alpha_{so} = 330$ meV·nm and $T = 2$K. Adapted from [102].

from Kubo formula (2.33) as

$$\sigma_{xy}(\omega) = \frac{e^2}{h}\left[\frac{1/4}{(\sqrt{N+1}-\sqrt{N})^2 + \left(\frac{1-i\omega\tau}{\sqrt{2}\Omega\tau}\right)^2} + \frac{1/4}{(\sqrt{N+1}+\sqrt{N})^2 + \left(\frac{1-i\omega\tau}{\sqrt{2}\Omega\tau}\right)^2}\right], \tag{5.210}$$

$$\sigma_{xx}(\omega) = \frac{e^2}{h}\frac{1-i\omega\tau}{\sqrt{2}\Omega\tau}\left[\frac{1/4}{(\sqrt{N+1}-\sqrt{N})\left[(\sqrt{N+1}-\sqrt{N})^2 + \left(\frac{1-i\omega\tau}{\sqrt{2}\Omega\tau}\right)^2\right]} + \frac{1/4}{(\sqrt{N+1}+\sqrt{N})\left[(\sqrt{N+1}+\sqrt{N})^2 + \left(\frac{1-i\omega\tau}{\sqrt{2}\Omega\tau}\right)^2\right]} + \sum_{n\geq N+1}\frac{1/2}{\sqrt{n+1}+\sqrt{n}}\frac{1}{(\sqrt{n+1}+\sqrt{n})^2 + \left(\frac{1-i\omega\tau}{\sqrt{2}\Omega\tau}\right)^2}\right]. \tag{5.211}$$

In Eq. (5.211) the first and second terms arise from the intraband $N \to N+1$ and interband $-N \to N+1$ transitions, respectively, while the third term accounts for the interband transitions from the hole LLs $n \leq -N-1$ to the particle LLs $n \geq N+1$. Such transitions contribute to σ_{xx} regardless of the presence (or absence) of the particle-hole symmetry.

For weak magnetic fields such that

$$N \approx \frac{\rho \Phi_0}{|B|} \gg 1, \tag{5.212}$$

the main contribution to Eqs. (5.210) and (5.211) comes from the intraband $N \to N+1$ transition, leading to the classical ac conductivities

$$\sigma_{xy} \approx \sigma_c \frac{\Omega_c \tau}{(\Omega_c \tau)^2 + (1 - i\omega\tau)^2}, \tag{5.213}$$

$$\sigma_{xx} \approx \sigma_c \frac{1 - i\omega\tau}{(\Omega_c \tau)^2 + (1 - i\omega\tau)^2}, \tag{5.214}$$

where Ω_c is the cyclotron frequency defined as

$$\Omega_c = \alpha_{so} \sqrt{\frac{2|eB|}{c\hbar}} (\sqrt{N+1} - \sqrt{N}) \approx \frac{|eB|\alpha_{so}}{c p_F}. \tag{5.215}$$

It corresponds to a resonant transition between the highest occupied (N) and lowest unoccupied ($N+1$) LLs for $\omega \to \Omega_c$ (see also Fig. 5.15). Unlike the parabolic-band case, Ω_c depends on the surface carrier density,

$$\Omega_c \propto \frac{1}{\sqrt{\rho}}, \tag{5.216}$$

due to the dependence of the Fermi momentum $p_F(\rho) = \hbar\sqrt{4\pi\rho}$.

Complemented by Maxwell's equations, the magnetotransport theory presented in this section lays the basis for magneto-optical spectroscopy of TIs. In particular, in the ac regime where $\omega\tau \gg 1$, but

$$\omega \ll \Omega(\sqrt{2(N+1)} - \sqrt{2N}), \tag{5.217}$$

the system is approximately on a QHE plateau with half-integer-quantized Hall conductivity (5.208). Condition (5.217) corresponds to adiabatically varying electromagnetic fields.

5.5 Emergent Axion Electrodynamics and Topological Magnetoelectric Effect

5.5.1 Emergent Axion Field: Modified Maxwell's Equations

Consider a 2DES in the plane at $z = 0$ in the QHE regime with current density

$$\boldsymbol{j} = -\sigma_{xy}\delta(z)\,\mathbf{z} \times \boldsymbol{E}, \tag{5.218}$$

where Hall conductivity σ_{xy} is given by Eq. (5.208). It is instructive to introduce an auxiliary field, $\theta(z)$, defined by the equation for its gradient

$$\nabla\theta = \frac{2\pi h}{e^2}\sigma_{xy}\delta(z)\mathbf{z}. \tag{5.219}$$

The current density can then be cast into the form

$$\boldsymbol{j} = -\frac{e^2}{2\pi h}\nabla\theta \times \boldsymbol{E}. \tag{5.220}$$

Using identity $\nabla\theta \times \boldsymbol{E} = \nabla \times (\theta\boldsymbol{E}) - \theta(\nabla \times \boldsymbol{E})$ and Faraday's law $\nabla \times \boldsymbol{E} = -\frac{1}{c}\partial_t \boldsymbol{B}$,[1] we can further transform Eq. (5.220) to

$$\boldsymbol{j} = -\frac{e^2}{2\pi h}\nabla \times (\theta\boldsymbol{E}) - \frac{e^2}{2\pi h}\frac{\theta}{c}\partial_t \boldsymbol{B}. \tag{5.221}$$

It has the form of a bound current density

$$\boldsymbol{j} = c\nabla \times \boldsymbol{M} + \partial_t \boldsymbol{P}, \tag{5.222}$$

with magnetization $\boldsymbol{M}$ and electric polarization $\boldsymbol{P}$ given by

$$\boldsymbol{M} = -\frac{e^2}{2\pi ch}\theta\boldsymbol{E} = -\frac{\alpha}{(2\pi)^2}\theta\boldsymbol{E}, \qquad \alpha = \frac{e^2}{c\hbar}, \tag{5.223}$$

$$\boldsymbol{P} = -\frac{e^2}{2\pi ch}\theta\boldsymbol{B} = -\frac{\alpha}{(2\pi)^2}\theta\boldsymbol{B}, \tag{5.224}$$

where $\alpha \approx 1/137$ is the fine-structure constant. The corresponding bound charge density can be written as

$$\rho = -\nabla \cdot \boldsymbol{P}. \tag{5.225}$$

[1] In the rest of this chapter, $\boldsymbol{B}$ denotes the magnetic component of a time-dependent electromagnetic field.

The electric and magnetic fields in Eqs. (5.223) and (5.224) should be found self-consistently, using Maxwell's equations

$$\nabla \times \boldsymbol{B} = \frac{4\pi}{c}(\boldsymbol{j}_0 + \boldsymbol{j}) + \frac{1}{c}\partial_t \boldsymbol{E}, \tag{5.226}$$

$$\nabla \cdot \boldsymbol{E} = 4\pi(\rho_0 + \rho), \tag{5.227}$$

where, for the sake of generality, we introduced extra current density $\boldsymbol{j}_0$ and charge density ρ_0. Using Eqs. (5.222) and (5.225), we arrive at modified Maxwell's equations

$$\nabla \times \boldsymbol{H} = \frac{4\pi}{c}\boldsymbol{j}_0 + \frac{1}{c}\partial_t \boldsymbol{D}, \tag{5.228}$$

$$\nabla \cdot \boldsymbol{D} = 4\pi\rho_0 \tag{5.229}$$

for magnetization and displacement fields $\boldsymbol{H} = \boldsymbol{B} - 4\pi\boldsymbol{M}$ and $\boldsymbol{D} = \boldsymbol{E} + 4\pi\boldsymbol{P}$. These are given explicitly by

$$\boldsymbol{H} = \boldsymbol{B} + \frac{\alpha\theta}{\pi}\boldsymbol{E}, \tag{5.230}$$

$$\boldsymbol{D} = \boldsymbol{E} - \frac{\alpha\theta}{\pi}\boldsymbol{B}. \tag{5.231}$$

We see that field θ generates a magnetoelectric effect characterized by a magnetic response to an electric field and vice versa. Analogous phenomena have been theorized in particle physics as the consequence of the hypothetical axion. This elusive particle couples to the electromagnetic field via the action (see [103, 104])

$$S_\theta = \frac{\alpha}{(2\pi)^2}\int \theta \boldsymbol{E} \cdot \boldsymbol{B} dt d^3\boldsymbol{r}, \tag{5.232}$$

where axion field θ is generally a dynamical variable. In our QHE system, the emergent axion field takes constant values in the materials above and below the plane $z = 0$. If, for example, $z = 0$ is the surface of a 3DTI occupying half-space $z > 0$ and facing an ordinary insulator at $z < 0$, then the 3DTI contains the axion field

$$\theta = \frac{2\pi h}{e^2}\sigma_{xy} = (2N + 1)\pi, \tag{5.233}$$

where we used Eqs. (5.208) and (5.219). That is, on a QHE plateau, the emerging axion field in a 3DTI is an odd integer of π. This is the topological magnetoelectric effect (TME) characterized by universal magnetoelectric corrections to $\boldsymbol{B}$ and $\boldsymbol{E}$ [104]:

$$\boldsymbol{H} = \boldsymbol{B} + \alpha(2N + 1)\boldsymbol{E}, \tag{5.234}$$

$$\boldsymbol{D} = \boldsymbol{E} - \alpha(2N + 1)\boldsymbol{B}. \tag{5.235}$$

It is worth noting that the emerging axion electrodymanics of 3DTIs is restricted to adiabatically varying electromagnetic fields, see Eq. (5.217).

5.5.2 Quantized Faraday Rotation due to Topological Magnetoelectric Effect

The TME can be observed in Faraday rotation experiments (see [104–106]). Here, we focus on the basic physical picture, leaving materials and apparatus specifics aside.

In a 3DTI, the magnetoelectric coupling results in magnetization and displacement fields, $\boldsymbol{H}$ and $\boldsymbol{D}$, that are generally non-collinear to vectors $\boldsymbol{B}$ and $\boldsymbol{E}$, respectively. This is, for example, the case if a 3DTI is subjected to a transverse electromagnetic wave in which $\boldsymbol{B}$ and $\boldsymbol{E}$ are orthogonal to each other and to the direction of incidence. The non-collinear orientation of the fields inside and outside the material causes a rotation of the $\boldsymbol{B}$- and $\boldsymbol{E}$-polarization planes in the transmitted wave, which is the Faraday effect.

To be more specific, we assume that the top and bottom surfaces of a 3DTI both contribute to the current density

$$\boldsymbol{j} = -[\sigma^t_{xy}\delta(z - d/2) + \sigma^b_{xy}\delta(z + d/2)]\mathbf{z} \times \boldsymbol{E}. \tag{5.236}$$

The use of the delta functions is justified if the localization lengths of the surface states are much smaller than the sample thickness d. The electromagnetic wave is incident normally at the bottom surface $z = -d/2$ and is monochromatic, with $\boldsymbol{B}$, $\boldsymbol{E} \propto e^{-i\omega t}$.

To find the electromagnetic field inside the material, we use the thin-film approximation

$$d \ll c/\omega, \tag{5.237}$$

which is the case for most epitaxially grown 3DTI films. Technically, the electric field is almost uniform inside the film and can be approximated by the average value

$$\langle \boldsymbol{E} \rangle = \frac{1}{d} \int_{-d/2}^{d/2} \boldsymbol{E}\, dz. \tag{5.238}$$

The equation for $\langle \boldsymbol{E} \rangle$ can be obtained by averaging the Maxwell equation

$$\nabla \times \boldsymbol{B} = \frac{4\pi}{c} \boldsymbol{j} - \frac{i\omega}{c} \boldsymbol{E} \tag{5.239}$$

over the film thickness. The resulting equation can be written as

$$-\frac{i\omega d}{c} \langle \boldsymbol{E} \rangle - \frac{4\pi}{c} \sigma_{xy} \mathbf{z} \times \langle \boldsymbol{E} \rangle = \mathbf{z} \times (\boldsymbol{B}_{+} - \boldsymbol{B}_{-}), \tag{5.240}$$

where $\sigma_{xy} = \sigma_{xy}^{t} + \sigma_{xy}^{b}$ is the total surface Hall conductivity.

The problem now reduces to finding the magnetic field components of the wave at the top and bottom surfaces, $\boldsymbol{B}_{+}$ and $\boldsymbol{B}_{-}$. For that, we need to solve Maxwell's equations outside the material,

$$\nabla \times \boldsymbol{B} = -\frac{i\omega}{c} \boldsymbol{E}, \qquad \nabla \times \boldsymbol{E} = \frac{i\omega}{c} \boldsymbol{B}. \tag{5.241}$$

For normal incidence, the solutions are

$$\boldsymbol{E} = \boldsymbol{E}_i e^{ik(z+d/2)} + \boldsymbol{E}_r e^{-ik(z+d/2)}, \tag{5.242}$$

$$\boldsymbol{B} = \mathbf{z} \times [\boldsymbol{E}_i e^{ik(z+d/2)} - \boldsymbol{E}_r e^{-ik(z+d/2)}], \quad z \leq -d/2, \tag{5.243}$$

and

$$\boldsymbol{E} = \boldsymbol{E}_t e^{ik(z-d/2)}, \quad \boldsymbol{B} = \mathbf{z} \times \boldsymbol{E}, \quad z \geq d/2, \tag{5.244}$$

where $\boldsymbol{E}_i$, $\boldsymbol{E}_r$ and $\boldsymbol{E}_t$ are the electric field amplitudes of the incident, reflected and transmitted waves; $k = \omega/c$. Hence,

$$\boldsymbol{B}_{+} = \mathbf{z} \times \boldsymbol{E}_t, \quad \boldsymbol{B}_{-} = \mathbf{z} \times (\boldsymbol{E}_i - \boldsymbol{E}_r). \tag{5.245}$$

Besides, for a thin film, we can set

$$\boldsymbol{E}_t \approx \langle \boldsymbol{E} \rangle, \quad \boldsymbol{E}_i + \boldsymbol{E}_r \approx \langle \boldsymbol{E} \rangle, \tag{5.246}$$

and exclude the reflection amplitude via

$$\boldsymbol{E}_r = \boldsymbol{E}_t - \boldsymbol{E}_i. \tag{5.247}$$

Inserting Eqs. (5.245)–(5.247) into Eq. (5.240) yields the following relation between amplitudes $\boldsymbol{E}_i$ and $\boldsymbol{E}_t$:

$$\left(1 - \frac{i\omega d}{2c}\right) \boldsymbol{E}_t - \frac{2\pi}{c} \sigma_{xy} \mathbf{z} \times \boldsymbol{E}_t = \boldsymbol{E}_i, \tag{5.248}$$

where term $i\omega d/(2c)$ can be neglected in thin-film approximation (5.237). To solve for transmission amplitude $\boldsymbol{E}_t$, we multiply Eq. (5.248) by $\mathbf{z}$ and combine the resulting equation with Eq. (5.248). After some algebra, we arrive at the result

$$\boldsymbol{E}_t = \frac{\boldsymbol{E}_i \cos\Theta + \mathbf{z} \times \boldsymbol{E}_i \sin\Theta}{\sqrt{1 + (2\pi\sigma_{xy}/c)^2}}, \tag{5.249}$$

where angle Θ is defined as

$$\Theta = \arcsin \frac{2\pi\sigma_{xy}/c}{\sqrt{1 + (2\pi\sigma_{xy}/c)^2}}. \tag{5.250}$$

The interpretation of the above results is pretty obvious. The polarization plane of the transmitted wave is rotated with respect to that of the incident wave by angle Θ. The latter is determined by total surface Hall conductivity

$$\sigma_{xy} = \sigma_{xy}^t + \sigma_{xy}^b = \frac{e^2}{h}(N_t + N_b + 1), \tag{5.251}$$

where N_t and N_b are the indices of the highest occupied LLs at top and bottom surfaces, respectively. Since these numbers are integer, rotation angle Θ takes quantized values

$$\Theta = \arcsin \frac{\alpha(N_t + N_b + 1)}{\sqrt{1 + \alpha^2(N_t + N_b + 1)^2}}, \qquad N_{t,b} = 0, 1, 2, \ldots \tag{5.252}$$

In particular, in a strong dc magnetic field such that $N_t = N_b = 0$, the rotation angle is close to the value of the fine-structure constant [104]:

$$\Theta \approx \alpha. \tag{5.253}$$

This is a magneto-optical manifestation of the TME discussed in Section 5.5.1. In experiments (e.g. in [105, 106]), more factors come into play such as the effects of substrates, disorder, frequency and polarization of the electromagnetic wave.

5.6 Problems

(1) Consider the Hamiltonian for a TI thin film with the Fermi level lying in the conduction band [see Eqs. (4.40), (4.41)

and Fig. 4.4]. Using the results of Sections 2.1.1 and 2.1.2, calculate Berry's phases for the upper and lower blocks $\hat{H}_{\mathbf{k}\pm}$ at the Fermi level.

Show that the elastic life-time τ in Eq. (5.83) can be expressed through the calculated Berry phases $\gamma_{\pm}$ as

$$\tau = \frac{\tau_0}{1 + (\gamma_+ - \pi)^2/\pi^2} = \frac{\tau_0}{1 + (\gamma_- - \pi)^2/\pi^2}. \qquad (5.254)$$

In the same way, prove the following expression for the backscattering rate $\tau_{\mathcal{M}}^{-1}$ in the Cooperons and WAL conductivity [see Eqs. (5.114) and (5.137)]:

$$\tau_{\mathcal{M}}^{-1} = \frac{2}{\tau}\left(\frac{\gamma_+ - \pi}{\pi}\right)^2 = \frac{2}{\tau}\left(\frac{\gamma_- - \pi}{\pi}\right)^2. \qquad (5.255)$$

Equations (5.254) and (5.255) demonstrate that the carrier backscattering occurs when the band Berry phase deviates from the universal value of π.

(2) Consider a 3DTI with an infinitely extended surface in the x-direction and a finite length L in the y-direction with the periodic boundary conditions at the ends. In this case, Eq. (5.155) for the WAL correction to the classical surface conductivity can be written as

$$\delta\sigma_{xx} = \frac{e^2 D}{\pi\hbar}\int \frac{dq_x}{2\pi L}\sum_n C^{00}(q_x, q_y^{(n)}), \qquad (5.256)$$

where the Cooperon $C^{00}(q_x, q_y^{(n)})$ is given by Eq. (5.150) with $q_y^{(n)}$ taking the discrete values $2\pi n/L$ with integer $n = 0, \pm 1, \pm 2, \ldots$. Calculate the integral over q_x and show that under the condition

$$\ell_\varphi^2 \gg \left(\frac{L}{2\pi}\right)^2 \qquad (5.257)$$

the WAL is effectively quasi-one-dimensional. Prove that for such a 3DTI "wire" the WAL correction (5.256) is

$$\delta\sigma_{xx} \approx \frac{e^2}{h}\frac{\ell_\varphi}{L}. \qquad (5.258)$$

Chapter 6

Unconventional Designer Superconductors

So far we have been dealing with single-particle properties of helical states in TIs. However, electron helicity can play a striking role in many-body phenomena as well. This chapter focuses on the interplay between helicity and pairing interaction among electrons in 3DTIs and related materials. This subject has attracted a great deal of interest for several reasons. First of all, pairing interaction results in unconventional superconductivity in TIs. Unlike typical superconductors, e.g. Pb, Al or Nb, where Cooper pairs have an ordinary s-wave symmetry, TIs can host odd-parity p-wave Cooper pairs. Further, in superconducting TI structures we encounter another type of topological states protected by TRS—helical Andreev bound states. Moreover, superconductivity in TIs can be induced through a proximity effect, by placing conventional superconducting metals, such as Al or Nb, on the surface of a TI material. This offers diverse possibilities for engineering novel superconducting states and devices.

Topological Quantum Materials: Concepts, Models, and Phenomena
Grigory Tkachov

ISBN 978-981-4968-12-6 (Hardcover), 978-1-003-26641-9 (eBook)
www.jennystanford.com

6.1 Introduction to Superconductivity

6.1.1 Electron–Electron Pairing: Mean-Field Theory

The theory of superconductivity developed by Bardeen, Cooper and Schrieffer [58] attributes this phenomenon to lattice-mediated attractive interactions between electrons in a metal. We adopt an insightful version of the BCS theory [107] in which an electron system is described by a real-space Hamiltonian with local pairwise interactions:

$$\hat{\mathcal{H}} = \int d\boldsymbol{r} \sum_{\alpha,\beta=\uparrow,\downarrow} \left[\psi_\alpha^\dagger(\boldsymbol{r}) h_{\alpha\beta}(\boldsymbol{r}) \psi_\beta(\boldsymbol{r}) + \frac{U}{2} \psi_\alpha^\dagger(\boldsymbol{r}) \psi_\beta^\dagger(\boldsymbol{r}) \psi_\beta(\boldsymbol{r}) \psi_\alpha(\boldsymbol{r}) \right], \tag{6.1}$$

where operator $\psi_\alpha^\dagger(\boldsymbol{r})$ ($\psi_\alpha(\boldsymbol{r})$) creates (annihilates) an electron at position $\boldsymbol{r}$ in spin state $\alpha =\uparrow, \downarrow$, $h_{\alpha\beta}(\boldsymbol{r})$ is the first-quantized form of the single-particle Hamiltonian, and $U < 0$ is the interaction constant. The electron motion is governed by equation $i\hbar\partial_t \psi_\alpha(\boldsymbol{r}, t) = \psi_\alpha(\boldsymbol{r}, t)\hat{\mathcal{H}} - \hat{\mathcal{H}}\psi_\alpha(\boldsymbol{r}, t)$ or, explicitly,

$$i\hbar\partial_t \psi_\alpha(\boldsymbol{r}, t) = \sum_\beta \left[h_{\alpha\beta}(\boldsymbol{r}) \psi_\beta(\boldsymbol{r}, t) - U \psi_\beta^\dagger(\boldsymbol{r}, t) \psi_\alpha(\boldsymbol{r}, t) \psi_\beta(\boldsymbol{r}, t) \right]. \tag{6.2}$$

In order to solve it, we employ the mean-field approximation for the interaction term:

$$\begin{aligned} U \psi_\beta^\dagger(\boldsymbol{r}, t) \psi_\alpha(\boldsymbol{r}, t) \psi_\beta(\boldsymbol{r}, t) &\to U \psi_\beta^\dagger(\boldsymbol{r}, t) \langle \psi_\alpha(\boldsymbol{r}, t) \psi_\beta(\boldsymbol{r}, t) \rangle \\ &= -\Delta_{\alpha\beta}(\boldsymbol{r}) \psi_\beta^\dagger(\boldsymbol{r}, t), \end{aligned} \tag{6.3}$$

where the pair of the annihilation operators is replaced by its expectation value denoted by brackets $\langle \ldots \rangle$, and $\Delta_{\alpha\beta}(\boldsymbol{r})$ is the resulting mean-field potential given by

$$\Delta_{\alpha\beta}(\boldsymbol{r}) = -U \langle \psi_\alpha(\boldsymbol{r}, t) \psi_\beta(\boldsymbol{r}, t) \rangle, \qquad \alpha, \beta =\uparrow, \downarrow. \tag{6.4}$$

Expectation value $\langle \psi_\alpha(\boldsymbol{r}, t) \psi_\beta(\boldsymbol{r}, t) \rangle$ is called the superconducting order parameter. The averaging is done with the statistical operator, using the mean-field Hamiltonian.

The nonzero order parameter indicates the specific character of the superconducting ground state which is formed by electron pairs

(Cooper pairs) rather than by independent particles. For this reason, $\Delta_{\alpha\beta}(\boldsymbol{r})$ is frequently called the pair potential or the superconducting gap function. It has four components that can be combined into a 2×2 matrix

$$\hat{\Delta}(\boldsymbol{r}) = \begin{bmatrix} \Delta_{\uparrow\uparrow}(\boldsymbol{r}) & \Delta_{\uparrow\downarrow}(\boldsymbol{r}) \\ \Delta_{\downarrow\uparrow}(\boldsymbol{r}) & \Delta_{\downarrow\downarrow}(\boldsymbol{r}) \end{bmatrix} = \Delta(\boldsymbol{r}) \begin{bmatrix} 0 & 1 \\ -1 & 0 \end{bmatrix} = \Delta(\boldsymbol{r}) i\sigma_y, \quad (6.5)$$

where we used anticommutation relation $\psi_\alpha(\boldsymbol{r}, t)\psi_\beta(\boldsymbol{r}, t) = -\psi_\alpha(\boldsymbol{r}, t)\psi_\beta(\boldsymbol{r}, t)$, or, explicitly,

$$\psi_\uparrow^2(\boldsymbol{r}, t) = \psi_\downarrow^2(\boldsymbol{r}, t) = 0, \quad \psi_\downarrow(\boldsymbol{r}, t)\psi_\uparrow(\boldsymbol{r}, t) = -\psi_\uparrow(\boldsymbol{r}, t)\psi_\downarrow(\boldsymbol{r}, t).$$

By $\Delta(\boldsymbol{r})$ we denote the orbital part of the gap function,

$$\Delta(\boldsymbol{r}) = -U \langle \psi_\uparrow(\boldsymbol{r}, t)\psi_\downarrow(\boldsymbol{r}, t) \rangle. \quad (6.6)$$

The gap function in Eq. (6.5) characterizes a conventional spin-singlet s-wave pairing. Indeed, the spin function of the electron pair, $i\sigma_y$, is antisymmetric under the exchange of spin states $\uparrow$ and $\downarrow$. Consequently, due to the Pauli principle, the orbital part of the pair potential, $\Delta(\boldsymbol{r})$, must be an even function of the particle coordinates, i.e. $\Delta(-\boldsymbol{r}) = \Delta(\boldsymbol{r})$. For a translationally invariant system, the pairing appears to be constant in space: $\Delta(\boldsymbol{r}) = const.$

6.1.2 Nambu Representation

From Eqs. (6.2)–(6.3) we obtain the mean-field equation of motion

$$i\hbar\partial_t\psi_\alpha(\boldsymbol{r}, t) = \sum_\beta \left[h_{\alpha\beta}(\boldsymbol{r})\psi_\beta(\boldsymbol{r}, t) + \Delta_{\alpha\beta}(\boldsymbol{r})\psi_\beta^\dagger(\boldsymbol{r}, t) \right]. \quad (6.7)$$

Since ψ and $\psi^\dagger$ are coupled, in order to solve Eq. (6.7) we also need its Hermitian conjugate:

$$i\hbar\partial_t\psi_\alpha^\dagger(\boldsymbol{r}, t) = \sum_\beta \left[-h_{\alpha\beta}^*(\boldsymbol{r})\psi_\beta^\dagger(\boldsymbol{r}, t) - \Delta_{\alpha\beta}^*(\boldsymbol{r})\psi_\beta(\boldsymbol{r}, t) \right]. \quad (6.8)$$

At this stage it is convenient to introduce the spinors

$$\psi(\boldsymbol{r}, t) = \begin{bmatrix} \psi_\uparrow(\boldsymbol{r}, t) \\ \psi_\downarrow(\boldsymbol{r}, t) \end{bmatrix}, \qquad \overline{\psi}(\boldsymbol{r}, t) = \begin{bmatrix} \psi_\uparrow^\dagger(\boldsymbol{r}, t) \\ \psi_\downarrow^\dagger(\boldsymbol{r}, t) \end{bmatrix}, \quad (6.9)$$

and write Eqs. (6.7) and (6.8) as

$$i\hbar\partial_t \begin{bmatrix} \psi(\boldsymbol{r},t) \\ \overline{\psi}(\boldsymbol{r},t) \end{bmatrix} = \begin{bmatrix} \hat{h}(\boldsymbol{r}) & \hat{\Delta}(\boldsymbol{r}) \\ -\hat{\Delta}^*(\boldsymbol{r}) & -\hat{h}^*(\boldsymbol{r}) \end{bmatrix} \begin{bmatrix} \psi(\boldsymbol{r},t) \\ \overline{\psi}(\boldsymbol{r},t) \end{bmatrix}, \tag{6.10}$$

where the hat over $h(\boldsymbol{r})$ indicates that the single-particle Hamiltonian is a 2×2 matrix in spin space. Further, we combine the spinors into a 4-component wave function

$$\Psi(\boldsymbol{r},t) = \begin{bmatrix} \psi(\boldsymbol{r},t) \\ \overline{\psi}(\boldsymbol{r},t) \end{bmatrix} = \begin{bmatrix} \psi_\uparrow(\boldsymbol{r},t) \\ \psi_\downarrow(\boldsymbol{r},t) \\ \psi_\uparrow^\dagger(\boldsymbol{r},t) \\ \psi_\downarrow^\dagger(\boldsymbol{r},t) \end{bmatrix}, \tag{6.11}$$

and introduce the 4×4 matrix Hamiltonian

$$\hat{H} = \begin{bmatrix} \hat{h}(\boldsymbol{r}) & \hat{\Delta}(\boldsymbol{r}) \\ -\hat{\Delta}^*(\boldsymbol{r}) & -\hat{h}^*(\boldsymbol{r}) \end{bmatrix}. \tag{6.12}$$

Then, Eq. (6.10) has the form of a free-particle equation of motion:

$$i\hbar\partial_t \Psi(\boldsymbol{r},t) = \hat{H}\,\Psi(\boldsymbol{r},t). \tag{6.13}$$

In order to obtain this important result we have used the particle-hole (also called Nambu) representation for field $\Psi(\boldsymbol{r},t)$. The name reflects the fact that spinors $\psi(\boldsymbol{r},t)$ and $\overline{\psi}(\boldsymbol{r},t)$ represent a particle- and a hole-like fermionic excitations, respectively. In the normal (non-interacting) metal, a particle-like excitation is created by adding an electron into an empty state above the Fermi sea, whereas a hole-like excitation is created by removing an electron from an occupied state of the Fermi sea. Since in normal metals these processes occur independently, a particle and a hole are described by decoupled equations of motion

$$\underbrace{i\hbar\partial_t \psi(\boldsymbol{r},t) = \hat{h}(\boldsymbol{r})\psi(\boldsymbol{r},t)}_{particle}, \quad \underbrace{i\hbar\partial_t \overline{\psi}(\boldsymbol{r},t) = -\hat{h}^*(\boldsymbol{r})\overline{\psi}(\boldsymbol{r},t)}_{hole}. \tag{6.14}$$

In the presence of a pairing interaction, particles and holes of the underlying non-interacting system are no longer valid elementary excitations. We can see this from Eq. (6.12) in which a particle and a hole are coupled by the pair potential. Now, the true elementary excitations are those that diagonalize Hamiltonian (6.12). Physically, a pairing interaction involves pairs of states which are related by time reversal. In order to see this, we switch to another Nambu basis defined by the unitary transformation

$$\Psi'(\boldsymbol{r}, t) = \hat{U}\,\Psi(\boldsymbol{r}, t), \quad \hat{H}' = \hat{U}\,\hat{H}\,\hat{U}^{\dagger}, \quad \hat{U} = \begin{bmatrix} \sigma_0 & 0 \\ 0 & i\sigma_y \end{bmatrix}, \tag{6.15}$$

or, explicitly,

$$\Psi'(\boldsymbol{r}, t) = \begin{bmatrix} \psi_{\uparrow}(\boldsymbol{r}, t) \\ \psi_{\downarrow}(\boldsymbol{r}, t) \\ \psi_{\downarrow}^{\dagger}(\boldsymbol{r}, t) \\ -\psi_{\uparrow}^{\dagger}(\boldsymbol{r}, t) \end{bmatrix}, \quad \hat{H}' = \begin{bmatrix} \hat{h}(\boldsymbol{r}) & \Delta(\boldsymbol{r})\sigma_0 \\ \Delta^*(\boldsymbol{r})\sigma_0 & -\mathbb{T}\hat{h}(\boldsymbol{r})\mathbb{T}^{-1} \end{bmatrix}. \tag{6.16}$$

In $\hat{H}'$, the hole is related to the particle by time-reversal operation $\mathbb{T} = i\sigma_y K$, while the pair potential becomes diagonal in spin space. That is, $\Delta(\boldsymbol{r})$ acts similar to a coupling between Kramers' partners of the underlying non-interacting system.

6.1.3 Bogoliubov–de Gennes equation: Particle-Hole Symmetry

In studies of superconducting systems (especially inhomogeneous ones) it is often convenient to deal with the wave functions instead of the field operators. In order to derive the equations for the wave functions we express the real-space field operators in the basis of new fermionic operators, γ_n and $\gamma_n^{\dagger}$,

$$\psi_{\alpha}(\boldsymbol{r}, t) = \sum_n [u_{\alpha n}(\boldsymbol{r}, t)\gamma_n + v_{\alpha n}^*(\boldsymbol{r}, t)\gamma_n^{\dagger}], \tag{6.17}$$

$$\psi_{\alpha}^{\dagger}(\boldsymbol{r}, t) = \sum_n [v_{\alpha n}(\boldsymbol{r}, t)\gamma_n + u_{\alpha n}^*(\boldsymbol{r}, t)\gamma_n^{\dagger}], \tag{6.18}$$

where the sum runs over the states with quantum numbers n that may include discrete and/or continuous indices. The position and time dependence is transferred from the operators to the wave functions $u_{\alpha n}(\boldsymbol{r}, t)$ and $v_{\alpha n}(\boldsymbol{r}, t)$. The corresponding transformation of the Nambu basis is

$$\begin{bmatrix} \psi(\boldsymbol{r}, t) \\ \overline{\psi}(\boldsymbol{r}, t) \end{bmatrix} = \sum_n \left(\begin{bmatrix} u_n(\boldsymbol{r}, t) \\ v_n(\boldsymbol{r}, t) \end{bmatrix} \gamma_n + \begin{bmatrix} v_n^*(\boldsymbol{r}, t) \\ u_n^*(\boldsymbol{r}, t) \end{bmatrix} \gamma_n^{\dagger} \right), \tag{6.19}$$

where we introduced the spinors

$$u_n(\boldsymbol{r}, t) = \begin{bmatrix} u_{\uparrow n}(\boldsymbol{r}) \\ u_{\downarrow n}(\boldsymbol{r}) \end{bmatrix} e^{-iE_n t/\hbar}, \qquad v_n(\boldsymbol{r}, t) = \begin{bmatrix} v_{\uparrow n}(\boldsymbol{r}) \\ v_{\downarrow n}(\boldsymbol{r}) \end{bmatrix} e^{-iE_n t/\hbar}, \tag{6.20}$$

describing a state with energy E_n. Inserting Eq. (6.19) into Eq. (6.10) and collecting the coefficients at γ_n and $\gamma_n^\dagger$, we arrive at two equations

$$E_n \begin{bmatrix} u_n(\boldsymbol{r}) \\ v_n(\boldsymbol{r}) \end{bmatrix} = \begin{bmatrix} \hat{h}(\boldsymbol{r}) & \hat{\Delta}(\boldsymbol{r}) \\ -\hat{\Delta}^*(\boldsymbol{r}) & -\hat{h}^*(\boldsymbol{r}) \end{bmatrix} \begin{bmatrix} u_n(\boldsymbol{r}) \\ v_n(\boldsymbol{r}) \end{bmatrix}, \tag{6.21}$$

and

$$-E_n \begin{bmatrix} v_n^*(\boldsymbol{r}) \\ u_n^*(\boldsymbol{r}) \end{bmatrix} = \begin{bmatrix} \hat{h}(\boldsymbol{r}) & \hat{\Delta}(\boldsymbol{r}) \\ -\hat{\Delta}^*(\boldsymbol{r}) & -\hat{h}^*(\boldsymbol{r}) \end{bmatrix} \begin{bmatrix} v_n^*(\boldsymbol{r}) \\ u_n^*(\boldsymbol{r}) \end{bmatrix}, \tag{6.22}$$

Equation (6.21) is a wave function analogue of Eq. (6.10), called the Bogolubov-de Gennes (BdG) equation. It comes in a pair with Eq. (6.22). This fact reflects a symmetry between the positive- and negative-energy states in a superconductor, known as the particle-hole symmetry. If, for example, $E_n > 0$, then for each positive-energy solution of Eq. (6.21) there is a negative-energy solution of Eq. (6.22). Mathematically, the particle hole-symmetry can be described by a conjugation operation

$$\mathbb{C} = \begin{bmatrix} 0 & \sigma_0 \\ \sigma_0 & 0 \end{bmatrix} K, \qquad \mathbb{C}^{-1} = \mathbb{C}. \tag{6.23}$$

where K denotes complex conjugation, and σ_0 is the unit matrix in spin space. It is easy to see that Eqs. (6.21) and (6.22) are related to each other by the transformation

$$\begin{bmatrix} v_n^*(\boldsymbol{r}) \\ u_n^*(\boldsymbol{r}) \end{bmatrix} = \mathbb{C} \begin{bmatrix} u_n(\boldsymbol{r}) \\ v_n(\boldsymbol{r}) \end{bmatrix}, \qquad \mathbb{C}\hat{H}\mathbb{C}^{-1} = -\hat{H}. \tag{6.24}$$

As an example illustrating the basic property of the superconducting state, let us calculate the energy spectrum of a superconductor with a spin-independent quadratic Hamiltonian,

$$\hat{h}(\boldsymbol{r}) = \left(-\frac{\hbar^2 \nabla^2}{2m} - \mu \right) \sigma_0, \tag{6.25}$$

and a uniform pairing, $\hat{\Delta}(\boldsymbol{r}) = const$; μ is the chemical potential. The system is translational invariant, i.e. quasiparticle momentum $\boldsymbol{p}$ is a good quantum number, and we seek the solutions to Eq. (6.21) in the form of plane waves $e^{i\boldsymbol{pr}/\hbar}$:

$$E_{\boldsymbol{p}} u_{\boldsymbol{p}} = \left(\frac{\boldsymbol{p}^2}{2m} - \mu \right) u_{\boldsymbol{p}} + \Delta i \sigma_y v_{\boldsymbol{p}} \tag{6.26}$$

$$E_{\boldsymbol{p}} v_{\boldsymbol{p}} = -\Delta^* i \sigma_y u_{\boldsymbol{p}} - \left(\frac{\boldsymbol{p}^2}{2m} - \mu \right) v_{\boldsymbol{p}}. \tag{6.27}$$

Another important symmetry is the invariance under spin rotation around the y axis. It allows us to choose the solutions to be the eigenstates of σ_y:

$$\sigma_y u_{s\boldsymbol{p}} = s u_{s\boldsymbol{p}}, \qquad \sigma_y v_{s\boldsymbol{p}} = s v_{s\boldsymbol{p}}, \tag{6.28}$$

with eigenvalues $s = \pm 1$. In this basis, the BdG equations become diagonal in spin space. Equating the determinant of the system to zero, we find the energy levels

$$E_{s\boldsymbol{p}} = \pm\sqrt{\left(\frac{\boldsymbol{p}^2}{2m} - \mu\right)^2 + |\Delta|^2}. \tag{6.29}$$

Note that they are degenerate with respect to spin projection s. Equation (6.29) shows that the minimum energy required to create an elementary excitation in a superconductor is $|\Delta|$. That is, the pairing interaction opens an energy gap between the ground state and excited states. Also, in accord with the generic particle-hole symmetry, for each positive-energy level there is a symmetric negative-energy state.

To conclude this introductory section, we note that the superconducting pair potential, Δ, must be determined self-consistently. We should first find wave functions $u_n(\boldsymbol{r})$ and $v_n(\boldsymbol{r})$ for given Δ from BdG equation (6.21). Then, we insert Eqs. (6.17) and (6.18) for the field operators into $\Delta = -U\langle\psi_\uparrow(\boldsymbol{r}, t)\psi_\downarrow(\boldsymbol{r}, t)\rangle$ and perform averaging with the statistical operator of the mean-field system. This yields an explicit equation for Δ as function of interaction constant U and other microscopic parameters. As it is a standard task of the BCS theory, we leave the self-consistent determination of Δ aside.

6.2 Induced Mixed-Parity Superconductivity in TIs

6.2.1 Classification of Superconducting Pairing: Mixed-Parity Order Parameter

The formalism described above can be used to study superconducting properties of TIs. We focus on a 3DTI with non-interacting surface carriers. The superconductivity in this system can be

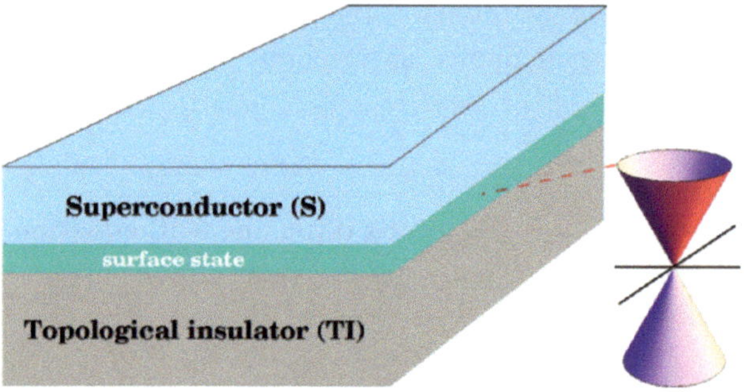

Figure 6.1 Schematic geometry of a superconductor (S)/topological insulator (TI) surface hybrid system. The surface state in the TI has a Dirac-cone band dispersion.

induced by placing a conventional, singlet s-wave superconductor on the TI surface, as depicted in Fig. 6.1. Such a hybrid system was first considered by Fu and Kane in Ref. [69]. In the presence of the superconductor, the TI surface layer can be described by a mean-field equation similar to Eq. (6.12):

$$i\hbar\partial_t \begin{bmatrix} c_{\boldsymbol{p}}(t) \\ \overline{c}_{-\boldsymbol{p}}(t) \end{bmatrix} = \underbrace{\begin{bmatrix} v\boldsymbol{\sigma}\cdot\boldsymbol{p}-\mu & i\sigma_y\Delta \\ -i\sigma_y\Delta^* & v\boldsymbol{\sigma}^*\cdot\boldsymbol{p}+\mu \end{bmatrix}}_{\hat{H}} \begin{bmatrix} c_{\boldsymbol{p}}(t) \\ \overline{c}_{-\boldsymbol{p}}(t) \end{bmatrix}, \qquad (6.30)$$

where $c_{\boldsymbol{p}}(t)$ and $\overline{c}_{-\boldsymbol{p}}(t)$ are the spinor field operators in the momentum ($\boldsymbol{p}$) representation

$$c_{\boldsymbol{p}}(t) = \begin{bmatrix} c_{\uparrow\boldsymbol{p}}(t) \\ c_{\downarrow\boldsymbol{p}}(t) \end{bmatrix}, \qquad \overline{c}_{-\boldsymbol{p}}(t) = \begin{bmatrix} c^{\dagger}_{\uparrow-\boldsymbol{p}}(t) \\ c^{\dagger}_{\downarrow-\boldsymbol{p}}(t) \end{bmatrix}, \qquad (6.31)$$

and $v\boldsymbol{\sigma}\cdot\boldsymbol{p}-\mu$ is the Hamiltonian for the surface electron (v is the Fermi velocity). The off-diagonal entries account for the induced singlet s-wave pairing with strength Δ. It should not be confused with the pair potential of an intrinsic superconductor introduced earlier in Eq. (6.6). The pair potential originates from an attractive electron-electron interaction and is proportional to coupling constant $U < 0$. In the TI we assume no attractive interactions, i.e. $U \geq 0$, which rules out the possibility of the intrinsic superconductivity. That is, the pair potential is identically zero in the normal system. This, however, does not apply to the superconducting order parameter $\langle \psi_\alpha(\boldsymbol{r}, t)\psi_\beta(\boldsymbol{r}, t)\rangle$. It can appear in the normal system without any interactions, due to the penetration of the Cooper pair condensate from the superconductor. This phenomenon is known as the superconducting proximity effect. Pairing energy Δ is therefore related to the proximity-induced order parameter in the TI. The proof of this fact requires a microscopic treatment of the coupling between the two systems, which will be reviewed later on.

As pointed out by Fu and Kane [69], the usual singlet s-wave pairing in Eq. (6.30) induces unconventional triplet p-wave pair correlations. This happens because spin-momentum locking $\boldsymbol{\sigma} \cdot \boldsymbol{p}$ violates the spin-rotation symmetry of the Hamiltonian in Eq. (6.30). In order to see this, and for a rigorous classification of the symmetry of the pair correlations, one has to define and calculate a suitable correlation function of the system. The standard way is to use Green's function of Eq. (6.30). It is defined as follows:

$$\hat{G}_{\boldsymbol{p}}(t,t') = \frac{1}{i\hbar}\left\langle \begin{bmatrix} c_{\boldsymbol{p}}(t) \\ \overline{c}_{-\boldsymbol{p}}(t) \end{bmatrix} \otimes \left[\overline{c}^T_{\boldsymbol{p}}(t')\; c^T_{-\boldsymbol{p}}(t')\right] \right\rangle \tag{6.32}$$

$$= \begin{bmatrix} \langle c_{\boldsymbol{p}}(t) \otimes \overline{c}^T_{\boldsymbol{p}}(t')\rangle & \langle c_{\boldsymbol{p}}(t) \otimes c^T_{-\boldsymbol{p}}(t')\rangle \\ \langle \overline{c}_{-\boldsymbol{p}}(t) \otimes \overline{c}^T_{\boldsymbol{p}}(t')\rangle & \langle \overline{c}_{-\boldsymbol{p}}(t) \otimes c^T_{-\boldsymbol{p}}(t')\rangle \end{bmatrix}, \tag{6.33}$$

where symbol $\otimes$ means the direct products of the spinors in Eq. (6.31) and the transposed spinors

$$\overline{c}^T_{\boldsymbol{p}}(t') = [c^\dagger_{\uparrow\boldsymbol{p}}(t'), c^\dagger_{\downarrow\boldsymbol{p}}(t')], \qquad c^T_{-\boldsymbol{p}}(t') = [c_{\uparrow-\boldsymbol{p}}(t'), c_{\downarrow-\boldsymbol{p}}(t')], \tag{6.34}$$

and the brackets $\langle \ldots \rangle$ denote averaging of the time-ordered operator pairs with the ground-state statistical operator. Equation (6.33) defines the Green function as a 2×2 matrix in the particle-hole space. For the matrix elements in the particle-hole space we introduce the notations

$$\hat{G}_{\boldsymbol{p}}(t,t') = \begin{bmatrix} \hat{G}_{11,\boldsymbol{p}}(t,t') & \hat{G}_{12,\boldsymbol{p}}(t,t') \\ \hat{G}_{21,\boldsymbol{p}}(t,t') & \hat{G}_{22,\boldsymbol{p}}(t,t') \end{bmatrix}, \tag{6.35}$$

where each entry is a 2×2 matrix in the spin space. The superconducting order parameter can be characterized by an off-diagonal Green function with equal time arguments, e.g.

$$\hat{G}_{12,\boldsymbol{p}}(t,t) = \frac{1}{i\hbar}\begin{bmatrix} \langle c_{\uparrow\boldsymbol{p}}(t)c_{\uparrow-\boldsymbol{p}}(t)\rangle & \langle c_{\uparrow\boldsymbol{p}}(t)c_{\downarrow-\boldsymbol{p}}(t)\rangle \\ \langle c_{\downarrow\boldsymbol{p}}(t)c_{\uparrow-\boldsymbol{p}}(t)\rangle & \langle c_{\downarrow\boldsymbol{p}}(t)c_{\downarrow-\boldsymbol{p}}(t)\rangle \end{bmatrix}. \tag{6.36}$$

This equation encodes all pairing scenarios allowed by the fermionic anti-commutation rules. Since the total spin of an electron pair can be either 0 (singlet state) or 1 (triplet state), Eq. (6.36) can be expanded in the basis of the singlet and triplet spin states:

$$\hat{G}_{12,\boldsymbol{p}}(t,t) = \frac{1}{i\hbar}\left[f_0(\boldsymbol{p})i\sigma_y + \boldsymbol{f}(\boldsymbol{p})\cdot\boldsymbol{\sigma}i\sigma_y\right]. \tag{6.37}$$

The first term is the singlet part. It consists of the antisymmetric spin function, $i\sigma_y$, and the symmetric (under $\boldsymbol{p} \to -\boldsymbol{p}$) orbital function

$$f_0(\boldsymbol{p}) = \frac{1}{2}\langle c_{\uparrow\boldsymbol{p}}(t)c_{\downarrow-\boldsymbol{p}}(t) - c_{\downarrow\boldsymbol{p}}(t)c_{\uparrow-\boldsymbol{p}}(t)\rangle. \tag{6.38}$$

The second term in Eq. (6.37) is the triplet part. Triplet spin functions $\sigma_j i\sigma_y$ (where $j = x, y$ or z) are symmetric under exchange of the spin indices. The corresponding orbital functions are combined into a vector $\boldsymbol{f}(\boldsymbol{p}) = [f_x(\boldsymbol{p}), f_y(\boldsymbol{p}), f_z(\boldsymbol{p})]$ defined by

$$f_x(\boldsymbol{p}) = -\frac{1}{2}\langle c_{\uparrow\boldsymbol{p}}(t)c_{\uparrow-\boldsymbol{p}}(t) - c_{\downarrow\boldsymbol{p}}(t)c_{\downarrow-\boldsymbol{p}}(t)\rangle, \tag{6.39}$$

$$f_y(\boldsymbol{p}) = -\frac{i}{2}\langle c_{\uparrow\boldsymbol{p}}(t)c_{\uparrow-\boldsymbol{p}}(t) + c_{\downarrow\boldsymbol{p}}(t)c_{\downarrow-\boldsymbol{p}}(t)\rangle, \tag{6.40}$$

$$f_z(\boldsymbol{p}) = \frac{1}{2}\langle c_{\uparrow\boldsymbol{p}}(t)c_{\downarrow-\boldsymbol{p}}(t) + c_{\downarrow\boldsymbol{p}}(t)c_{\uparrow-\boldsymbol{p}}(t)\rangle. \tag{6.41}$$

All three triplet orbital functions are odd in momentum $\boldsymbol{p}$. Unlike the singlet state, the triplet pairing is nodal in both momentum and coordinate spaces, e.g. vanishing for two electrons at the same position, as illustrated in Fig. 6.2. Both the singlet and triplet pairings are consistent with the fermionic anti-commutation rules and the Pauli exclusion principle. It is worth noting that the states with $f_{x,y}(\boldsymbol{p})$ are formed by Cooper pairs with nonzero (± 1) projections of the total spin, whereas $f_z(\boldsymbol{p})$ corresponds to a state with zero total spin projection.

In order to identify the pairing symmetry in the TI we need to calculate $\hat{G}_{12,\boldsymbol{p}}(t,t)$ explicitly and compare it with the expansion in

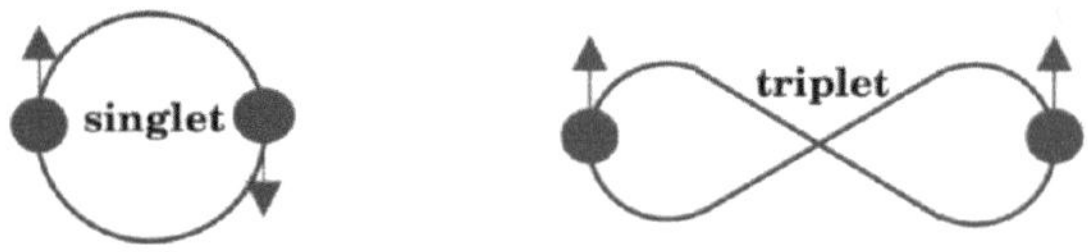

Figure 6.2 Illustration of the singlet and triplet pairing (see also the text).

Eq. (6.37). We start with the equation of motion for Green's function (6.32):

$$i\hbar\partial_t \hat{G}_{\boldsymbol{p}}(t,t') - \hat{H}\hat{G}_{\boldsymbol{p}}(t,t') = \hat{I}\delta(t-t'). \tag{6.42}$$

It follows from Eqs. (6.30) and (6.32) by differentiating $\hat{G}_{\boldsymbol{p}}(t,t')$ with respect to the first time argument ($\hat{I}$ is the 4×4 unit matrix). Using the energy representation

$$\hat{G}_{\boldsymbol{p}}(t,t') = \int \frac{dE}{h}\hat{G}_{\boldsymbol{p}E}\, e^{-iE(t-t')/\hbar}, \tag{6.43}$$

and Eqs. (6.30) and (6.35) for $\hat{H}$ and $\hat{G}_{\boldsymbol{p}E}$, respectively, we rewrite Eq. (6.42) as

$$\left\{E\hat{I} - \begin{bmatrix} v\boldsymbol{\sigma}\cdot\boldsymbol{p}-\mu & i\sigma_y\Delta \\ -i\sigma_y\Delta^* & v\boldsymbol{\sigma}^*\cdot\boldsymbol{p}+\mu \end{bmatrix}\right\}\begin{bmatrix} \hat{G}_{11,\boldsymbol{p}E} & \hat{G}_{12,\boldsymbol{p}E} \\ \hat{G}_{21,\boldsymbol{p}E} & \hat{G}_{22,\boldsymbol{p}E} \end{bmatrix} = \hat{I}. \tag{6.44}$$

It splits into two pairs of coupled equations

$$(E+\mu - v\boldsymbol{\sigma}\boldsymbol{p})\hat{G}_{11,\boldsymbol{p}E} - \Delta i\sigma_y\hat{G}_{21,\boldsymbol{p}E} = \sigma_0, \tag{6.45}$$

$$\Delta^* i\sigma_y\hat{G}_{11,\boldsymbol{p}E} + (E-\mu - v\boldsymbol{\sigma}^*\boldsymbol{p})\hat{G}_{21,\boldsymbol{p}E} = 0, \tag{6.46}$$

and

$$(E+\mu - v\boldsymbol{\sigma}\boldsymbol{p})\hat{G}_{12,\boldsymbol{p}E} - \Delta i\sigma_y\hat{G}_{22,\boldsymbol{p}E} = 0, \tag{6.47}$$

$$\Delta^* i\sigma_y\hat{G}_{12,\boldsymbol{p}E} + (E-\mu - v\boldsymbol{\sigma}^*\boldsymbol{p})\hat{G}_{22,\boldsymbol{p}E} = \sigma_0. \tag{6.48}$$

To illustrate the solution procedure let us consider the first pair of the equations. Applying $\sigma_y(E-\mu-v\boldsymbol{\sigma}^*\boldsymbol{p})\sigma_y$ to Eq. (6.45) from the left and excluding $\hat{G}_{21,\boldsymbol{p}E}$ with the help of Eq. (6.46), we obtain a closed equation for $\hat{G}_{11,\boldsymbol{p}E}$

$$\left[(E-\mu+v\boldsymbol{\sigma}\boldsymbol{p})(E+\mu-v\boldsymbol{\sigma}\boldsymbol{p}) - |\Delta|^2\right]\hat{G}_{11,\boldsymbol{p}E} = E-\mu+v\boldsymbol{\sigma}\boldsymbol{p}. \tag{6.49}$$

Then, applying $(E+\mu+v\boldsymbol{\sigma}\boldsymbol{p})(E-\mu-v\boldsymbol{\sigma}\boldsymbol{p}) - |\Delta|^2$ from the left, we arrive after some algebra at the solution

$$\hat{G}_{11,\boldsymbol{p}E} = \frac{(E+\mu+v\boldsymbol{\sigma p})[(E-\mu)^2 - v^2p^2] - |\Delta|^2(E-\mu+v\boldsymbol{\sigma p})}{\left[E^2-(vp-\mu)^2-|\Delta|^2\right]\left[E^2-(vp+\mu)^2-|\Delta|^2\right]}, \tag{6.50}$$

where $p \equiv |\boldsymbol{p}|$. The function $\hat{G}_{21,\boldsymbol{p}E}$ is then obtained from Eq. (6.46) as

$$\hat{G}_{21,\boldsymbol{p}E} = \frac{-\Delta^* i\sigma_y\,(E^2-\mu^2-v^2p^2-|\Delta|^2-2\mu v\boldsymbol{\sigma p})}{\left[E^2-(vp-\mu)^2-|\Delta|^2\right]\left[E^2-(vp+\mu)^2-|\Delta|^2\right]}. \tag{6.51}$$

Similarly, the solutions of Eqs. (6.47) and (6.48) are

$$\hat{G}_{22,\boldsymbol{p}E} = \frac{(E-\mu+v\boldsymbol{\sigma}^*\boldsymbol{p})[(E+\mu)^2 - v^2p^2] - |\Delta|^2(E+\mu+v\boldsymbol{\sigma}^*\boldsymbol{p})}{\left[E^2-(vp-\mu)^2-|\Delta|^2\right]\left[E^2-(vp+\mu)^2-|\Delta|^2\right]}, \tag{6.52}$$

$$\hat{G}_{12,\boldsymbol{p}E} = \frac{(\overbrace{E^2-\mu^2-v^2p^2-|\Delta|^2}^{singlet}-\overbrace{2\mu v\boldsymbol{\sigma p}}^{triplet})\,\Delta i\sigma_y}{\left[E^2-(vp-\mu)^2-|\Delta|^2\right]\left[E^2-(vp+\mu)^2-|\Delta|^2\right]}. \tag{6.53}$$

The latter equation provides the information about the character of the pair correlations in the TI. Comparing Eqs. (6.37) and (6.53), we see that the induced order parameter has both singlet ($\propto i\sigma_y$) and triplet ($\propto \sigma_j i\sigma_y$) parts, with corresponding orbital functions $f_0(\boldsymbol{p}, E)$ and $\boldsymbol{f}(\boldsymbol{p}, E)$ given by

$$f_0(\boldsymbol{p}, E) = \frac{\Delta(E^2-\mu^2-v^2p^2-|\Delta|^2)}{\left[E^2-(vp-\mu)^2-|\Delta|^2\right]\left[E^2-(vp+\mu)^2-|\Delta|^2\right]}, \tag{6.54}$$

$$\boldsymbol{f}(\boldsymbol{p}, E) = -\frac{2\Delta\mu v\boldsymbol{p}}{\left[E^2-(vp-\mu)^2-|\Delta|^2\right]\left[E^2-(vp+\mu)^2-|\Delta|^2\right]}. \tag{6.55}$$

Comparing these results with the case of an ordinary s-wave superconductor (see Section 6.1), we conclude that the induced superconductivity in the TI has a mixed s- and p-wave character. The p-wave correlations occur as the combined effect of the induced s-wave pairing and spin-momentum locking $\boldsymbol{\sigma}\cdot\boldsymbol{p}$, which breaks the spin-rotation symmetry. As expected, the orbital triplet pairing is odd in momentum $\boldsymbol{p}$. It has only two nonzero components, f_x and f_y. The third, f_z, is absent, as it is not allowed by TRS. Besides, the triplet pairing vanishes for $\mu = 0$, i.e. at the charge neutrality point of the surface state.

6.2.2 Theory of the Mixed-Parity Proximity Effect

Proximity structures of conventional superconductors and TIs are an example of noncentrosymmetric superconductors (NCSs). Such superconductors (called "proximity NCSs" below) generically exhibit an antisymmetric SOC which mixes even-parity (spin-singlet) and odd-parity (spin-triplet) Cooper pairs, producing a mixed-parity superconducting order parameter. It is unconventional because it does not fit into traditional classification of superconducting states which invokes definite (even or odd) spatial parity of the Cooper-pair wave function. Unambiguous verification of the mixed-parity superconducting order is one of the outstanding challenges in the NCS research.

For intrinsic NCSs, candidate order parameters can be classified according to their behavior under the symmetry elements (space group) of the crystal. We shall take a different route and derive the mixed-parity order parameter, using a microscopic model for a proximity NCS. It assumes that superconductivity is induced in a 2D SOC system by an overlying s-wave singlet superconductor,[1] as depicted in Fig. 6.3. In such proximity structures, the SOC forces the electron spins in a tunneling singlet Cooper pair to follow the electron momentum in the plane of the normal system, which causes a spin flip, hence an admixture of spin-triplet odd-parity Cooper pairs. Since each spin in a singlet pair can be flipped, both up- and down-spin pairs are induced with no net spin magnetization. This phenomenology is behind many microscopic studies of the proximity effect in systems with broken spin rotation symmetry, such as Rashba 2DES and TIs.

More insight can be gained from the tunneling model of the superconducting proximity effect used earlier for conventional

[1]Surface superconductivity in 3DTIs is a special case of this model.

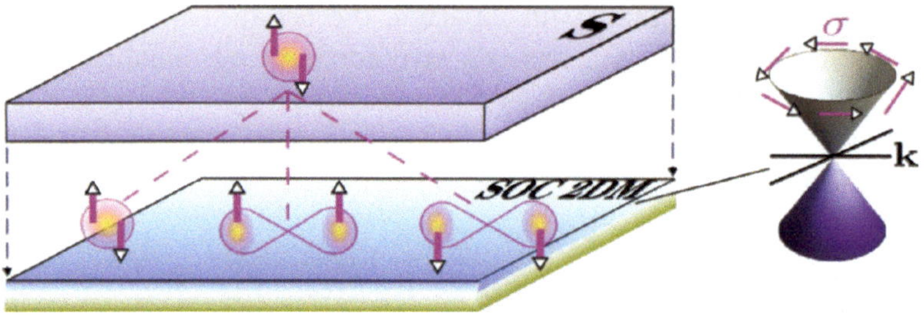

Figure 6.3 Schematic of a mixed-parity proximity effect in a SOC 2D material (2DM) contacted by a conventional (s-wave singlet) superconductor (S). Spin-momentum locking facilitates conversion of singlet Cooper pairs into a mixture of singlet and triplet states in the SOC 2DM. Adapted from [60].

2DESs (see e.g. [109–113]) and later on for TIs (see e.g. [114–117]). In this model, the proximity of the superconductor is accounted for by a tunneling self-energy $\hat{\Sigma}_{\mathcal{T}}$ in the equation of motion for Green's function of the normal system $\hat{\mathcal{G}}_{\boldsymbol{k}}$:

$$[E - \hat{H}^{(0)}_{\boldsymbol{k}} - \hat{\Sigma}_{\mathcal{T}}]\hat{\mathcal{G}}_{\boldsymbol{k}} = \hat{I}, \quad \hat{H}^{(0)}_{\boldsymbol{k}} = \begin{bmatrix} \hat{h}_{\boldsymbol{k}} & 0 \\ 0 & -\hat{h}^*_{-\boldsymbol{k}} \end{bmatrix}, \tag{6.56}$$

where $\hat{H}^{(0)}_{\boldsymbol{k}}$ is the bare Hamiltonian of the normal system in the Nambu representation, and $\hat{I}$ is the corresponding unit matrix. The self-energy is a matrix in the Nambu space with the following structure:

$$\hat{\Sigma}_{\mathcal{T}} = \begin{bmatrix} -i\Gamma(E) & \Delta(E)i\sigma_y \\ -\Delta^*(E)i\sigma_y & -i\Gamma(E) \end{bmatrix}.$$

Its off-diagonal entries yield the induced singlet pair potential, while the diagonal elements account for the shift of the spectrum due to the tunneling:

$$\Gamma(E) = \Gamma_0 g_s(E) = \Gamma_0 \frac{E}{\sqrt{E^2 - \Delta_s^2}}, \qquad \Gamma_0 = \pi\mathcal{T}^2\rho_s,$$

$$\Delta(E) = i\Gamma_0 f_s(E) = i\Gamma_0 \frac{\Delta_s}{\sqrt{E^2 - \Delta_s^2}}.$$

Here, $g_s(E)$ and $f_s(E)$ are the momentum-integrated quasiparticle and condensate Green functions of the overlying superconductor which has the gap energy Δ_s. The energy Γ_0 is determined by the single-particle tunneling rate depending on the normal-state density of the states in the superconducting metal, ρ_s, and the tunneling matrix element $\mathcal{T}$. At low energies $E \ll \Delta_s$, in the main approximation $\Gamma = 0$ and $\Delta = \Gamma_0$, hence the effective BdG Hamiltonian

$$\hat{H}_{\boldsymbol{k}} = \hat{H}^{(0)}_{\mathbf{k}} + \hat{\Sigma}_{\mathcal{T}} = \begin{bmatrix} \hat{h}_{\boldsymbol{k}} & \Delta i\sigma_y \\ -\Delta i\sigma_y & -\hat{h}^*_{-\boldsymbol{k}} \end{bmatrix}. \tag{6.57}$$

The normal system is described by a 2D Hamiltonian

$$\hat{h}_{\boldsymbol{k}} = \boldsymbol{\sigma} \cdot \boldsymbol{\gamma}_{\boldsymbol{k}} - \mu,$$

with an antisymmetric SOC field $\boldsymbol{\gamma}_{\boldsymbol{k}}$.

Similar to the preceding section, the induced order parameter is a 2×2 spin matrix of the form

$$\left\langle \begin{matrix} c_{\uparrow\boldsymbol{k}}(t)c_{\uparrow-\boldsymbol{k}}(t) & c_{\uparrow\boldsymbol{k}}(t)c_{\downarrow-\boldsymbol{k}}(t) \\ c_{\downarrow\boldsymbol{k}}(t)c_{\uparrow-\boldsymbol{k}}(t) & c_{\downarrow\boldsymbol{k}}(t)c_{\downarrow-\boldsymbol{k}}(t) \end{matrix} \right\rangle = f_0(t, \boldsymbol{k}) i\sigma_y \tag{6.58}$$

$$+ \boldsymbol{f}(t, \boldsymbol{k}) \cdot \boldsymbol{\sigma} i\sigma_y. \tag{6.59}$$

As usual, brackets $\langle \ldots \rangle$ denote the ground-state expectation value. Also, we use the singlet-triplet basis, with the singlet pair amplitude $f_0(t, \boldsymbol{k})$ and the triplet vector

$$\boldsymbol{f}(t, \boldsymbol{k}) = \left[\frac{f_{\downarrow\downarrow} - f_{\uparrow\uparrow}}{2}, \frac{f_{\uparrow\uparrow} + f_{\downarrow\downarrow}}{2i}, f_{\uparrow\downarrow+\downarrow\uparrow} \right], \tag{6.60}$$

combining the amplitudes $f_{\uparrow\uparrow}(t, \boldsymbol{k})$, $f_{\downarrow\downarrow}(t, \boldsymbol{k})$ and $f_{\uparrow\downarrow+\downarrow\uparrow}(t, \boldsymbol{k})$ of the triplet pair states with the total spin projections $S_z = 1, -1$ and 0. In proximity NCSs, the role of the $\boldsymbol{f}$ vector is similar to that of the Balian–Wertheimer $\boldsymbol{d}$ vector in intrinsic NCSs.

All pairing amplitudes can be obtained from Green's function of Eq. (6.56), which is the Nambu matrix[2]

$$\mathcal{G}(E, \boldsymbol{k}) = \begin{bmatrix} G(E, \boldsymbol{k}) & F(E, \boldsymbol{k}) \\ F^\dagger(E, \boldsymbol{k}) & \overline{G}(E, \boldsymbol{k}) \end{bmatrix}.$$

Here, each entry is a 2×2 matrix in spin space: $G(E, \boldsymbol{k})$ and $\overline{G}(E, \boldsymbol{k})$ are the quasiparticle Green functions, while $F(E, \boldsymbol{k})$ is the anomalous (condensate) Green function given by

$$F(E, \boldsymbol{k}) = [f_0(E, \boldsymbol{k}) + \boldsymbol{f}(E, \boldsymbol{k}) \cdot \boldsymbol{\sigma}] i\sigma_y,$$

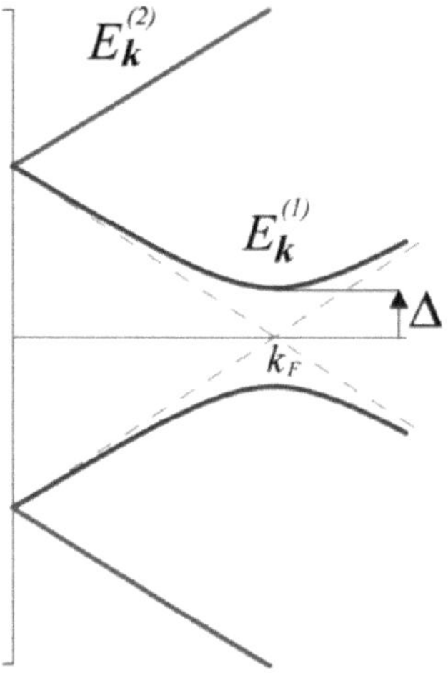

Figure 6.4 Energy spectrum of a proximity NCS, see Eq. (6.64). Adapted from [60].

where $f_0(E, \boldsymbol{k})$ and $\boldsymbol{f}(E, \boldsymbol{k})$ are the orbital amplitudes for the singlet and triplet pairing at given energy. From Eqs. (6.56) and (6.57) one readily finds (see Appendices B.1 and B.2)

$$f_0(E, \boldsymbol{k}) = \frac{\Delta}{\Pi(E, \boldsymbol{k})}(E^2 - \mu^2 - \Delta^2 - \gamma_{\boldsymbol{k}}^2), \tag{6.61}$$

$$\boldsymbol{f}(E, \boldsymbol{k}) = -\frac{2\mu\Delta}{\Pi(E, \boldsymbol{k})}\gamma_{\boldsymbol{k}}, \tag{6.62}$$

$$\Pi(E, \boldsymbol{k}) = \left[E^2 - (\mu - |\gamma_{\boldsymbol{k}}|^2 - \Delta^2\right]\left[E^2 - (\mu + |\gamma_{\boldsymbol{k}}|)^2 - \Delta^2\right]. \tag{6.63}$$

Noteworthy is the information about the energy spectrum and the order parameter of a proximity NCS. The energy spectrum is given by the roots of $\Pi(E, \boldsymbol{k})$ and consists of two spin-split BCS-like branches

$$E_{\boldsymbol{k}}^{(1)} = \pm\sqrt{(\mu - |\gamma_{\boldsymbol{k}}|)^2 + \Delta^2},\ E_{\boldsymbol{k}}^{(2)} = \pm\sqrt{(\mu + |\gamma_{\boldsymbol{k}}|)^2 + \Delta^2}. \tag{6.64}$$

These are plotted in Fig. 6.4. As for the order parameter, it has a mixed parity, with an even singlet $f_0(E, \boldsymbol{k})$ (6.61) and an odd triplet

[2]Hereafter, we adopt more compact notations for the Green functions.

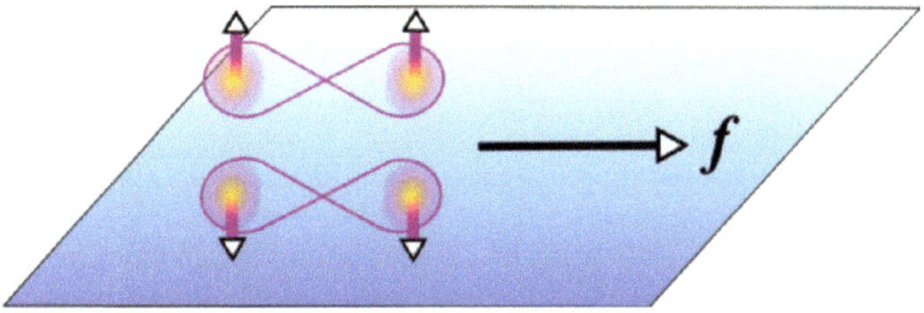

Figure 6.5 Schematic of the triplet pairing in a 2D proximity NCS. There are two species of opposite-spin pairs in orbital $(k_x \pm ik_y)$ states. Adapted from [60].

$f(E, \boldsymbol{k})$ (6.62) component. The triplet admixture is proportional to the induced singlet pair potential, Δ, and the SOC vector, $\boldsymbol{\gamma_k}$. That is, the SOC converts some of the tunneling singlet pairs into triplets, as depicted in Fig. 6.3. For example, for the Rashba SOC with

$$\boldsymbol{\gamma_k} = \alpha_{\text{so}}(\boldsymbol{k} \times \boldsymbol{z}) = \alpha_{\text{so}}[k_y, -k_x, 0]$$

the $\boldsymbol{f}$ vector lies in the basal plane, describing the equal-spin triplets with the orbital $(k_x \pm ik_y)$ symmetries (see also Fig. 6.5):

$$f_{\uparrow\uparrow,\downarrow\downarrow} \propto \alpha_{\text{so}}(k_x \mp ik_y), \qquad f_{\uparrow\downarrow+\downarrow\uparrow} = 0,$$

where α_{so} is the SOC constant. As before, the $S_z = 0$ triplet is absent by time-reversal symmetry. We note that Eq. (6.62) is not limited to the linear—in—$\boldsymbol{k}$ SOC. It can, for example, include a cubic warping term, which is relevant for the surface states of tetradymite compounds. Its role in the proximity-induced superconductivity has been addressed theoretically in Ref. [118].

The role of disorder deserves separate comment. In dirty systems, p-wave component (6.62) is suppressed relative to the s-wave pairing when the elastic mean-free path is much smaller than the superconducting coherence length [116]. The suppression is due to the generic nonlocality of the odd-parity Cooper pairs, which makes them sensitive to the electron mean-free path limited by disorder. In cleaner systems, however, the p-wave component can be comparable to the s-wave one and should therefore be observable despite the presence of disorder (e.g. random potential of ionized impurities). Hugdal *et al* [119] have developed a quasiclassical theory of the proximity effect in impure Dirac materials. Non-equilibrium Eilenberger and Usadel equations were derived to first

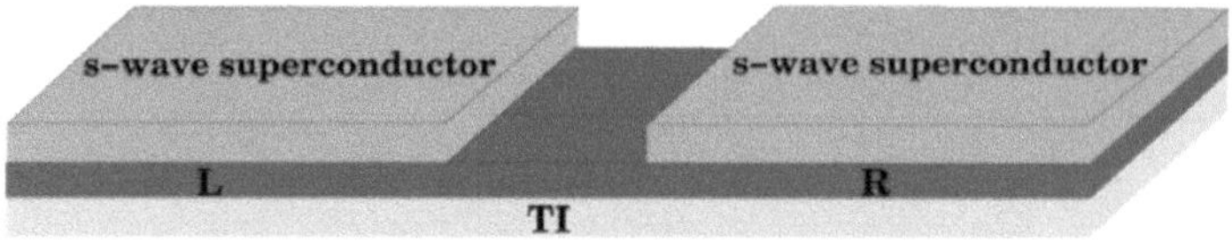

Figure 6.6 Schematic of a superconducting TI junction.

order in quantities small compared to the Fermi energy for Dirac edge and surface states with spin-momentum locking.

6.2.3 Helical Andreev Bound States

At this stage, one might ask: what are the potentially observable phenomena that could serve as evidence for the unconventional superconductivity in TIs? Here we address this question, following a general idea of the paper of Fu and Kane, Ref. [69]. They proposed that the unconventional superconductivity in TIs should manifest itself in the formation of topological bound states at the boundaries between superconducting and normal TI phases. A typical example of such a hybrid system is a Josephson junction created by depositing two spatially separated superconductors (Ss) on the surface of a 3DTI, as sketched in Fig. 6.6. Junctions of this type have been realized experimentally (see e.g. [120–127] and Fig. 6.7). We shall see that the S/TI/S junctions support protected states that, at the same time, are capable of carrying electric current without dissipation.

In the junction, the pairing Δ is non-uniform in space. It vanishes in the normal region and is nonzero in the left (L) and right (R) S terminals

$$\Delta|_{L,R} = \Delta_0 e^{i\phi_{L,R}}, \quad \phi_L = 0, \quad \phi_R \equiv \phi. \tag{6.65}$$

The pairing term is assumed to have different phases $\phi_{L,R}$ and the same absolute value $\Delta_0 = |\Delta|_{L,R}$ in terminals L and R. The junction length defined as distance between the S terminals is assumed much smaller than the superconducting coherence length $\sim \hbar v/\Delta_0$. For example, for Nb/TI junctions with typical parameters $\Delta_0 \approx 0.2$ meV and $\hbar v \approx 250 \div 350$ meV·nm [127], the junction length should be smaller than $1.25 \div 1.75$ μm, which is easily achievable experimentally. In the limit of zero length, the junction can be

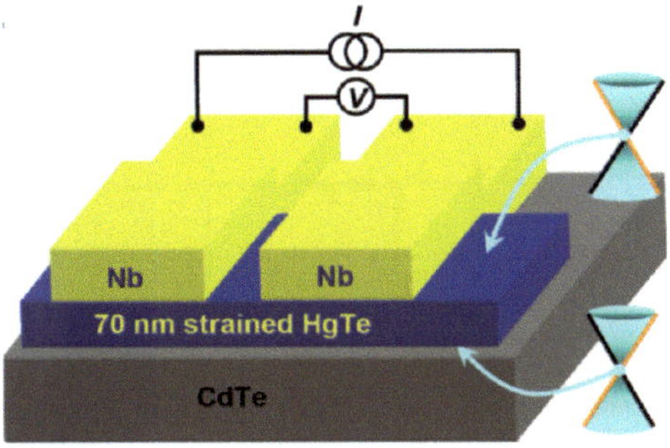

Figure 6.7 Schematic experimental realization of a superconducting junction with a 70-nm-thick strained HgTe film as a 3DTI and two Nb contacts (adapted from [127]).

described by the BdG equation

$$\left[E\hat{I} - \begin{pmatrix} -i\hbar v\boldsymbol{\sigma}\nabla + U(x) - \mu & i\sigma_y\Delta_0 e^{i\phi_{L,R}} \\ -i\sigma_y\Delta_0 e^{-i\phi_{L,R}} & -i\hbar v\boldsymbol{\sigma}^*\nabla - U(x) + \mu \end{pmatrix}\right]\begin{bmatrix} u(\boldsymbol{r}) \\ v(\boldsymbol{r}) \end{bmatrix}, \tag{6.66}$$

or, explicitly,

$$\begin{bmatrix} E+\mu-U(x) & \hbar v(i\partial_x+\partial_y) & 0 & -\Delta_0 e^{i\phi_{L,R}} \\ \hbar v(i\partial_x-\partial_y) & E+\mu-U(x) & \Delta_0 e^{i\phi_{L,R}} & 0 \\ 0 & \Delta_0 e^{-i\phi_{L,R}} & E-\mu+U(x) & \hbar v(i\partial_x-\partial_y) \\ -\Delta_0 e^{-i\phi_{L,R}} & 0 & \hbar v(i\partial_x+\partial_y) & E-\mu+U(x) \end{bmatrix} \times \begin{bmatrix} u_\uparrow(\boldsymbol{r}) \\ u_\downarrow(\boldsymbol{r}) \\ v_\uparrow(\boldsymbol{r}) \\ v_\downarrow(\boldsymbol{r}) \end{bmatrix} = 0. \tag{6.67}$$

To account for the possibility of electron scattering in the junction we introduce a potential $U(x) = U\delta(x)$ between the terminals. The precise form of the scattering potential is not of crucial importance here because, as we shall see now, the main interesting effects originate from Klein tunneling which takes place independently of the barrier shape [102, 128, 129].

We are interested in the particular solutions of the BdG equations—Andreev bound states (ABSs)—that have energies smaller than the induced excitation gap, Δ_0. The mechanism of the formation of the bound states—Andreev reflection—is illustrated in

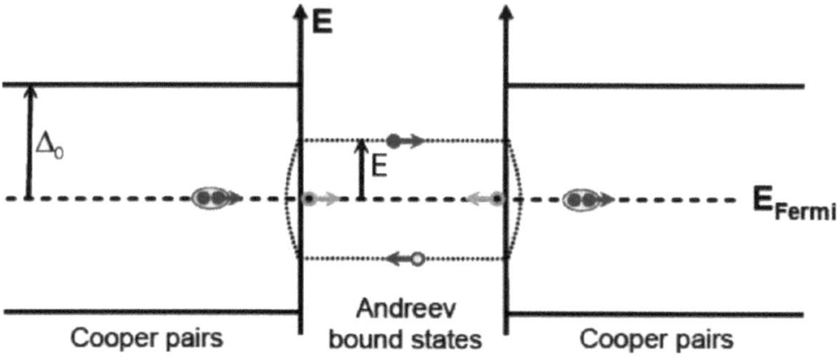

Figure 6.8 Illustration of Andreev reflection and formation of Andreev bound states in a superconductor/normal metal/superconductor junction (adapted from [127]).

Fig. 6.8. Consider a particle with an energy $E < \Delta_0$ in the normal region that propagates toward the superconductor on the right. Since below the gap there are no single-particle states, a particle entering the superconductor must be accompanied by a partner with energy $-E$ to create a Cooper pair in the ground state at the Fermi level. This leaves a hole at energy $-E$ in the Fermi sea of the normal system, and, by TRS, the hole must retrace the trajectory of the particle. This is the Andreev reflection process. In order to enter the superconductor on the left, the hole undergoes another Andreev reflection, being emitted back into the normal region as a particle with energy E. In this way a bound state is formed between the two superconductors.

We seek the ABS solutions as the sum of independent modes with different k_y wave numbers:

$$\Phi(\boldsymbol{r}) \equiv \begin{bmatrix} u_\uparrow(\boldsymbol{r}) \\ u_\downarrow(\boldsymbol{r}) \\ v_\uparrow(\boldsymbol{r}) \\ v_\downarrow(\boldsymbol{r}) \end{bmatrix} = \sum_{k_y} \Phi_{k_y}(x) \frac{e^{ik_y y}}{\sqrt{W}}, \tag{6.68}$$

where W is the junction width, and $\Phi_{k_y}(x)$ is the evanescent spinor function decaying for $x \to \pm\infty$. For a large Fermi energy

$$\mu \gg \Delta_0, \tag{6.69}$$

the spinors $\Phi_{k_y}(x)$ are readily obtained from Eq. (6.67) as

$$\Phi_{k_y}(x>0) = \begin{bmatrix} 1 \\ e^{i\theta} \\ -e^{i\theta}a_R^+ \\ a_R^+ \end{bmatrix} C_R^> e^{ixk_F\cos\theta - x/\xi} + \begin{bmatrix} 1 \\ -e^{-i\theta} \\ e^{-i\theta}a_R^- \\ a_R^- \end{bmatrix} C_R^< e^{-ixk_F\cos\theta - x/\xi}, \tag{6.70}$$

and

$$\Phi_{k_y}(x<0) = \begin{bmatrix} 1 \\ e^{i\theta} \\ -e^{i\theta}a_L^- \\ a_L^- \end{bmatrix} C_L^> e^{ixk_F\cos\theta + x/\xi} + \begin{bmatrix} 1 \\ -e^{-i\theta} \\ e^{-i\theta}a_L^+ \\ a_L^+ \end{bmatrix} C_L^< e^{-ixk_F\cos\theta + x/\xi}, \tag{6.71}$$

where $C_{R,L}^{>,<}$ are constants, ξ is the length of the penetration of the ABSs into the superconductors:

$$\xi = \frac{\hbar v}{\sqrt{\Delta_0^2 - E^2}}, \tag{6.72}$$

and other parameters and notations are explained below:

$$a_{R,L}^{\pm}(E) = \frac{\Delta_0 e^{-i\phi_{R,L}}}{E \pm i\sqrt{\Delta_0^2 - E^2}}, \tag{6.73}$$

$$\cos\theta = \sqrt{1 - (k_y/k_F)^2}. \tag{6.74}$$

Here $a_{R,L}^{\pm}(E)$ are the amplitudes of Andreev reflection from superconductors R and L, the angle θ indicates the propagation direction, spanning the range from $-\pi/2$ to $\pi/2$ around the x-axis, as shown in Fig. 6.9. The allowed values of θ are determined by the condition that the transverse wave number k_y is smaller than the Fermi wave number k_F. The matching condition for spinors (6.70)

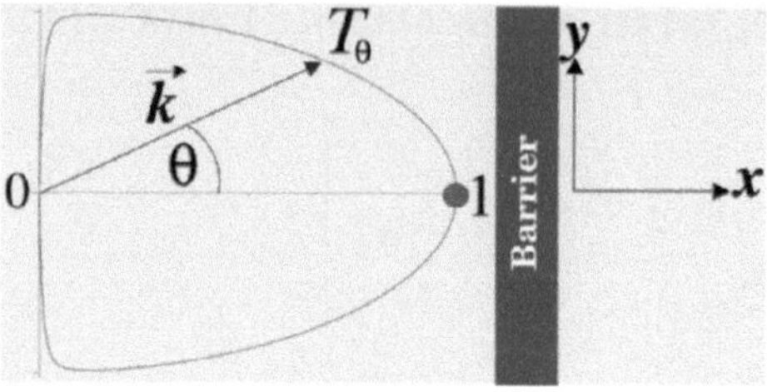

Figure 6.9 In-plane scattering geometry and polar plot of barrier transparency T_θ as a function of incidence angle θ [see also Eq. (6.77)].

and (6.71) is obtained by integrating Eq. (6.66) around $x = 0$, which yields

$$\Phi_{k_y}(-0) = \begin{bmatrix} 1 + iZ\sigma_x & 0 \\ 0 & 1 - iZ\sigma_x \end{bmatrix} \Phi_{k_y}(+0), \quad Z = \frac{U}{\hbar v}, \tag{6.75}$$

where the dimensionless parameter Z characterizes the barrier strength. Inserting (6.70) and (6.71) into (6.75) we arrive at four linear algebraic equations for the four constants $C_R^>$, $C_R^<$, $C_L^>$ and $C_L^<$. Equating then the determinant of the system to zero, we arrive after some algebra the following equation for the ABSs:

$$\begin{aligned} &[a_R^+(E) - a_R^-(E)][a_L^+(E) - a_L^-(E)] \\ &= T_\theta\, [a_R^+(E) - a_L^+(E)][a_R^-(E) - a_L^-(E)]. \end{aligned} \tag{6.76}$$

Here the carrier helicity is accounted for by the angle-dependent transmission coefficient $T(\theta)$ (see also Fig. 6.9):

$$T_\theta = \frac{\cos^2(\theta)}{1 - \sin^2(\theta)/(1 + Z^2)}. \tag{6.77}$$

For the perpendicular propagation with $\theta = 0$, the transparency T_0 is 1 independently of the barrier strength. This is the familiar Klein tunneling of Dirac fermions. In the following, we discuss how this phenomenon manifests itself in the energy spectrum of the ABSs.

Using Eq. (6.73) for the Andreev amplitudes, we obtain from (6.76) the ABS spectrum as a function of the superconducting phase difference ϕ and the propagation angle θ:

$$E_\theta^\pm(\phi) = \pm\Delta_0\sqrt{1 - T_\theta \sin^2\frac{\phi}{2}}. \tag{6.78}$$

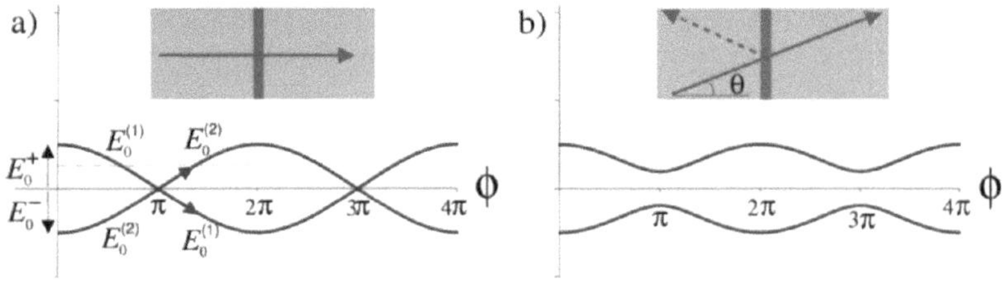

Figure 6.10 Andreev bound states (6.78) versus phase difference ϕ for (a) normal incidence at potential barrier with $\theta = 0$ and (b) oblique incidence with $\theta = \pi/4$. Arrows indicate the 4π-periodic branches (6.80). Dashed line corresponds to fixed energy (6.81).

Figure 6.10 shows the ABSs for two distinct cases: normal incidence at the potential barrier with $\theta = 0$ (panel a) and oblique incidence with $\theta \neq 0$ (panel b). As the perpendicular propagation is reflectionless, the ABSs for $\theta = 0$ are given by

$$E_0^{\pm}(\phi) = \pm\Delta_0 \left|\cos\frac{\phi}{2}\right|. \tag{6.79}$$

For the phase difference $\phi = \pi$ the two $\pm$ states have zero energy, i.e. they are gapless. We recall that gapless ABSs also occur in ballistic junctions with conventional s-wave superconductors. However, unlike the latter, the gapless solutions (6.79) are protected by TRS. In order to prove this we notice that the $\pm$ states in Eq. (6.79) are equivalent to two 4π-periodic branches crossing in the middle of the gap:

$$E_0^{(1,2)}(\phi) = \pm\Delta_0 \cos\frac{\phi}{2}, \tag{6.80}$$

where we introduced notation $E_0^{(1,2)}(\phi)$ to distinguish the 4π-periodic solutions (6.80) from (6.79). Let us now examine their wave functions at the fixed energy [shown by the dashed line in Fig. 6.10(a)], which implies

$$E_0^{(1)}(\phi) = E_0^{(2)}(2\pi - \phi) = E. \tag{6.81}$$

The wave functions of these states [see e.g. Eq. (6.70) for $\theta = 0$] appear to consist only of one-way propagating states:

$$\Phi_0^{(1)}(x) = \begin{bmatrix} 1 \\ 1 \\ -a_R^+(\phi) \\ a_R^+(\phi) \end{bmatrix} \frac{e^{ik_F x - x/\xi}}{2\sqrt{\xi}}, \tag{6.82}$$

$$\Phi_0^{(2)}(x) = \begin{bmatrix} 1 \\ -1 \\ a_R^-(2\pi - \phi) \\ a_R^-(2\pi - \phi) \end{bmatrix} \frac{e^{-ik_F x - x/\xi}}{2\sqrt{\xi}}, \tag{6.83}$$

where $1/2\sqrt{\xi}$ is the normalization factor. Note that for the given energy (6.81) the Andreev amplitude a_R^- in $\Phi_0^{(2)}(x)$ has the phase $2\pi - \phi$. It is now easy to see that states (6.82) and (6.83) are connected by the TRS operation:

$$\Phi_0^{(2)}(x) = \begin{bmatrix} i\sigma_y & 0 \\ 0 & i\sigma_y \end{bmatrix} \Phi_0^{(1)*}(x). \tag{6.84}$$

Moreover, both particle and hole components of $\Phi_0^{(1,2)}(x)$ are the eigenstates of the spin projection σ_x, which can be expressed in the matrix form as

$$\begin{bmatrix} \sigma_x & 0 \\ 0 & -\sigma_x \end{bmatrix} \Psi_0^{(1,2)}(x) = \pm\Psi_0^{(1,2)}(x). \tag{6.85}$$

Thus, the spins of these states are tied to the momentum direction along the x-axis. As a consequence of Eq. (6.84), helical ABSs (6.82) and (6.83) are orthogonal to each other and, therefore, protected against TR-preserving perturbations.

Clearly, the protection by the TRS does not hold for the ABSs with oblique incidence at the potential barrier since in this case the incident and reflected trajectories are not related by the time reversal [see Fig. 6.10(b)]. The reflection from the potential barrier generates a minigap $\propto \sqrt{1 - T_\theta}$ at $\phi = \pi, 3\pi, \ldots,$ similar to the ABSs in conventional Josephson junctions.

Although for $\theta \neq 0$ the ABSs are not protected from potential scattering, they still feature the spin-momentum locking and, therefore, can be called helical in the same sense as the normal-state 2D surface carriers. It is worth noting that despite the dependence on the normal-state quantity T_θ the helical ABSs reflect the nature of the induced superconducting condensate, i.e. its mixed s- and p-wave character. In Eq. (6.67) the s- and p-wave correlations are explicitly accounted for by the couplings between the opposite-spin (e.g. $u_\uparrow$ and $v_\downarrow$) and the same-spin (e.g. $u_\uparrow$ and $v_\uparrow$) particle and hole components, respectively. Alternatively, one can examine the spin structure of the Green function of the ABSs. This Green function can be constructed from the eigenspinors and energies of the ABSs by means of the Hilbert-Schmidt expansion. For our purpose it suffices to consider the contributions of the normally propagating states (6.82) and (6.83). For $E \to E_0^{(1)}(\phi)$, the contribution of the state (6.82) is obtained by making the direct product of the spinor $\Phi_0^{(1)}(x)$ with its conjugate $\tilde{\Phi}_0^{(1)}(x') = \left[1, 1, -a_R^{+*}, a_R^{+*}\right] \mathrm{e}^{-ik_F x' - x'/\xi}/(2\sqrt{\xi})$ as follows:

$$G_0^{(1)}(x, x') \approx \frac{\Phi_0^{(1)}(x) \otimes \tilde{\Phi}_0^{(1)}(x')}{E - E_0^{(1)}(\phi)} =$$
$$\begin{bmatrix} 1 & 1 & -a_R^{+*} & a_R^{+*} \\ 1 & 1 & -a_R^{+*} & a_R^{+*} \\ -a_R^{+} & -a_R^{+} & 1 & -1 \\ a_R^{+} & a_R^{+} & -1 & 1 \end{bmatrix} \times \frac{\mathrm{e}^{ik_F(x-x')-(x+x')/\xi}}{4\xi\,[E - E_0^{(1)}(\phi)]} =$$
$$\begin{bmatrix} \sigma_0 + \sigma_x & (\sigma_0 + \sigma_x) i\sigma_y a_R^{+*} \\ -i\sigma_y(\sigma_0 + \sigma_x) a_R^{+} & \sigma_0 - \sigma_x \end{bmatrix} \times \frac{\mathrm{e}^{ik_F(x-x')-(x+x')/\xi}}{4\xi\,[E - E_0^{(1)}(\phi)]}. \tag{6.86}$$

The spin structure of the diagonal and off-diagonal matrix elements of Eq. (6.86) is similar to that of the Green functions in Eqs. (6.50)–(6.53) for $\boldsymbol{p}||x$. This proves that ABS (6.82) is of the mixed $s-$ and p-wave type. Similar calculation for the counter-propagating state (6.83) yields

$$G_0^{(2)}(x, x') \approx \frac{\Psi_0^{(2)}(x) \otimes \tilde{\Psi}_0^{(2)}(x')}{E - E_0^{(2)}(\phi)} =$$
$$\begin{bmatrix} \sigma_0 - \sigma_x & (\sigma_0 - \sigma_x) i\sigma_y a_R^{+} \\ -i\sigma_y(\sigma_0 - \sigma_x) a_R^{+*} & \sigma_0 + \sigma_x \end{bmatrix} \times \frac{\mathrm{e}^{-ik_F(x-x')-(x+x')/\xi}}{4\xi\,[E - E_0^{(2)}(\phi)]}. \tag{6.87}$$

The spin matrices in Eqs. (6.86) and (6.87) are related to each other by the TRS.

6.2.4 Emergent Majorana Fermions

We have seen that in S/TI/S junctions the TRS guaranties the existence of a pair of gapless ABSs regardless of the presence or absence of a potential barrier in the junction. We also know that there is another generic symmetry of the system—the particle-hole conjugation—which allows one to obtain any hole state by acting with the conjugation operator $\mathbb{C}$ (6.24) on the particle one or vice versa. This has an interesting implication for the states with an exactly zero energy. Due to the particle-hole symmetry, a zero-energy particle state must be identical to its hole counterpart and vice versa. Such zero-energy states are the condensed matter analogues of Majorana fermions of relativistic field theory, where those appear as particles and anti-particles at the same time. The physical origin of Majorana-like fermions in S/TI/S junctions and their properties are discussed in detail below (see also Ref. [69]).

Let us first understand why the zero-energy states appear only when the superconducting phase difference is an odd-integer of π,

$$\phi = \pi, 3\pi, \ldots \tag{6.88}$$

For simplicity, we consider the strictly one-dimensional junction in the x-direction, which is a particular case of the model presented in the preceding section. Under condition (6.88) the superconducting paring potential [see Eq. (6.65)] has the opposite signs in the left and the right S terminals:

$$\Delta|_L = \Delta_0, \qquad \Delta|_R = -\Delta_0. \tag{6.89}$$

The sign of Δ determines the relative ordering of the particle and hole spectral branches in a superconductor in the same manner as the sign of the band gap determines the band sequence in an insulator. Therefore, under conditions (6.88) and (6.89) the ordering of the bands in one of the superconductors is inverted with respect to the other. This situation is illustrated in Fig. 6.11 where the particle (hole) branch corresponds to positive (negative) energies on the left and negative (positive) energies on the right. The inversion of the band structure in space implies that the gap between the particle and hole states must close in the middle of the junction. This has an important consequence. Due to the absence of the superconducting gap, there is no energy cost for adding unpaired

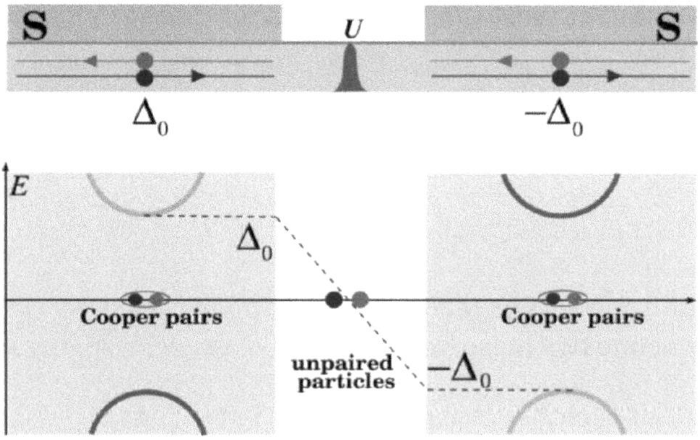

Figure 6.11 Schematic of the formation of zero-energy (midgap) states in an S/TI/S junction (see also text).

particles to the ground state of the junction, whereas in the S terminals the ground state consists of Cooper pairs and is separated from the excited single-particle states by a minimum energy of Δ_0.

The states accommodating the unpaired particles in the middle of the junction are indeed very special. First, they have zero energy with respect to the chemical potential and, for this reason, are related by the particle-hole conjugation. Second, the zero-energy states form a Kramers' doublet and, therefore, remain unaffected by a potential barrier in the junction. The particle-hole and TR symmetries impose the constraints on the zero-energy states, due to which the number of independent degrees of freedom is reduced to a pair of orthogonal Majorana fermions. To prove this let us take a closer look at the zero-energy solutions (6.82) and (6.83),

$$\Phi_0^{(1)}(x) = \begin{bmatrix} 1 \\ 1 \\ -i \\ i \end{bmatrix} \frac{e^{ik_F x - x/\xi}}{2\sqrt{\xi}}, \tag{6.90}$$

$$\Phi_0^{(2)}(x) = \begin{bmatrix} 1 \\ -1 \\ -i \\ -i \end{bmatrix} \frac{e^{-ik_F x - x/\xi}}{2\sqrt{\xi}}. \tag{6.91}$$

The two states transform into each other under the particle-hole conjugation

$$\mathbb{C}\Phi_0^{(1)}(x) = \Phi_0^{(2)}(x), \qquad \mathbb{C}\Phi_0^{(2)}(x) = \Phi_0^{(1)}(x), \tag{6.92}$$

where the $\mathbb{C}$ operation is defined, for convenience, as

$$\mathbb{C} = \begin{bmatrix} 0 & -i\sigma_0 \\ -i\sigma_0 & 0 \end{bmatrix} C, \tag{6.93}$$

which differs from Eq. (6.23) only by the allowed phase factor $-i$. Let us now define two linear combinations of the zero-energy solutions as

$$\gamma_+(x) = \frac{\Phi_0^{(1)}(x) + \Phi_0^{(2)}(x)}{2}, \qquad \gamma_-(x) = \frac{\Phi_0^{(1)}(x) - \Phi_0^{(2)}(x)}{2i}. \tag{6.94}$$

From Eqs. (6.92) and (6.94) it is clear that $\gamma_\pm(x)$ are invariant under the charge conjugation:

$$\mathbb{C}\gamma_\pm(x) = \gamma_\pm(x). \tag{6.95}$$

In other words, $\gamma_\pm(x)$ are the particle and the hole states at the same time. In this sense, the self-conjugated solutions defined by Eq. (6.94) mimic the Majorana fermions of relativistic field theory where the role of $\mathbb{C}$ is played by the charge-conjugation operation of the Dirac equation. Importantly, in view of the TR relation $\Phi_0^{(2)}(x) = \mathbb{T}\Phi_0^{(1)}(x)$ between the zero-energy states, $\gamma_\pm(x)$ are orthogonal,

$$\langle\gamma_+|\gamma_-\rangle = 0, \tag{6.96}$$

and, hence, independent states. We note that the expressions for $\gamma_\pm$ and $\mathbb{C}$ depend on the chosen basis, whereas the self-conjugation property (6.95) is generic. The emergent Majorana fermions are therefore a generic manifestation of the unconventional superconductivity in TIs.

In order to realize a "π-junction" with zero-energy states, one should be able to control the Josephson phase difference ϕ. A way to achieve this is to incorporate an S/TI/S junction into a superconducting ring with a magnetic flux Φ, as depicted in Fig. 6.12. If the radius of the ring is much larger than its thickness, the phase difference depends on the magnetic flux through a simple formula

$$\phi(\Phi) = 2\pi\frac{\Phi}{\Phi_0}, \tag{6.97}$$

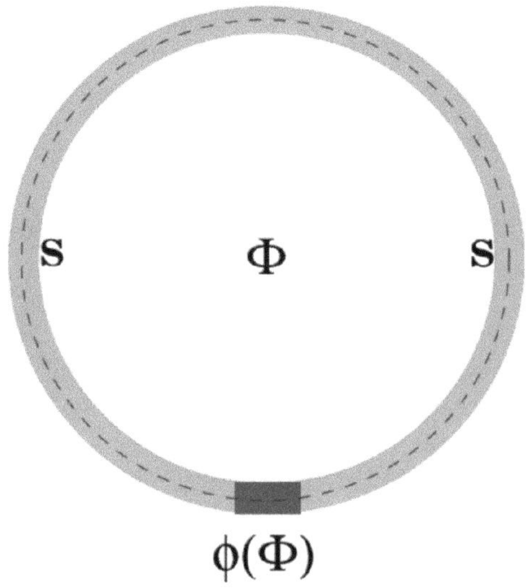

Figure 6.12 Scheme to control the Josephson phase difference ϕ at an S/TI/S junction in a superconducting ring with a magnetic flux Φ.

where $\Phi_0 = ch/2|e|$ is the magnetic flux quantum. Thus, for the half-integer magnetic flux $\Phi = \Phi_0/2, 3\Phi_0/2, \ldots$ the Josephson phase difference will be odd-integer of π.

Equation (6.97) can be derived from the principle of gauge invariance. The latter requires that in a magnetic field the local phase of the superconducting order parameter has the form

$$\phi(\boldsymbol{r}) = \phi_0(\boldsymbol{r}) - \frac{2e}{c\hbar}\int^{\boldsymbol{r}} \boldsymbol{A}(\boldsymbol{r}')\cdot d\boldsymbol{r}', \tag{6.98}$$

where $\phi_0(\boldsymbol{r})$ is the phase in the absence of the field, $\boldsymbol{A}(\boldsymbol{r})$ is the vector potential of the external magnetic field, and $2e$ is the charge of a Cooper pair. Let us take the phase gradient

$$\nabla\phi(\boldsymbol{r}) = \nabla\phi_0(\boldsymbol{r}) - \frac{2e}{c\hbar}\boldsymbol{A}(\boldsymbol{r}), \tag{6.99}$$

and integrate it along the ring in the anti-clockwise direction. Since the phase drop occurs mainly across the nonsuperconducting part of the junction, the integral of the left-hand-side of Eq. (6.99) is $\phi_R - \phi_L = \phi$, whereas the right-hand-side yields

$$\oint \nabla\phi_0(\boldsymbol{r})\cdot d\boldsymbol{r} + 2\pi\frac{\Phi}{\Phi_0}. \tag{6.100}$$

For the uniquely defined order parameter the first term vanishes, which brings us to Eq. (6.97). Tuning the Josephson phase difference with the magnetic flux in a ring is one of the experimental tools to measure the Josephson current-phase relation.

6.2.5 Josephson Effect: Superconducting Klein Tunneling

The Josephson effect is a macroscopic quantum phenomenon that consists in the appearance of the electric current between two superconductors in the absence of the bias voltage. This current, called the supercurrent, is carried by Cooper pairs. The measurement of the supercurrent is one of the accessible tools to probe the symmetry of the superconducting order parameter. We may therefore ask the question of how the unconventional superconductivity and the helical ABSs in TIs manifest themselves in the Josephson effect.

In a short junction the supercurrent can be calculated from the spectrum of the ABSs according to the thermodynamic formula:

$$I(\phi) = -\frac{eN}{\hbar} \int_0^{\pi/2} d\theta \cos\theta \, \frac{\partial E_\theta^+(\phi)}{\partial \phi} \tanh \frac{E_\theta^+(\phi)}{2\Xi}, \tag{6.101}$$

where $N = k_F W/\pi$ is the number of the transport channels in the system, Ξ is the temperature in energy units, and the integral in Eq. (6.101) accounts for all possible propagation directions in the junction plane. For the ABSs in Eq. (6.78) the current (6.101) is

$$\frac{I(\phi)}{I_0} = \frac{\sin\phi}{2} \int_0^{\pi/2} d\theta \cos\theta \frac{T_\theta}{\left[1 - T_\theta \sin^2 \frac{\phi}{2}\right]^{1/2}} \tanh \frac{\left[1 - T_\theta \sin^2 \frac{\phi}{2}\right]^{1/2}}{2\Xi/\Delta_0} \tag{6.102}$$

where $I_0 = eN\Delta_0/2\hbar$. It is instructive to consider in parallel a reference S/N/S junction, where N is a conventional 2D system with a spin-degenerate parabolic dispersion (6.25). We assume that the reference system has the same geometry and the same (delta-like) potential in the middle. In this case, the ABSs are also given by Eq. (6.78), but with the angle-independent transmission probability $T' = 1/(1 + Z^2)$ for the nonhelical carriers in the N. Consequently,

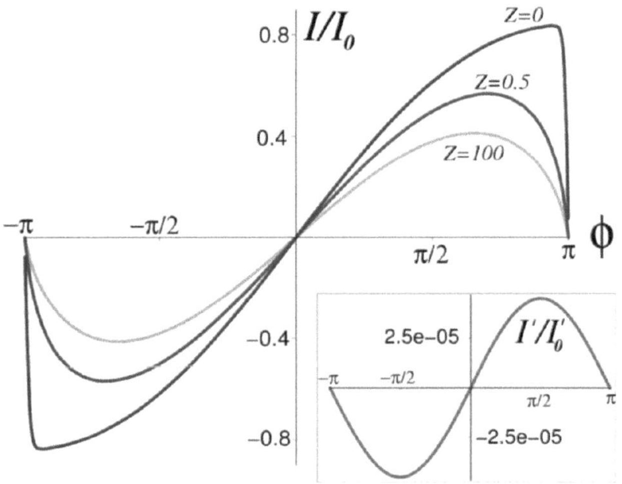

Figure 6.13 Current-phase relation $I(\phi)$ (6.102) in S/TI/S junction for different barrier strengths Z. Inset: $I'(\phi)$ (6.103) for a conventional reference S/N/S junction for strong barrier with $Z = 100$. In both cases $\Xi = 0.01\Delta_0$ (adapted from [129]).

for the reference supercurrent, $I'(\phi)$, Eq. (6.101) gives

$$\frac{I'(\phi)}{I'_0} = \frac{\sin\phi}{2} \frac{T'}{\left[1 - T' \sin^2 \frac{\phi}{2}\right]^{1/2}} \tanh \frac{\left[1 - T' \sin^2 \frac{\phi}{2}\right]^{1/2}}{2\Xi/\Delta_0}, \quad (6.103)$$

where $I'_0 = 2I_0$ includes the factor of two due to the spin degeneracy.

The equilibrium current-phase relation (6.102) is 2π-periodic. It is plotted in Fig. 6.13 for different values of the barrier strengths Z. The most interesting behavior is observed in the case of large Z. In this case, the conventional current-phase relation (6.103) is sinusoidal,

$$I'(\phi) \propto \sin\phi, \quad (6.104)$$

as shown for comparison in the inset of Fig. 6.13. In sharp contrast to Eq. (6.104) for the S/TI/S the current $I(\phi)$ remains nonsinusoidal and is by orders of magnitude larger than the conventional current $I'(\phi)$. The reason for this drastic difference is gapless ABSs (6.78) whose contribution is $\propto \sin(\phi/2)\text{sgn}(\cos(\phi/2))$ and independent of the barrier strength. Thus, the first signature of the unconventional

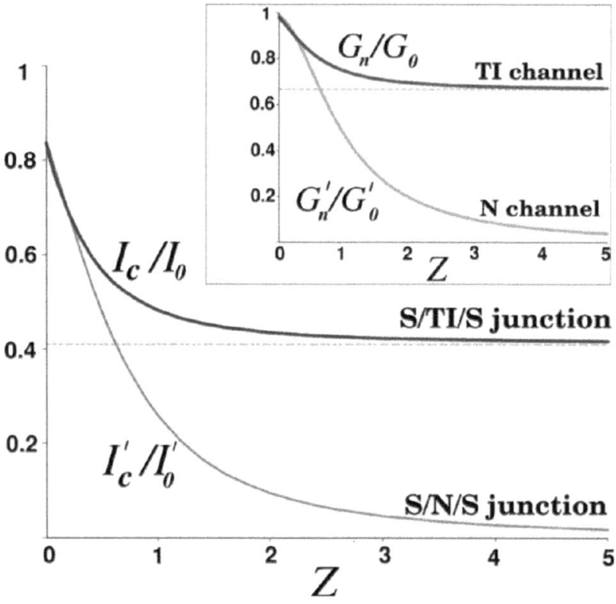

Figure 6.14 Normalized critical current versus barrier strength Z for S/TI/S and conventional S/N/S junctions. Inset: Normal-state conductance versus barrier strength Z for TI and N channels (see also text); $\Xi = 0.01\Delta_0$ (adapted from [129]).

superconductivity and the helical ABSs in TIs is a nonsinusoidal Josephson current-phase relation.

Next we analyze numerically the critical supercurrent I_c. It is defined as the maximum of $I(\phi)$ (6.102) at fixed parameters Z and Ξ. Figure 6.14 shows the normalized I_c as a function of the barrier strength Z. Initially decreasing with Z, the critical current saturates at a constant value for $Z > 3$. This is again in sharp contrast to the conventional S/N/S junction in which $I_c'(Z)$ is strongly suppressed for $Z > 3$. The large supercurrent despite the strong barrier is a manifestation of the helical ABSs (6.78) which are able to transport the Cooper pairs through the barrier without reflection for $\theta = 0$. This is a superconducting analogue of the normal-state Klein tunneling. For comparison, the inset in Fig. 6.14 shows the normal-state conductance of the TI:

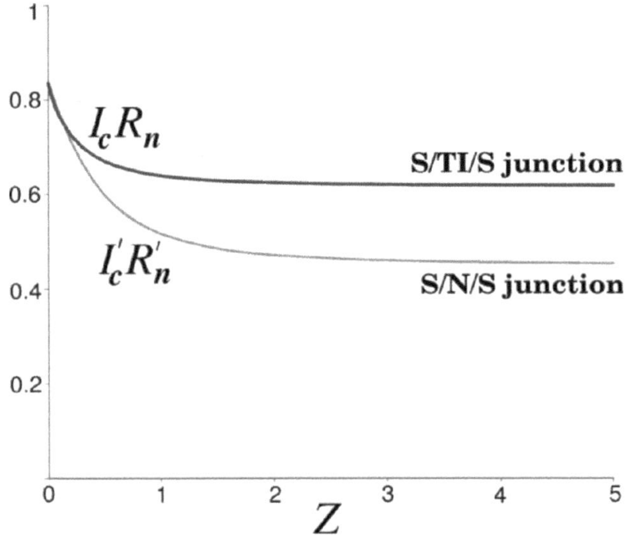

Figure 6.15 $I_c R_n$ product in units of $\pi \Delta_0/e$ versus barrier strength Z for S/TI/S and conventional S/N/S junctions; $\Xi = 0.01\Gamma_0$ (adapted from [129]).

$$G_n = G_0 \int_0^{\pi/2} d\theta \cos\theta \, T_\theta, \quad G_0 = \frac{e^2 N}{h}, \tag{6.105}$$

along with that of the conventional N system $G_n' = G_0' T'$, with $G_0' = 2G_0$. Klein tunneling with $T_0 = 1$ for $\theta = 0$ leads to the saturation of the conductance G_n at large Z, which clearly correlates with the behavior of the critical current I_c. Figure 6.15 shows another experimentally relevant characteristic, the product of the critical current I_c and the normal-state resistance $R_n = 1/G_n$ (6.105) normalized by $\pi \Delta_0/e$. Due to the Klein tunneling, the $I_c R_n$ product is larger for the S/TI/S junction than for the conventional S/N/S system.

Summarizing the results of this section, we can conclude that the Josephson current-phase relation, critical current and the $I_c R_n$ product all show the signs of the reflectionless passage of Cooper pairs through the junction. This is a superconducting analogue of the single-particle Klein tunneling. The effect is directly linked to the helical ABSs and, therefore, can be used to detect them and to establish an unconventional character of superconductivity in TIs.

6.3 Induced Triplet Superconductivity in TIs

6.3.1 Odd-Frequency Triplet Superconductivity

Breaking the TRS enriches the unconventional proximity effect in topological materials. An informative generalization of the model in Section 6.2.2 is achieved by adding an exchange (or Zeeman) spin field $\boldsymbol{h}$:

$$\boldsymbol{\gamma}'_{\boldsymbol{k}} = \boldsymbol{\gamma}_{\boldsymbol{k}} + \boldsymbol{h}.$$

The total spin field $\boldsymbol{\gamma}'_{\boldsymbol{k}}$ is no longer antisymmetric in momentum:

$$\boldsymbol{\gamma}'_{-\boldsymbol{k}} \neq \boldsymbol{\gamma}'_{\boldsymbol{k}},$$

which leads to the following generalization of the triplet $\boldsymbol{f}$ vector [cf. Eq. (6.62)]:

$$\boldsymbol{f}(E, \boldsymbol{k}) = \frac{\Delta}{\Pi(E, \boldsymbol{k})}[-\mu(\boldsymbol{\gamma}'_{\boldsymbol{k}} - \boldsymbol{\gamma}'_{-\boldsymbol{k}}) + E(\boldsymbol{\gamma}'_{\boldsymbol{k}} + \boldsymbol{\gamma}'_{-\boldsymbol{k}}) + i\boldsymbol{\gamma}'_{\boldsymbol{k}} \times \boldsymbol{\gamma}'_{-\boldsymbol{k}}], \tag{6.106}$$

or, explicitly,

$$\boldsymbol{f}(E, \boldsymbol{k}) = \frac{2\Delta}{\Pi(E, \boldsymbol{k})}[-\mu\boldsymbol{\gamma}_{\boldsymbol{k}} + E\boldsymbol{h} + i\boldsymbol{\gamma}_{\boldsymbol{k}} \times \boldsymbol{h}]. \tag{6.107}$$

The calculation details can be found in Appendices B.1 and B.2.

We now have two new contributions. One is linear in energy E. It corresponds to odd Matsubara frequencies and, for this reason, such type of pairing is called odd-frequency triplet pairing (see recent review [130]). Interestingly, the odd-frequency component is even in momentum space and, consequently, odd-frequency Cooper pairs have an even spatial symmetry. They are, therefore, more robust to disorder and other spatial inhomogeneities, thus offering both new interesting physics and application potential.

The triplet component in Eq. (6.107) was calculated in Ref. [115] for a TI surface state with the linear Dirac spectrum. After that, a growing body of work has been dealing with different aspects of the odd-frequency superconductivity in hybrid structures involving 3DTIs, 2DTIs and related Rashba materials. In particular, Cayao *et al* [131] have looked into the emergence of an odd-frequency s-wave pairing at the edge of a 2DTI without any magnetism.

The odd-frequency proximity effects have been studied for superconductors of various symmetry classes and other relevant findings have been discussed in literature (see also review [132]). While the literature on the odd-frequency superconductivity is abundant, the emergence and physical consequences of the imaginary term in Eqs. (6.106) and (6.107) have gone largely unnoticed. This type of pairing is an analogue of the paradigmatic non-unitary pairing in triplet superfluids [133] with a complex vector order parameter. We elaborate on this point below.

6.3.2 Non-Unitary Triplet Superconductivity: Charge-Spin Conversion

As mentioned above, the standard classification of the triplet Cooper pairing which invokes a Balian–Wertheimer $\boldsymbol{d}$ vector is not applicable to proximity NCSs where no pairing interaction takes place. For that purpose, we employ the matrix condensate Green function

$$F(E, \boldsymbol{k}) = [f_0(E, \boldsymbol{k}) + \boldsymbol{f}(E, \boldsymbol{k}) \cdot \boldsymbol{\sigma}]i\sigma_y,$$

which has proved useful in diverse proximity structures and driven superconductors.

Following Leggett [133], we call the pairing non-unitary if product $\hat{F}\hat{F}^\dagger$ is not proportional to a unit spin matrix. A simple calculation yields

$$\hat{F}\hat{F}^\dagger = (|f_0|^2 + \boldsymbol{f} \cdot \boldsymbol{f}^*)\sigma_0 + (f_0^* \boldsymbol{f} + f_0 \boldsymbol{f}^*) \cdot \boldsymbol{\sigma} + (i\,\boldsymbol{f} \times \boldsymbol{f}^*) \cdot \boldsymbol{\sigma}, \quad (6.108)$$

where arguments E and $\boldsymbol{k}$ are omitted for brevity. The second and third terms above indicate the non-unitary pairing due to the lack of inversion and time-reversal, respectively. We are interested in the latter case where, by analogy with triplet superfluids, Cooper pairs have a net spin polarization:

$$\langle \boldsymbol{S} \rangle \propto i\,\boldsymbol{f} \times \boldsymbol{f}^*.$$

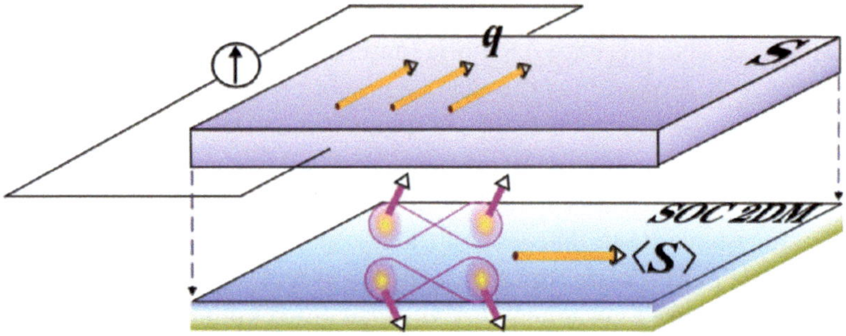

Figure 6.16 Sketch of charge-spin conversion in a SOC 2DM proximitized by a current-biased conventional superconductor. An applied electric current generates in-plane spin polarization $\langle \boldsymbol{S} \rangle \propto i\boldsymbol{f} \times \boldsymbol{f}^*$ reflecting the triplet pairing with total spin projection $S_z = 0$ on the SOC plane. This can be interpreted as tilting of pair spins ↑↑ and ↓↓ such that they acquire a common in-plane projection [see also Eqs. (6.110) and (6.111)]. Adapted from [60].

For example, for a 2D NCS in a perpendicular spin field $\boldsymbol{h}$, Eq. (6.107) yields the following result for the axial vector $i\boldsymbol{f} \times \boldsymbol{f}^*$ at the Fermi level:

$$i\boldsymbol{f}(0, \boldsymbol{k}) \times \boldsymbol{f}^*(0, \boldsymbol{k}) = \frac{8\mu\Delta^2\gamma_{\boldsymbol{k}}^2}{\Pi^2(0, \boldsymbol{k})}\boldsymbol{h}. \tag{6.109}$$

As expected, the pair spin polarization is parallel to the spin field $\boldsymbol{h}$, indicating an imbalance between equal-spin triplets ↑↑ and ↓↓.

In the above example, the $\boldsymbol{f}$ vector formalism allows us to extend the notion of the non-unitary pairing beyond its original context [133], viz. to treat proximiy-induced superconductivity. Furthermore, the non-unitary pairing does not generally require the spin field $\boldsymbol{h}$. The pair spin polarization $i\boldsymbol{f} \times \boldsymbol{f}^*$ can be induced just by an electric current via charge-spin conversion [134, 135]. To illustrate this point let us consider the BdG Hamiltonian

$$\hat{H}_{\boldsymbol{k}} = \begin{bmatrix} \boldsymbol{\sigma} \cdot \boldsymbol{\gamma}_{\boldsymbol{k}+\boldsymbol{q}} - \mu & \Delta i\sigma_y \\ -\Delta i\sigma_y & -(\boldsymbol{\sigma} \cdot \boldsymbol{\gamma}_{-\boldsymbol{k}+\boldsymbol{q}} - \mu)^* \end{bmatrix},$$

where wave vector shift $\boldsymbol{q}$ accounts for the presence of a superconducting phase gradient associated with a dissipationless electric current. The current is applied to the overlying superconductor and is weak enough to disregard the depairing effects in Δ. The corresponding triplet $\boldsymbol{f}$ vector is obtained in Appendix B.2. It is given by

$$\boldsymbol{f}(E, \boldsymbol{k}) = \frac{2\Delta}{\Pi(E, \boldsymbol{k})}[-\mu\boldsymbol{\gamma}_k + E\,\boldsymbol{\gamma}_q + i\,\boldsymbol{\gamma}_k \times \boldsymbol{\gamma}_q], \qquad (6.110)$$

where we used $\boldsymbol{\gamma}_{\pm k+q} = \pm\boldsymbol{\gamma}_k + \boldsymbol{\gamma}_q$ for any linear SOC. The combined effect of the SOC and the supercurrent produces a Zeeman-like field $\boldsymbol{\gamma}_q$, thereby generating both odd-frequency and non-unitary pairing akin to $\boldsymbol{h}$. Generally, the direction of $\boldsymbol{\gamma}_q$ depends on the type of the structural or lattice asymmetry behind the SOC, so docs thc pair spin polarization

$$i\,\boldsymbol{f}(0, \boldsymbol{k}) \times \boldsymbol{f}^*(0, \boldsymbol{k}) = -\frac{8\mu\Delta^2}{\Pi^2(0, \boldsymbol{k})}\boldsymbol{\gamma}_k \times (\boldsymbol{\gamma}_k \times \boldsymbol{\gamma}_q). \qquad (6.111)$$

For the Rashba SOC, the spin polarization is carried by the triplet state with total spin projection $S_z = 0$ on the SOC plane. Loosely speaking, the supercurrent causes a tilting of pair spins $\uparrow\uparrow$ and $\downarrow\downarrow$ such that they acquire a common in-plane projection, as sketched in Fig. 6.16. Importantly, the spin polarization does not vanish upon averaging over the directions of wave vector $\boldsymbol{k}$. Using $\boldsymbol{\gamma}_k = \alpha_{\text{so}}(\boldsymbol{k} \times \boldsymbol{z})$ and the results of Appendix B.2, one finds

$$\overline{i\,\boldsymbol{f}(0, \boldsymbol{k}) \times \boldsymbol{f}^*(0, \boldsymbol{k})} = \frac{4\mu\Delta^2\alpha_{\text{so}}^3\boldsymbol{k}^2}{\Pi^2(0, \boldsymbol{k})}(\boldsymbol{q} \times \boldsymbol{z}), \qquad (6.112)$$

where the bar means the average value. This result just means that an unpolarized charge current is converted into spin magnetization of the superconducting condensate, a form of the magnetoelectric effect in NCSs [136]. Here, the magnetoelectric effect refers to the spin magnetization induced by a phase gradient of the order parameter, while in the inverse magnetoelectric effect the magnetic polarization causes charge and spin flows. In both cases, the magnetoelectric coupling is characterized by SOC constant α_{so}.

Equation (6.112) resembles the current-induced thermodynamic magnetization of a Rashba NCS [136]. The averaged spin polarization retains the dependence on the structural or lattice asymmetry. For the Rashba SOC, the average polarization direction is perpendicular to an applied supercurrent in the SOC plane (see

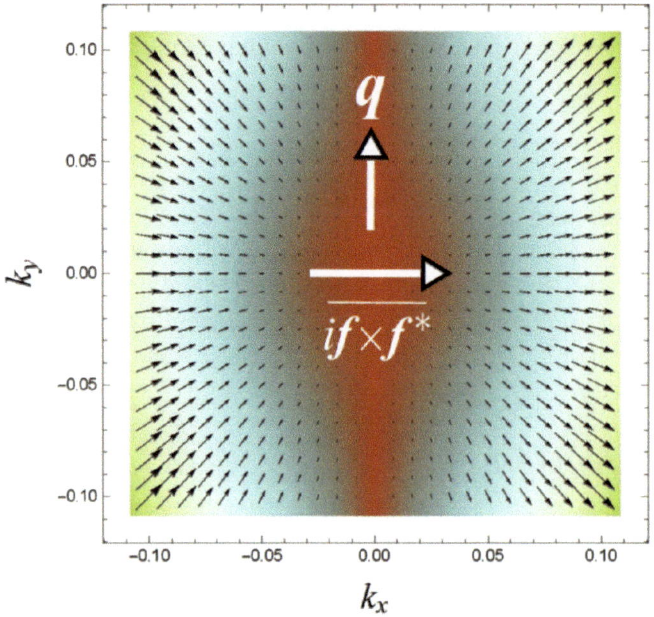

Figure 6.17 Vector plot of the pair spin polarization (6.111) for the Rashba SOC. Vector $\overline{i\boldsymbol{f} \times \boldsymbol{f}^*}$ shows the average polarization (6.112). Adapted from [60].

also Fig. 6.17). For NCSs of the cubic crystal group we expect a different result. In this case, the SOC vector is simply parallel to the momentum, $\boldsymbol{\gamma_k} = \alpha_{\text{so}}\boldsymbol{k} = \alpha_{\text{so}}[k_x, k_y, k_z]$, and Eq. (6.111) yields the following result:

$$\overline{i\,\boldsymbol{f}(0,\boldsymbol{k}) \times \boldsymbol{f}^*(0,\boldsymbol{k})} = \frac{16\mu\Delta^2\alpha_{\text{so}}^3\boldsymbol{k}^2}{3\Pi^2(0,\boldsymbol{k})}\,\boldsymbol{q}. \tag{6.113}$$

As we see, in cubic NCSs the pair spin polarization is locked parallel to the applied current.

The charge-spin conversion is an indicator of the unconventional, mixed-parity order parameter. Still, the direct magnetoelectric effect has not been verified experimentally yet despite a diverse range of other observed properties of NCSs. The theory reviewed in this section may help in planning a decisive experiment.

6.4 Problems

(1) Consider the Hamiltonian

$$\hat{\mathcal{H}} = \int d\boldsymbol{r} \sum_{\alpha,\beta=\uparrow,\downarrow} \left[\psi_\alpha^\dagger(\boldsymbol{r},t) h_{\alpha\beta}(\boldsymbol{r}) \psi_\beta(\boldsymbol{r},t) \right. \\ \left. + \frac{1}{2}\psi_\alpha^\dagger(\boldsymbol{r},t)\Delta_{\alpha\beta}\psi_\beta^\dagger(\boldsymbol{r},t) + \frac{1}{2}\psi_\alpha(\boldsymbol{r},t)[\Delta_{\alpha\beta}]^\dagger\psi_\beta(\boldsymbol{r},t)\right], \tag{6.114}$$

where $[\Delta_{\alpha\beta}]^\dagger$ denotes the Hermitian conjugation

$$[\Delta_{\alpha\beta}]^\dagger = \Delta^*_{\beta\alpha}. \tag{6.115}$$

Derive the equations of motion for the operators $\psi_\alpha(\boldsymbol{r},t)$ and $\psi_\alpha^\dagger(\boldsymbol{r},t)$. Prove that they coincide with Eqs. (6.7) and (6.8) provided that $\Delta_{\alpha\beta}$ is antisymmetric under the exchange of the spin indices: $\Delta_{\alpha\beta} = -\Delta_{\beta\alpha}$. Equation (6.114) is, therefore, the mean-field Hamiltonian of a spin singlet superconductor.

(2) Using the results of Section 6.1.3, calculate the density of states (DOS)

$$N_S(E) = \sum_{s\boldsymbol{p}} \delta(E - E_{s\boldsymbol{p}}) \tag{6.116}$$

for a conventional superconductor with the $s-$wave pairing and spin-independent quadratic normal-state spectrum. Show that for a large Fermi energy $\mu \gg |\Delta|$ the DOS is given by

$$N_S(E) = N\frac{|E|}{\sqrt{E^2 - |\Delta|^2}}\Theta(|E| - |\Delta|), \tag{6.117}$$

where N is the normal-state DOS at the Fermi level.

(3) Calculate the Nambu Green function of a conventional superconductor from the matrix equation

$$\left\{E\hat{I} - \begin{bmatrix} \frac{\boldsymbol{p}^2}{2m} - \mu & i\sigma_y\Delta \\ -i\sigma_y\Delta^* & -\frac{\boldsymbol{p}^2}{2m} + \mu \end{bmatrix}\right\} \begin{bmatrix} \hat{G}_{11,\boldsymbol{p}E} & \hat{G}_{12,\boldsymbol{p}E} \\ \hat{G}_{21,\boldsymbol{p}E} & \hat{G}_{22,\boldsymbol{p}E} \end{bmatrix} = \hat{I}. \tag{6.118}$$

Show that for a large Fermi energy $\mu \gg |\Delta|$ the Green function is given by

$$\hat{G}_{\boldsymbol{p}E} = \begin{bmatrix} \hat{G}_{11,\boldsymbol{p}E} & \hat{G}_{12,\boldsymbol{p}E} \\ \hat{G}_{21,\boldsymbol{p}E} & \hat{G}_{22,\boldsymbol{p}E} \end{bmatrix} = \frac{\begin{bmatrix} \sigma_0\left(E + \frac{\boldsymbol{p}^2}{2m} - \mu\right) & i\sigma_y \Delta \\ -i\sigma_y \Delta^* & \sigma_0\left(E - \frac{\boldsymbol{p}^2}{2m} + \mu\right) \end{bmatrix}}{E^2 - \left(\frac{\boldsymbol{p}^2}{2m} - \mu\right)^2 - |\Delta|^2}. \tag{6.119}$$

(4) The knowledge of the Green function allows us to calculate various thermodynamic characteristics of the superconductor. For example, its DOS can be related to the imaginary part of the quasi-particle Green function $\hat{G}_{11,\boldsymbol{p}E}$ as

$$N_S(E) = -\frac{1}{\pi}\sum_{\boldsymbol{p}} \mathrm{Tr}(\Im \hat{G}_{11,\boldsymbol{p}E+i0}), \tag{6.120}$$

where Tr denotes the trace operation in spin space, while $\Im$ the imaginary part. Changing the integration variable from $\boldsymbol{p}$ to $\xi_{\boldsymbol{p}} = \mathbf{p}^2/2m - \mu$, derive Eq. (6.117) for the DOS.

(5) The BdG Hamiltonian (6.12) can be used to describe superconductors with broken TRS. An illustrative example is a superconductor carrying a dissipationless current. For the particles with a spin-independent quadratic spectrum the current-carrying superconducting state is described by Eq. (6.12) in which the Hamiltonian $\hat{h}$ is given by

$$\hat{h}(\boldsymbol{r}) = \left[\frac{(-i\hbar\nabla + \boldsymbol{q}/2)^2}{2m} - \mu\right]\sigma_0, \tag{6.121}$$

where the vector $\boldsymbol{q}$ has the meaning of the momentum of the Cooper-pair condensate. Write the corresponding BdG equations in the momentum representation [cf. Eqs. (6.26) and (6.27)]. Assuming $\mu \gg \boldsymbol{q}^2/8m$, prove that the excitation spectrum of the superconductor is

$$E_{s\boldsymbol{p}} = -\frac{\boldsymbol{p}\cdot\boldsymbol{q}}{2m} \pm \sqrt{\left(\frac{\boldsymbol{p}^2}{2m} - \mu\right)^2 + |\Delta|^2}. \tag{6.122}$$

What happens when $\frac{|\boldsymbol{p}\cdot\boldsymbol{q}|}{2m} = |\Delta|$?

(6) Starting from Eq. (6.37) for the condensate correlation function, obtain the classification of the superconducting pairing in the Nambu basis defined in Eq. (6.16). What are the spin functions of the singlet and triplet Cooper pairs in this basis?

Chapter 7

Topological Superconductors and Majorana Modes

As argued in the preceding chapter, pairing interaction in TIs generates a triplet order parameter with an orbital $p_x + ip_y$ symmetry. Under appropriate conditions, such $p_x + ip_y$ superconductors can be topological as, for example, in the model of Read and Green [63]. It has illustrated the concept of a topological superconductor as a system with a bulk pairing gap and a gapless Majorana mode at the boundary or in a vortex core. Emergent Majorana modes in solids obey the exchange statistics of non-Abelian anyons [137, 138], offering a route to braiding-based topological quantum computation [62]. In this chapter, we shall discuss how topological $p_x + ip_y$ superconductivity can emerge in hybrid structures of quantum materials and how it can be detected by means of magnetotransport and a fractional Josephson effect. A special focus will be on models and properties of Majorana modes in 0D and 1D cases.

Topological Quantum Materials: Concepts, Models, and Phenomena
Grigory Tkachov

ISBN 978-981-4968-12-6 (Hardcover), 978-1-003-26641-9 (eBook)
www.jennystanford.com

7.1 Chiral Superconductors: Majorana Edge Modes

7.1.1 Topological $p_x + ip_y$ Superconductivity: Phenomenology

We recall that superconductivity originates from attractive pairwise interactions among electrons in a metal and, as is common in many-body physics, the pairing interaction can be treated in the mean-field approximation. In particular, a superconducting state can be described by the Bogoliubov–de Gennes (BdG) Hamiltonian of the form

$$\hat{H}_{BdG} = \begin{bmatrix} h_{\boldsymbol{k}} & \Delta_{\boldsymbol{k}} \\ \Delta^{\dagger}_{\boldsymbol{k}} & -h^{*}_{-\boldsymbol{k}} \end{bmatrix}.$$

Above, $h_{\boldsymbol{k}}$ is a single-particle Hamiltonian of the normal (non-superconducting) state, while the off-diagonal entry $\Delta_{\boldsymbol{k}}$ and its Hermitian conjugate $\Delta^{\dagger}_{\boldsymbol{k}}$ account for the pairing interaction. The BdG Hamiltonian acts on a pseudospinor,

$$\Psi_{\boldsymbol{k}} = \begin{bmatrix} u_{\boldsymbol{k}} \\ v_{\boldsymbol{k}} \end{bmatrix},$$

which has two (Nambu) components, $u_{\boldsymbol{k}}$ and $v_{\boldsymbol{k}}$, being a particle- and a hole-like wave function, respectively.

Coming from the pairing interaction, the matrix structure of the BdG Hamiltonian is crucial for emergent topological superconductivity. A paradigmatic example is a 2D superconductor with an odd-parity gap function of the form [63, 137]

$$\Delta_{\boldsymbol{k}} = \Delta'(k_x + ik_y),$$

where Δ' is a constant (assumed real and positive throughout). Such a gap function describes Cooper pairs in the orbital p-wave state with quantum number $m_\ell = 1$, which defines a $p_x + ip_y$ superconductor. It has a partner with the orbital $p_x - ip_y$ symmetry. These are the 2D versions of the A-phase of superfluid ^{3}He [133]. All electrons are assumed to have the same spin, so the system is a superconducting analogue of a 2D Chern insulator. As we shall see later, it is a "building block" for a topological superconductor emerging in hybrid structures of QAHIs and conventional superconductors.

As for the normal-state Hamiltonian, here we choose the simplest, parabolic band conductor with

$$h_{\boldsymbol{k}} = \frac{\boldsymbol{k}^2}{2m} - \mu,$$

where μ is the chemical potential. The BdG Hamiltonian is then given by

$$\hat{H}_{BdG} = \begin{bmatrix} \frac{\boldsymbol{k}^2}{2m} - \mu & \Delta'(k_x + ik_y) \\ \Delta'(k_x - ik_y) & -(\frac{\boldsymbol{k}^2}{2m} - \mu) \end{bmatrix}, \qquad (7.1)$$

or, in the basis of Pauli matrices $\boldsymbol{\tau} = [\tau_1, \tau_2, \tau_3]$,

$$\hat{H}_{BdG} = \boldsymbol{d_k} \cdot \boldsymbol{\tau}, \quad \boldsymbol{d_k} = \left[\Delta' k_x, -\Delta' k_y, \frac{\boldsymbol{k}^2}{2m} - \mu\right].$$

Up to notations, this is a particle-hole-symmetric model of a Chern insulator studied in Section 2.3.2. In this case, vector $\boldsymbol{d_k}$ defines the texture of Nambu pseudospin $\boldsymbol{\tau}/2$ in the 2D momentum space. Again, the topology of the pseudospin texture can be characterized by the winding number

$$n = \frac{1}{4\pi} \int \boldsymbol{u_k} \cdot \left(\frac{\partial \boldsymbol{u_k}}{\partial k_x} \times \frac{\partial \boldsymbol{u_k}}{\partial k_y}\right) dk_x dk_y, \quad \boldsymbol{u_k} = \frac{\boldsymbol{d_k}}{|\boldsymbol{d_k}|}.$$

It acquires nontrivial values $n = \pm 1$ under condition

$$\mu m > 0.$$

It is nothing else as the criterion for an inverted band structure, when the configuration of $\boldsymbol{d_k}$ is topologically equivalent to a skyrmion (cf. Fig. 2.11). That is, for $\mu m > 0$ our $p_x + ip_y$ superconductor is topological. If, however,

$$\mu m < 0,$$

the superconducting state is topologically trivial. It is an analogue of an ordinary band insulator in which no winding of $\boldsymbol{d_k}$ takes place (cf. Fig. 2.12) and, therefore, $n = 0$.

Further, using the results of Section 2.3.2, we find the Chern number for our $p_x + ip_y$ superconductor as

$$n = \frac{1}{2}[\text{sgn}(-\mu) - \text{sgn}(m)].$$

This analytical result confirms the geometric analysis of the winding of $\boldsymbol{d}_k$ in momentum space. As in Chern insulators, nontrivial number n indicates a chiral gapless state at the edge of the system (see also Fig. 1.14). Furthermore, due to the generic particle-hole symmetry

$$\mathbb{C}\hat{H}_{BdG}(x, y)\mathbb{C}^{\dagger} = -\hat{H}_{BdG}(x, y),$$

the edge mode mimics a Majorana fermion of the relativistic quantum theory. Above, $\hat{H}_{BdG}(x, y)$ is the real-space BdG Hamiltonian, and $\mathbb{C}$ is the particle-hole conjugation operation. In nontopological superconductors, $\mathbb{C}$ converts a particle into a hole and vice versa. In contrast to that, the Majorana mode, $\Psi_M(x, y, t)$, transforms to itself under $\mathbb{C}$:

$$\Psi_M^{\mathbb{C}}(x, y, t) = \Psi_M(x, y, t),$$

that is, the emergent Majorana fermion embodies both a particle and a hole. Its wave function, $\Psi_M(x, y, t)$, can be chosen real. Indeed, for BdG Hamiltonian (7.1), the conjugation operation is given by $\mathbb{C} = \tau_1 K$. If we choose the Majorana mode to be the eigenstate of τ_1 with eigenvalue $+1$, then $\mathbb{C}$ is just complex conjugation:

$$\Psi_M^{*}(x, y, t) = \Psi_M(x, y, t).$$

Using the formalism of Section 2.3.2, we readily prove that such a real edge mode indeed exists:

$$\Psi_M(x, y, t) = \begin{bmatrix} 1 \\ 1 \end{bmatrix} \sum_{k_x} \mathcal{N}_{k_x} \left[e^{\varkappa_+(k_x)y} - e^{\varkappa_-(k_x)y} \right] \times \cos\left[k_x x - E(k_x)t/\hbar\right]. \tag{7.2}$$

It is assumed that the superconductor occupies half-space $y \leq 0$ and the edge mode propagates along x. It has a linear dispersion

$$E(k_x) = \Delta' k_x,$$

and is localized on the scale given by

$$\varkappa_{\pm} = m\Delta' \pm \sqrt{m^2\Delta'^2 + k_x^2 - k_F^2}.$$

The coefficients, $\mathcal{N}_{k_x}$, of the sum in Eq. (7.2) are the normalizing factors. The normalizability requires $|k_x| < k_F$.

7.1.2 Topological $p_x + ip_y$ Superconductivity in Hybrid QAHI Structures

The model above has illustrated the principal possibility and essential attributes of topological chiral p-wave superconductivity. We now consider a concrete realization of a topological superconductor (TS) in TI structures with competing superconducting and magnetic orders. Since the work of Fu and Kane [69], TI materials with induced superconducting and magnetic interactions have become a fertile ground for theoretical modeling of TSs. In particular, in the series of papers [70, 71] and [72], a topological $p_x + ip_y$ state with a chiral Majorana edge mode has been identified theoretically in QAHI/superconductor structures. Its expected transport signatures are Majorana backscattering and a half-integer longitudinal electric conductance. In the following, these ideas are discussed in more detail.

The QAHI in question is a magnetic thin film described by the normal-state Hamiltonian

$$\hat{\mathcal{H}} = \sum_{\boldsymbol{k}} c^{\dagger}_{\boldsymbol{k}} \hat{H}_{\boldsymbol{k}} c_{\boldsymbol{k}}, \qquad c_{\boldsymbol{k}} = \begin{bmatrix} c^{t}_{\boldsymbol{k}\uparrow} \\ c^{t}_{\boldsymbol{k}\downarrow} \\ c^{b}_{\boldsymbol{k}\uparrow} \\ c^{b}_{\boldsymbol{k}\downarrow} \end{bmatrix}$$

where $\hat{H}_{\boldsymbol{k}}$ is a four-band Hamiltonian [72]:

$$\hat{H}_{\boldsymbol{k}} = A(\sigma_x k_y - \sigma_y k_x)\nu_3 + M_{\boldsymbol{k}}\sigma_0\nu_1 + \lambda\sigma_z\nu_0 - \mu\sigma_0\nu_0. \qquad (7.3)$$

Operator $c^{t,b}_{\boldsymbol{k}\sigma}$ annihilates an electron with momentum $\boldsymbol{k}$ and spin $\sigma = \uparrow, \downarrow$, and superscripts t and b refer to the top and bottom surfaces of the film. Pauli matrices σ_i (with $i = x, y, z$) and ν_j (with $j = 1, 2, 3$) represent spin and surface degrees of freedom, respectively (σ_0 and ν_0 are the corresponding unit matrices).

The first term in Eq. (7.3) is the Hamiltonian of the two surfaces, where constant A determines the surface velocity. The second term, $M_{\boldsymbol{k}}\sigma_0\nu_1$, introduces the coupling between the surfaces, opening a hybridization gap at the Γ point ($\boldsymbol{k} = 0$). The hybridization energy is

$$M_{\boldsymbol{k}} = M_0 + M_1(k_x^2 + k_y^2),$$

where M_0 yields the half of the gap between the conduction and valence bands, while M_1 accounts for the band curvature. The third term, $\lambda\sigma_z\tau_0$, is the mean-field exchange Hamiltonian due to the ferromagnetic ordering, with λ being the exchange energy. Up to the choice of the basis states and notations, Hamiltonian (7.3) is the model of a QAHI studied in Section 3.2. Note also that notation Δ is reserved for the superconducting gap energy.

Using the formalism of Section 3.2, we find that Hamiltonian (7.3) decouples into two Chern subsystems with Dirac "masses" $\lambda \pm M_{\boldsymbol{k}}$, where signs $\pm$ are dictated by TRS. The corresponding Chern number is the sum of the Chern numbers for the two subsystems. For the purpose of further analysis, we denote it by C and define as follows:

$$C = \frac{1}{2}\sum_{\nu=\pm}[\mathrm{sgn}(\lambda + \nu M_0) - \mathrm{sgn}(\nu M_1)]$$
$$= \frac{1}{2}[\mathrm{sgn}(\lambda + M_0) + \mathrm{sgn}(\lambda - M_0)].$$

It is worth emphasizing that a similar topological number is encountered in Haldane's model [6]. Taking for simplicity $M_0 > 0$ and $\lambda > 0$, we see that the system undergoes a topological phase transition from an ordinary insulator with $C = 0$ for $\lambda < M_0$ to a QAHI with $C = 1$ for $\lambda > M_0$:

$$C = \begin{cases} 0, \ \lambda < M_0, \\ 1, \ \lambda > M_0. \end{cases}$$

In the latter case, there is a chiral edge mode which realizes a gapless Dirac fermion with a four-component wave function.

We now consider the magnetic film in contact with an overlying superconductor which has a singlet s-wave order parameter. Save for the normal system, the type of the hybrid structure is the same as in Fig. 6.3. As has been shown in Section 6.2.2, the proximity of the superconductor results in a pairing interaction in the normal system, which can be described at low energies by the BdG Hamiltonian

$$\hat{H}_{BdG} = \begin{bmatrix} \hat{H}_{\boldsymbol{k}} & \hat{\Delta} \\ \hat{\Delta}^{\dagger} & -\hat{H}^{*}_{-\boldsymbol{k}} \end{bmatrix},$$

where $\hat{\Delta}$ is the induced singlet pair potential. For a thin film, it can be written as

$$\hat{\Delta} = \begin{bmatrix} \Delta_t i\sigma_y & 0 \\ 0 & \Delta_b i\sigma_y \end{bmatrix},$$

where Δ_t and Δ_b are the induced pair potentials at the top and bottom surfaces of the film. As found in Ref. [72], the essential condition for an emerging TS is to have unequal pair potentials $\Delta_t \neq \Delta_b$. This point is best illustrated in the special case where

$$\Delta_t = \Delta, \qquad \Delta_b = -\Delta, \qquad \mu = 0,$$

and Δ is real. In this case, the BdG Hamiltonian can be written in a compact matrix form

$$\hat{H}_{BdG} = A(\Sigma_x k_y - \Sigma_y k_x) + (\lambda + M_{\boldsymbol{k}}\mathcal{V}_1 + \Delta\mathcal{T}_1)\Sigma_z, \tag{7.4}$$

where Σ_x, Σ_y, Σ_z, $\mathcal{V}_1$ and $\mathcal{T}_1$ are 8×8 matrices defined by

$$\Sigma_x = \tau_3\nu_3\sigma_x, \quad \Sigma_y = \tau_3\nu_3\sigma_y, \quad \Sigma_z = \tau_0\nu_0\sigma_z,$$
$$\mathcal{V}_1 = \tau_0\nu_1\sigma_z, \quad \mathcal{T}_1 = \tau_1\nu_0\sigma_z.$$

In Eq. (7.4), the sum

$$\lambda + M_{\boldsymbol{k}}\mathcal{V}_1 + \Delta\mathcal{T}_1$$

is the matrix of Dirac masses. Furthermore, $\mathcal{V}_1$ and $\mathcal{T}_1$ commute with each other and with any of Σ_i, so in the basis of the common eigenstates of $\mathcal{V}_1$ and $\mathcal{T}_1$ the mass matrix has a diagonal structure with the entries

$$\lambda + \nu M_{\boldsymbol{k}} + \tau\Delta, \qquad \nu, \tau = \pm 1.$$

ν and τ are the eigenvalues of $\mathcal{V}_1$ and $\mathcal{T}_1$, respectively. In other words, the BdG model for the QAHI decouples into four Chern subsystems or, equivalently, into four species of the $p_x + ip_y$ superconductor. Accordingly, the total Chern number is

$$\begin{aligned} N &= \frac{1}{2}\sum_{\nu,\tau=\pm}[\mathrm{sgn}(\lambda + \nu M_0 + \tau\Delta) - \mathrm{sgn}(\nu M_1)] \\ &= \frac{1}{2}[\mathrm{sgn}(\lambda + M_0 + \Delta) + \mathrm{sgn}(\lambda - M_0 + \Delta) \\ &\quad + \mathrm{sgn}(\lambda + M_0 - \Delta) + \mathrm{sgn}(\lambda - M_0 - \Delta)]. \end{aligned}$$

To distinguish the superconducting case, we use here notation N instead of C (cf. Ref. [72]).

A finite energy scale of the pairing interaction, Δ, enables additional phase transitions that are absent in the normal case (see Fig. 7.1). For low-T_c superconductor structures, we can safely

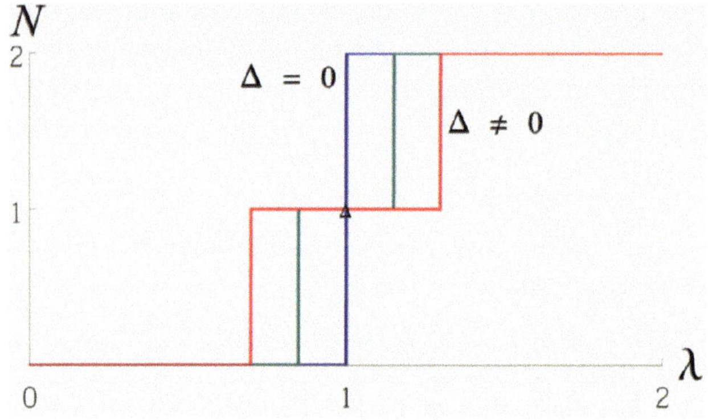

Figure 7.1 Chern number of a QAHI/superconductor hybrid, N (7.5), as function of exchange energy λ in units of M_0. For a finite pairing energy Δ, a plateau develops at $N = 1$, which corresponds to a chiral $p_x + ip_y$ superconductor with a single Majorana edge mode (7.2). Adapted from [60].

assume $\Delta < M_0$. Then, for positive parameters, the possible values of the Chern number are

$$N = \begin{cases} 0 & \lambda < M_0 - \Delta, \\ 1 & M_0 - \Delta < \lambda < M_0 + \Delta, \\ 2, & \lambda > M_0 + \Delta. \end{cases} \tag{7.5}$$

In other words, an increasing exchange field λ induces a series of phase transitions from an ordinary superconductor with $N = 0$ to the topological phases with $N = 1$ and $N = 2$. The phase with $N = 2$ has two chiral Majorana edge states making up a single Dirac mode, so the $N = 2$ phase matches the QAHI with $C = 1$. The truly new phase is that with odd Chern number $N = 1$ (7.5). This is nothing else as a $p_x + ip_y$ TS with a single chiral Majorana edge mode, as discussed earlier in Section 7.1.1. Note that by reversing magnetization $\lambda \to -\lambda$, one can also access the opposite-chirality phases with $N = -1$ and $N = -2$.

As argued in Ref. [72], the $N = 1$ phase can be identified by a half-integer plateau $0.5\,e^2/h$ in the longitudinal electric conductance as a function of the exchange field. The proposal relies on the backscattering of Majorana edge modes [71] which is expected in a QAHI/superconductor device at the topological transition to the $N = 1$ phase. The basic setup consists of a magnetic thin film and a superconducting bar placed across it, as depicted in Fig. 7.2. For a large enough exchange field,

$$\lambda > M_0 + \Delta,$$

see also Fig. 7.2a, the normal regions are the $C = 1$ QAHI with a Dirac edge state propagating along the sample boundary, while the superconducting region is the $N = 2$ phase supporting two Majorana edge modes. Since a Dirac mode is made up of two Majorana ones, we can think of two incident Majorana edge modes which match those in the superconducting region, getting transmitted almost perfectly through the device. Consequently, the longitudinal conductance between the left and right contacts reaches the quantum of e^2/h:

$$\sigma_{xx} = \frac{e^2}{h}.$$

Upon lowering the exchange field to

$$M_0 < \lambda < M_0 + \Delta,$$

the state of the superconducting region switches to the chiral TS with $N = 1$, such that one of the paired Majorana edge modes vanishes, while the normal regions are still the $C = 1$ QAHI (see Fig. 7.2b). In this regime, only one of the incident Majorana edge modes can be transmitted to the superconducting region, whereas the other Majorana mode is almost perfectly reflected. This is what Wang *et al* [72] called the separation of the two Majorana modes at the superconducting boundary, resulting in a half-integer plateau in the longitudinal electric conductance:

$$\sigma_{xx} = 0.5\frac{e^2}{h}.$$

Following this theoretical prediction, experimental work [73] reported an observation of the half-integer longitudinal conductance plateaus. They were seen close to the coercive fields during magnetization reversals in an external out-of-plane magnetic field. These findings have stimulated further theoretical and experimental research, including alternative interpretations of the half-integer longitudinal conductance which invoke specific contact physics, disorder and other factors typical for mesoscopic systems.

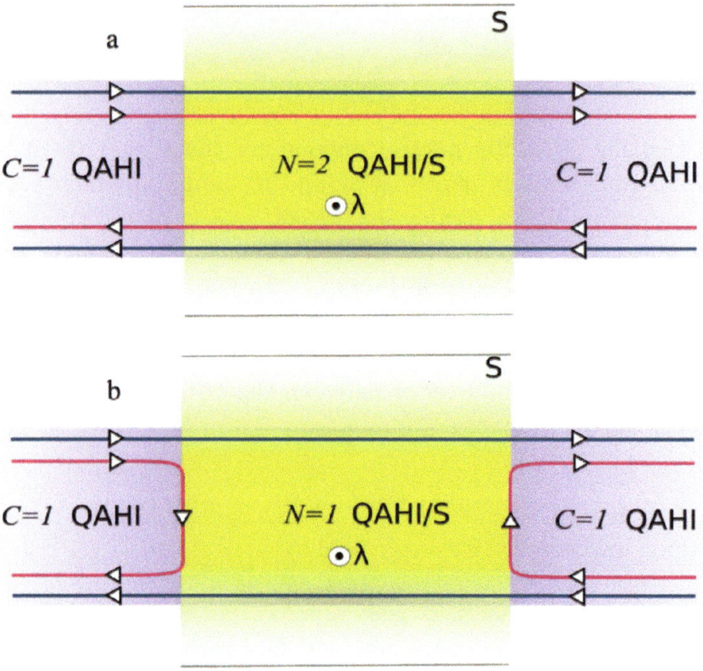

Figure 7.2 Schematic of the chiral edge modes and Majorana backscattering in QAHI/superconductor (S) structure (after [72, 73]). Adapted from [60].

7.2 Topological Josephson Junctions

7.2.1 Majorana Zero Modes and Ground State Degeneracy

Striking manifestations of topological superconductivity are expected to occur in weak links between low-dimensional systems supporting Majorana modes. Two illustrative examples of such systems are 1D *p*-wave superconductors [138, 139] and superconductor/semiconductor SOC wires [140, 141]. The boundaries of 1D TSs host a pair of Majorana zero modes (MZMs). These are the 0D cousins of the chiral Majorana edge mode which appear at the midgap energy, i.e. at exactly zero energy relative to the Fermi level.

An MZM can be described by a self-adjoint fermion operator

$$\gamma = \gamma^{\dagger}$$

which squares to 1:

$$\gamma^2 = 1.$$

It is called a Majorana operator (see end-of-chapter problems for more details).

An interesting implication of MZMs is the degeneracy of the ground state. Consider a pair of MZMs described by anticommuting operators γ_1 and γ_2. We can define a Hermitian operator $i\gamma_1\gamma_2$. Its eigenvalues ± 1 correspond to two distinct ground states. Combining the Majorana operators into a usual fermion operator

$$c = \frac{1}{2}(\gamma_1 + i\gamma_2),$$

we find that the two ground states are distinguished by the occupation number

$$c^{\dagger}c = \frac{1}{2}(1 + i\gamma_1\gamma_2) = \begin{cases} 1 \\ 0 \end{cases}. \tag{7.6}$$

In other words, they have different fermion parities. One practical example is a Josephson junction (JJ) of two TSs brought into electric contact. In this case, γ_1 and γ_2 represent the MZMs in the contact, while $c^{\dagger}c$ its occupation number, meaning that the JJ has two ground states with different fermion parities [138].

However, MZMs are not readily available in solid-state devices, and much effort has been put into engineering and detecting them in accessible materials and structures. Josephson effects (JEs) play a particular role in this quest.

The JE diagnostics of topological weak links relies on the so-called fractional JE associated with the ground state degeneracy. First proposed for model p-wave superconductors [138, 139] the fractional JE is also achievable in hybrid structures of conventional superconductors and TI materials, which has caused a surge of interest in this and related phenomena recently. In the following, we aim to learn more about MZMs and the fractional JE.

7.2.2 Majorana Zero Modes in a p_x Superconductor

As an exemplary model, we can choose a 1D version of the p_x+ip_y TS with BdG Hamiltonian (7.1). Alternatively, we can think of the chiral TS in the QAHI/superconductor structures at the transition to the $N = 1$ state [see Eqs. (7.4) and (7.5)]. The reported observability of such transitions [73, 74] gives us an extra reason for this choice. With the x-axis parallel to the system, we can write

$$\hat{H}_{BdG} = \begin{bmatrix} \frac{k_x^2}{2m} - \mu & \Delta' e^{i\varphi} k_x \\ \Delta' e^{-i\varphi} k_x & -(\frac{k_x^2}{2m} - \mu) \end{bmatrix} = \tau_3 \left(\frac{k_x^2}{2m} - \mu \right) + \tau_1 \Delta' e^{-i\tau_3\varphi} k_x,$$

where $k_x = -i\partial_x$ and φ is the phase of the order parameter. It is convenient to make a unitary transformation of the BdG wave function,

$$\Psi(x) \to e^{i\tau_3\varphi/2}\Psi(x), \tag{7.7}$$

bringing the Hamiltonian to the form

$$\hat{H}_{BdG} \to e^{-i\tau_3\varphi/2} \hat{H}_{BdG} e^{i\tau_3\varphi/2} = \tau_3 \left(\frac{k_x^2}{2m} - \mu \right) + \tau_1 \Delta' k_x. \tag{7.8}$$

It maps to a 1D Dirac fermion model with a mass term $\frac{k_x^2}{2m} - \mu$. Furthermore, akin to Jackiw–Rebbi model [75] the system possesses a chiral symmetry,

$$\tau_2 \hat{H}_{BdG} \tau_2 = -\hat{H}_{BdG}.$$

This ensures the existence of MZMs as eigenstates of the chirality matrix, τ_2:

$$\tau_2 \Psi(x) = \tau \Psi(x), \qquad \tau_2 \Psi^{\mathbb{C}}(x) = \tau \Psi^{\mathbb{C}}(x),$$

where $\tau = \pm 1$ are the eigenvalues. The eigenstates of τ_2 are self-adjoint, satisfying the equation $\hat{H}_{BdG}\Psi(x) = 0$ or, explicitly,

$$(\partial_x^2 + 2\mu m - 2\tau_2 \Delta' m \partial_x)\Psi(x) = 0.$$

This is an ordinary differential equation with simple solutions in the half space (say, $x \geq 0$) for boundary condition $\Psi(0) = 0$. Substitution $\Psi(x) \propto e^{-\varkappa x}$ yields two solutions for the decay constant

$$\varkappa_{\pm} = -\tau \Delta' m \pm \sqrt{(\Delta' m)^2 - 2\mu m}.$$

Since $\Delta' m > 0$, normalizability condition $\varkappa_{\pm} > 0$ is met for eigenvalue $\tau = -1$ and for

$$\mu m > 0.$$

This defines the topological regime with a single MZM at $x = 0$:

$$\Psi_M(x) = \mathcal{N} \begin{bmatrix} 1 \\ -i \end{bmatrix} \left(e^{-\varkappa_+ x} - e^{-\varkappa_- x} \right), \tag{7.9}$$

where $\mathcal{N}$ is the normalizing factor. By the same token, the other eigenstate of τ_2 would correspond to an MZM localized at the opposite end of the TS.

7.2.3 Majorana Zero Modes in a QSHI Channel

Our next example is a helical edge channel in a QSHI. Albeit very different from a p_x TS, this system also possesses a chiral symmetry. This symmetry is an indicator of MZMs regardless of other specifics of the system.

The system consists of a QSHI and a conventional superconductor placed over the edge of the QSHI. In the normal state, a QSHI edge channel can be described by the Hamiltonian

$$\hat{h} = (\upsilon \sigma_z + \alpha \sigma_y) k_x + \lambda \sigma_z,$$

where υ yields the edge velocity, α is a Rashba SOC constant, λ is, depending on the context, an exchange or Zeeman energy, and $k_x = -i\partial_x$. To describe the proximity-induced superconductivity in the channel, we use the BdG Hamiltonian

$$\begin{aligned} \hat{H}_{BdG} &= \begin{bmatrix} (\upsilon \sigma_z + \alpha \sigma_y) k_x + \lambda \sigma_z & \Delta \\ \Delta & -(\upsilon \sigma_z + \alpha \sigma_y) k_x + \lambda \sigma_z \end{bmatrix} \\ &= \tau_3 (\upsilon \sigma_z + \alpha \sigma_y) k_x + \lambda \tau_0 \sigma_z + \Delta \tau_1 \sigma_0, \end{aligned} \tag{7.10}$$

with a real Δ. The Hamiltonian has the chiral symmetry

$$\tau_3 \sigma_x \hat{H}_{BdG} \tau_3 \sigma_x = -\hat{H}_{BdG},$$

so we expect MZMs in this system as well.

Indeed, let us look at the zero energy solutions of the BdG equation:

$$\hat{H}_{BdG} \Psi(x) = 0, \qquad \Psi(x) = \begin{bmatrix} u(x) \\ v(x) \end{bmatrix},$$

where $u(x)$ and $v(x)$ are spinors satisfying two coupled equations:

$$[(\upsilon\sigma_z + \alpha\sigma_y)k_x + \lambda\sigma_z]u(x) + \Delta v(x) = 0,$$

$$\Delta u(x) + [-(\upsilon\sigma_z + \alpha\sigma_y)k_x + \lambda\sigma_z]v(x) = 0.$$

Excluding the hole spinor, $v(x)$, yields

$$[(\upsilon\sigma_z + \alpha\sigma_y)k_x - \lambda\sigma_z][(\upsilon\sigma_z + \alpha\sigma_y)k_x + \lambda\sigma_z]u(x) + \Delta^2 u(x) = 0,$$

and, after some algebra, we arrive at the second-order differential equation

$$[(\upsilon^2 + \alpha^2)\partial_x^2 - 2\alpha\lambda\sigma_x\partial_x + \lambda^2 - \Delta^2]u(x) = 0.$$

The required solution is a bound state of the form

$$u(x) = ue^{-\varkappa x},$$

where u is a position-independent spinor chosen to be an eigenstate of σ_x:

$$\sigma_x u = \sigma u, \qquad \sigma = \pm 1,$$

and $\varkappa$ is a decay constant, satisfying the quadratic equation

$$(\upsilon^2 + \alpha^2)\varkappa^2 + 2\alpha\lambda\sigma\varkappa + \lambda^2 - \Delta^2 = 0.$$

Its roots are

$$\varkappa_\pm = -\frac{\alpha\lambda\sigma}{\upsilon^2 + \alpha^2} \pm \sqrt{\frac{\alpha^2\lambda^2}{(\upsilon^2 + \alpha^2)^2} - \frac{\lambda^2 - \Delta^2}{\upsilon^2 + \alpha^2}}.$$

Under condition

$$|\lambda| > |\Delta|,$$

we find one normalizable solution at each boundary. For example, at $x = 0$ there is an MZM with spin projection $\sigma = -\text{sgn}(\alpha\lambda)$, which decays in half-space $x \geq 0$ with $\varkappa_\pm > 0$. If, for concreteness, $\alpha\lambda > 0$, the MZM wave function is given by

$$u_M(x) = \mathcal{N}\begin{bmatrix} 1 \\ -1 \end{bmatrix} \left(e^{-\varkappa_+ x} - e^{-\varkappa_- x}\right). \tag{7.11}$$

Similar to the MZM of a p_x TS, solution (7.11) vanishes at the edge. The hole spinor, $v_M(x)$, has an analogous structure.

7.2.4 Fractional Josephson Effect

The fractional JE is caused by the coupling of the MZMs across a weak link between two TSs. The phenomenology is rather independent of the details of the TSs and can be illustrated by an effective junction Hamiltonian

$$\hat{H}_J = \begin{bmatrix} \hat{H}_R & \hat{V}^\dagger \\ \hat{V} & \hat{H}_L \end{bmatrix},$$

where $\hat{H}_R$ and $\hat{H}_L$ are the Hamiltonians of the right and left TSs, whereas the operator

$$\hat{V} = \begin{bmatrix} \hat{V}_0 & 0 \\ 0 & -\hat{V}_0^* \end{bmatrix}$$

models the coupling between them. $\hat{V}$ is a diagonal Nambu matrix where $\hat{V}_0$ is the normal-state coupling (generally complex-valued, which would break TRS). The unitary transformation (7.7) in each TS yields

$$\hat{H}_J \to \begin{bmatrix} \hat{H}_R & \hat{V}^\dagger e^{-i\tau_3\phi/2} \\ \hat{V} e^{i\tau_3\phi/2} & \hat{H}_L \end{bmatrix}, \qquad \phi = \varphi_R - \varphi_L,$$

where the coupling acquires the dependence on the phase difference, ϕ, between the TSs, whereas transformed $\hat{H}_R$ and $\hat{H}_L$ are both phase-independent akin to the Hamiltonians in Eqs. (7.8) and (7.10).

The energy spectrum of the JJ is obtained from the BdG equation

$$E\Psi = \hat{H}_J \Psi, \qquad \Psi = \begin{bmatrix} \Psi_R \\ \Psi_L \end{bmatrix},$$

where Ψ_R and Ψ_L are the BdG wave functions of the right and left TSs, satisfying two coupled equations,

$$E\Psi_R = \hat{H}_R \Psi_R + \hat{V}^\dagger e^{-i\tau_3\phi/2} \Psi_L,$$

$$E\Psi_L = \hat{V} e^{i\tau_3\phi/2} \Psi_R + \hat{H}_L \Psi_L,$$

together with normalization conditions

$$\langle \Psi_{L,R} | \Psi_{L,R} \rangle = 1.$$

In the lowest order in $\hat{V}$, the energy levels can be expressed through the MZMs by putting

$$\hat{H}_R \Psi_R = 0, \qquad \hat{H}_L \Psi_L = 0,$$

and projecting the BdG equations on the bra states $\langle \Psi_{R,L}|$. Using notation E_+ for the solution, we have

$$E_+ = \langle \Psi_L|\hat{V} e^{i\tau_3\phi/2}|\Psi_R\rangle = \langle \Psi_R|\hat{V}^\dagger e^{-i\tau_3\phi/2}|\Psi_L\rangle. \tag{7.12}$$

Since

$$\langle \Psi_R|\hat{V}^\dagger e^{-i\tau_3\phi/2}|\Psi_L\rangle = \langle \Psi_L|\hat{V} e^{i\tau_3\phi/2}|\Psi_R\rangle^*,$$

the second equality in Eq. (7.12) just means that the solution is real.

The particle-hole symmetry ensures the existence of another solution, E_-, irrespective of other symmetries of the system. Replacing $\Psi_L \to \Psi_L^{\mathbb{C}}$ and $\Psi_R \to \Psi_R^{\mathbb{C}}$, we have

$$E_- = \langle \Psi_L^{\mathbb{C}}|V e^{i\tau_3\phi/2}|\Psi_R^{\mathbb{C}}\rangle,$$

where the matrix element can be evaluated as follows:

$$\begin{aligned}\langle \Psi_L^{\mathbb{C}}|V e^{i\tau_3\phi/2}|\Psi_R^{\mathbb{C}}\rangle &= \langle \Psi_L|\mathbb{C}^\dagger V e^{i\tau_3\phi/2}\mathbb{C}|\Psi_R\rangle \\ &= \langle \Psi_L|\mathbb{C}^\dagger V \mathbb{C} e^{i\tau_3\phi/2}|\Psi_R\rangle \\ &= -\langle \Psi_L|V e^{i\tau_3\phi/2}|\Psi_R\rangle.\end{aligned}$$

We use the particle-hole symmetry $\mathbb{C}^\dagger V \mathbb{C} = -V$, due to which the second level comes with the opposite sign

$$E_- = -\langle \Psi_L|V e^{i\tau_3\phi/2}|\Psi_R\rangle. \tag{7.13}$$

Physically, $E_\pm$ are the levels of the Andreev bound states (ABSs) formed by the two hybridized MZMs. The corresponding wave functions $\Psi_\pm$ are the linear combinations of the right and left MZMs. Remarkably, the topological ABSs are 4π-periodic in the Josephson phase difference ϕ. This is a qualitative distinction from usual JJs where the ABSs are 2π-periodic. For a TR-symmetric coupling

$$V = V_0 \tau_3,$$

the terms $\propto \sin(\phi/2)$ vanish because of the orthogonality of the MZMs, so the phase dependence of the energy levels is reduced to

$$E_\pm(\phi) = \pm\langle \Psi_L|V|\Psi_R\rangle \cos(\phi/2), \tag{7.14}$$

see also Fig. 7.3. It is worth noting that microscopic calculations assuming a potential barrier in the JJ (e.g. in [139]) give a similar

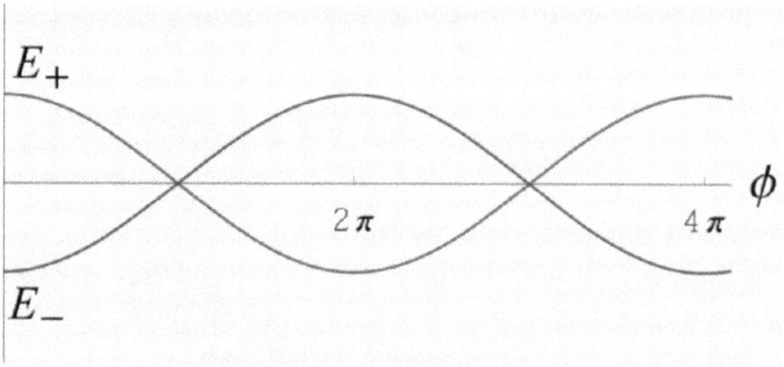

Figure 7.3 4π-periodic ABSs from hybridization of two MZMs [see Eq. (7.14)]. Adapted from [60].

result for the ABS spectrum

$$E_{\pm}(\phi) = \pm(\Delta' k_F)\sqrt{\mathcal{T}}\cos(\phi/2),$$

where $\mathcal{T}$ is the barrier transparency.

Two aspects of the topological ABSs merit special attention. First, their 4π-periodicity harbors the topological degeneracy due to the underlying MZMs. A phase translation $\phi \to \phi + 2\pi$ brings the JJ into a state with the same energy

$$E_+(\phi + 2\pi) = E_-(\phi),$$

but with the opposite fermion parity, since the relative sign of Ψ_R and Ψ_L changes under the unitary transformation in Eq. (7.7) and the occupation number in Eq. (7.6) switches between 0 and 1. Second, each ABS carries a 4π-periodic supercurrent. Since the phase and the particle number are conjugate dynamical variables, a phase-dependent coupling energy gives rise to a current flow between the systems. In sufficiently short JJs, a major contribution to the current-phase relation (CPR) comes from the ABSs. At equilibrium, the CPR is given by the thermodynamic formula

$$J(\phi) = J_+(\phi) + J_-(\phi),$$

$$J_{\pm}(\phi) = \frac{e}{\hbar}\frac{\partial E_{\pm}(\phi)}{\partial \phi} n[E_{\pm}(\phi)],$$

where $J_{\pm}(\phi)$ are the contributions of the two ABS levels occupied according to the Fermi distribution $n(E)$. Using Eq. (7.14) and specially chosen occupations $n[E_+(\phi)] = 0$ and $n[E_-(\phi)] = 1$, one has

$$J(\phi) = J_-(\phi) = \frac{e\langle\Psi_L|V|\Psi_R\rangle}{2\hbar}\sin(\phi/2). \tag{7.15}$$

This example illustrates the fractional JE which is characterized by a subharmonic CPR with frequency 1/2. As proposed in [139], CPR (7.15) should be observable in voltage-biased JJs. In this case, the Josephson phase difference evolves with time as

$$\phi(t) = 2eUt/\hbar,$$

where U is a voltage drop at the JJ. Consequently, the current is an oscillating function of time,

$$J(t) = \frac{e\langle\Psi_L|V|\Psi_R\rangle}{2\hbar}\sin(eUt/\hbar). \tag{7.16}$$

It oscillates at frequency $eU/\hbar$ which is only half of the usual Josephson frequency $2eU/\hbar$ observed in nontopological JJs. This is quite remarkable for the following reason. The frequency $2eU/\hbar$ indicates that the current is carried by electron pairs with charge $2e$. Therefore, the fractional frequency $eU/\hbar$ means that in topological JJs the Josephson current is single-electron one. It is carried by a single-fermion ABS due to the overlap of the MZMs of the right and left TSs. That is, the fractional JE is an indicator of the MZMs, each of them being, loosely speaking, half a fermion.

A number of recent theories have revisited the fractional JE in the context of MZMs in hybrid structures combining conventional superconductors with TIs or semiconductor SOC nanowires.

7.3 Non-Abelian Exchange Statistics and Braiding

Systems supporting MZMs hold promise for braiding-based topological quantum computation in which computing operations are performed by unitary transformations within a degenerate set of ground states [62]. The proposal exploits a non-Abelian exchange statistics of MZMs in which pairwise exchanges of MZMs are realized by unitary operators.

To conclude this chapter, we illustrate the idea of the non-Abelian statistics for a pair of MZMs described by Majorana operators γ_1 and γ_2 (see also end-of-chapter problems). Let us define the operator

$$\hat{B} = \frac{1}{\sqrt{2}}(1 + \gamma_1\gamma_2)$$

and its Hermitian conjugate

$$\hat{B}^\dagger = \frac{1}{\sqrt{2}}(1 + \gamma_2\gamma_1).$$

First, we prove that operator $\hat{B}$ is unitary, which is straightforward:

$$\begin{aligned}\hat{B}\hat{B}^\dagger &= \frac{1}{2}(1 + \gamma_1\gamma_2)(1 + \gamma_2\gamma_1)\\ &= \frac{1}{2}(1 + \gamma_1\gamma_2 + \gamma_2\gamma_1 + \gamma_1\gamma_2^2\gamma_1)\\ &= 1,\end{aligned}$$

where we use the anticommutation of γ_1 and γ_2 and the fact that they square to 1, i.e. $\gamma_1^2 = \gamma_2^2 = 1$. Since $\hat{B}^\dagger$ is equal to $\hat{B}^{-1}$, the operator is indeed unitary.

We then look at the unitary transformations of γ_1:

$$\begin{aligned}\hat{B}\gamma_1\hat{B}^\dagger &= \frac{1}{2}(1 + \gamma_1\gamma_2)\gamma_1(1 + \gamma_2\gamma_1)\\ &= \frac{1}{2}(1 + \gamma_1\gamma_2)(\gamma_1 - \gamma_2)\\ &= \frac{1}{2}(\gamma_1 - \gamma_2 - \gamma_2 - \gamma_1)\\ &= -\gamma_2,\end{aligned}$$

where, again, the anticommutation and $\gamma_1^2 = \gamma_2^2 = 1$ rules are used. By the same token,

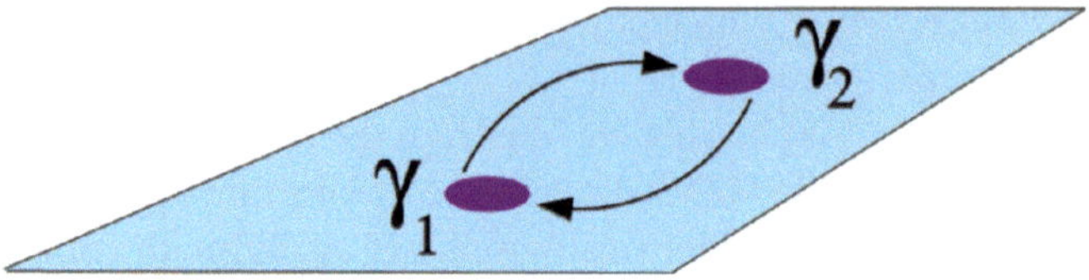

Figure 7.4 Exchange of two MZMs as unitary operation in Eq. (7.17).

$$
\begin{aligned}
\hat{B}\gamma_2\hat{B}^\dagger &= \frac{1}{2}(1+\gamma_1\gamma_2)\gamma_2(1+\gamma_2\gamma_1) \\
&= \frac{1}{2}(1+\gamma_1\gamma_2)(\gamma_2+\gamma_1) \\
&= \frac{1}{2}(\gamma_2+\gamma_1+\gamma_1-\gamma_2) \\
&= \gamma_1,
\end{aligned}
$$

That is, a pair γ_1, γ_2 turns to $-\gamma_2, \gamma_1$, respectively. This can be expressed as the unitary operation

$$\hat{B}\gamma_1\gamma_2\hat{B}^\dagger = -\gamma_2\gamma_1. \tag{7.17}$$

It is just an exchange of two fermions, as depicted in Fig. 7.4. However, this exchange is performed by a unitary operation, which is a characteristic of a non-Abelian statistics.

Furthermore, a double exchange of γ_1 and γ_2 is also nontrivial:

$$\hat{B}^2\gamma_1(\hat{B}^\dagger)^2 = -\hat{B}\gamma_2\hat{B}^\dagger = -\gamma_1, \tag{7.18}$$

$$\hat{B}^2\gamma_2(\hat{B}^\dagger)^2 = \hat{B}\gamma_1\hat{B}^\dagger = -\gamma_2. \tag{7.19}$$

A double exchange of two MZMs is equivalent to a round trip of one of them about the other, as sketched in Fig. 7.5. We do not recover

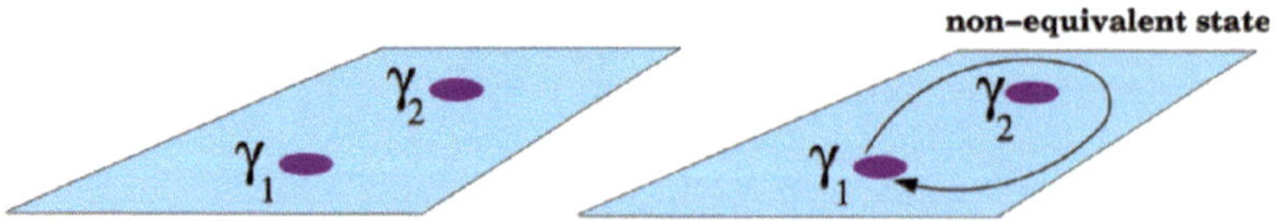

Figure 7.5 Double exchange of two MZMs as round trip of one of them about the other [see Eqs. (7.18) and (7.19)].

the original state of γ_1 and γ_2, as each MZM acquires a phase of π. Only a quadruple exchange leads to the original state of γ_1 and γ_2:

$$\hat{B}^4\gamma_1(\hat{B}^\dagger)^4 = \gamma_1, \qquad \hat{B}^4\gamma_2(\hat{B}^\dagger)^4 = \gamma_2.$$

The unitary operations performed by $\hat{B}$ above illustrate the basic idea of braiding of MZMs. It lays the basis for quantum computing with MZMs, which, however, requires more than one pair of them [62].

Currently, non-Abelian excitations and chiral superconductivity in TI materials are the subjects of intense experimental and theoretical effort. The experimental quest for chiral superconductivity in both intrinsic and proximity-induced superconductors continues. Speaking of hybrid proximity structures, we have seen that chiral superconductivity can be understood in terms of the duality between $p_x + ip_y$ superconductors and Chern insulators. Is there more new physics beyond this intricate duality? The answer to this question is not only of theoretical interest, but may also uncover hitherto unexplored routes towards topological quantum computation and dissipationless electronics.

7.4 Problems

(1) Consider the following representation for fermion creation and annihilation operators:

$$c_j = \frac{1}{2}(\gamma_{1j} + i\gamma_{2j}), \qquad c_j^\dagger = \frac{1}{2}(\gamma_{1j} - i\gamma_{2j}),$$

where γ_{1j} and γ_{2j} are self-adjoint (Majorana) operators defined as follows:

$$\gamma_{1j}^\dagger = \gamma_{1j}, \qquad \gamma_{2j}^\dagger = \gamma_{2j}.$$

Using the commutation relations for c_j and $c_j^\dagger$, obtain the commutation relations for Majorana operators:

$$\gamma_{1i}\gamma_{2j} = -\gamma_{2j}\gamma_{1i}, \quad \gamma_{1i}\gamma_{1j} + \gamma_{1j}\gamma_{1i} = 2\delta_{ij},$$
$$\gamma_{2i}\gamma_{2j} + \gamma_{2j}\gamma_{2i} = 2\delta_{ij}.$$

Prove the following expression for the fermion occupation number:

$$c_j^\dagger c_j = \frac{1}{2}(1 + i\gamma_{1j}\gamma_{2j}).$$

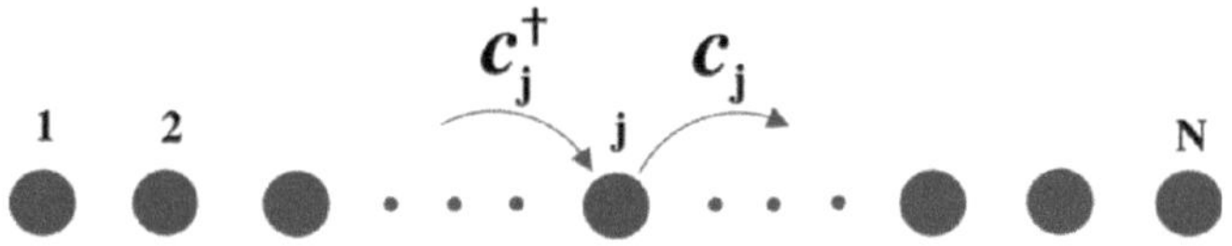

Figure 7.6 Electron hopping on a 1D lattice in tight-binding model (7.20).

(2) A 1D conductor is described by the tight-binding Hamiltonian

$$\hat{h}_{t.b.} = V\sum_i c_i^\dagger c_i + t\sum_{\langle ij\rangle}(c_i^\dagger c_j + c_j^\dagger c_i), \tag{7.20}$$

where $\langle ij\rangle$ means the nearest-neighbor sites of a 1D lattice shown in Fig. 7.6. Obtain the following representation in terms of Majorana operators:

$$\hat{h}_{t.b.} = \sum_i\left(\frac{V}{2}+t\right) + \frac{iV}{2}\sum_i \gamma_{1i}\gamma_{2i} + it\sum_{\langle ij\rangle}\gamma_{1i}\gamma_{2j}.$$

(3) A 1D p-wave superconductor can be described by a tight-binding Hamiltonian of the form [138]:

$$\hat{h}_{t.b.} = V\sum_i c_i^\dagger c_i + \sum_{\langle ij\rangle}[t(c_i^\dagger c_j + c_j^\dagger c_i) + \Delta(c_i c_j + c_j^\dagger c_i^\dagger)].$$

Obtain its representation in terms of Majorana operators.

(4) The Hamiltonian of the Jackiw–Rebbi model [75] can be written as

$$\hat{h} = c\,\sigma_x(-i\partial_x) + m(x)\sigma_y,$$

where $m(x)$ changes the sign at $x = x_0$. The Hamiltonian has a zero-energy eigenstate localized at the node $x = x_0$. Consider the BdG Hamiltonian

$$\hat{H}_{BdG} = \begin{bmatrix} c\,\sigma_x(-i\partial_x) + m(x)\sigma_y & \Delta \\ \Delta & c\,\sigma_x(+i\partial_x) + m(x)\sigma_y \end{bmatrix}.$$

Check whether it has a chiral symmetry. Identify the topological regime in which $\hat{H}_{BdG}$ has an MZM. Obtain an explicit solution for it.

Appendix A

Weak Antilocalization in Topological Insulators

A.1 Equations (5.94) for Cooperon Amplitudes

In this appendix we outline the derivation of Eq. (5.94) for the Cooperon amplitudes $C^{ij}(\omega, \boldsymbol{q})$. Inverting the expansion in Eq. (5.92), we have

$$C^{ij}(\omega, \boldsymbol{q}) = \sum_{\alpha\beta\alpha'\beta'} \Psi^{i*}_{\alpha\beta} \Psi^{j}_{\alpha'\beta'} C_{\alpha\beta\alpha'\beta'}(\omega, \boldsymbol{q}). \tag{A.1}$$

Inserting this into Eq. (5.91) yields the equation

$$C^{ij}(\omega, \boldsymbol{q}) = \frac{\tau^2}{\tau_0}\delta_{ij} + \zeta \sum_{s} \int \frac{d\boldsymbol{k}}{(2\pi)^2} \tag{A.2}$$
$$\times \left[\sum_{\alpha\beta\gamma'\delta'} \Psi^{i*}_{\alpha\beta} G^{R}_{\alpha\gamma'}(\boldsymbol{k}, E + \hbar\omega) G^{A}_{\beta\delta'}(\boldsymbol{q} - \boldsymbol{k}, E) \Psi^{s}_{\gamma'\delta'} \right] C^{sj}(\omega, \boldsymbol{q}).$$

It can be rewritten in a simpler form with the help of the identity

$$\sum_{\alpha\beta\gamma'\delta'} \Psi^{i*}_{\alpha\beta} G^{R}_{\alpha\gamma'}(\boldsymbol{k}, E + \hbar\omega) G^{A}_{\beta\delta'}(\boldsymbol{q} - \boldsymbol{k}, E) \Psi^{s}_{\gamma'\delta'} =$$
$$\sum_{\delta'\beta\alpha\gamma'} \left[G^{A^T}(\boldsymbol{q} - \boldsymbol{k}, E) \right]_{\delta'\beta} \left[\Psi^{i\dagger} \right]_{\beta\alpha} G^{R}_{\alpha\gamma'}(\boldsymbol{k}, E + \hbar\omega) \Psi^{s}_{\gamma'\delta'}$$
$$= \mathrm{Tr}\left[\hat{G}^{A^T}(\boldsymbol{q} - \boldsymbol{k}, E) \Psi^{i\dagger} \hat{G}^{R}(\boldsymbol{k}, E + \hbar\omega) \Psi^{s} \right], \tag{A.3}$$

where Tr and T denote the trace and transposition operations, respectively, in $\boldsymbol{\sigma}$ space. We therefore have

$$C^{ij}(\omega, \boldsymbol{q}) = \frac{\tau^2}{\tau_0}\delta_{ij} + \zeta \sum_s \mathrm{Tr}\Big[\int \frac{d\boldsymbol{k}}{(2\pi)^2}$$

$$\times \hat{G}^{A^T}(\boldsymbol{q}-\boldsymbol{k}, E)\Psi^{i\dagger}\hat{G}^R(\boldsymbol{k}, E+\hbar\omega)\Psi^s\Big]C^{sj}(\omega, \boldsymbol{q}). \qquad \text{(A.4)}$$

For the chosen orthonormal spin basis $\Psi^j = \sigma_j\sigma_y/\sqrt{2}$ [see also Eq. (5.93)] Eq. (A.4) assumes the following form:

$$C^{ij}(\omega, \boldsymbol{q}) = \frac{\tau^2}{\tau_0}\delta_{ij} + \frac{\zeta}{2} \sum_s \mathrm{Tr}\Big[\int \frac{d\boldsymbol{k}}{(2\pi)^2}$$

$$\times \widetilde{\hat{G}^A}(\boldsymbol{q}-\boldsymbol{k}, E)\sigma_i\hat{G}^R(\boldsymbol{k}, E+\hbar\omega)\sigma_s\Big]C^{sj}(\omega, \boldsymbol{q}). \qquad \text{(A.5)}$$

Here the tilde denotes the time-reversal operation,

$$\widetilde{\hat{G}^A} = \sigma_y\left[\hat{G}^A\right]^T\sigma_y, \qquad \text{(A.6)}$$

which flips the spin: $\widetilde{\boldsymbol{\sigma}} = \sigma_y\boldsymbol{\sigma}^T\sigma_y = -\boldsymbol{\sigma}$. As expected, for $\boldsymbol{q} \to 0$ Eq. (A.5) describes interference between the state with $\boldsymbol{k}, \boldsymbol{\sigma}$ and its "time-reversed" partner with $-\boldsymbol{k}, -\boldsymbol{\sigma}$.

Next, we evaluate the $\boldsymbol{k}$ integral in Eq. (A.5). For that we again use the sharpness of the Green functions (5.80) near the Fermi surface, $|\boldsymbol{k}| \approx k_F$, under condition (5.38) as well as the smallness of the Cooperon momentum

$$|\boldsymbol{q}| \ll k_F. \qquad \text{(A.7)}$$

We expand the denominator of $\widetilde{\hat{G}^A}_{\boldsymbol{q}-\boldsymbol{k}}$ [see Eq. (5.80)] up to the linear order in $\boldsymbol{q}$, i.e. $\xi_{\boldsymbol{q}-\boldsymbol{k}} \approx \xi_{\boldsymbol{k}} - \hbar v\boldsymbol{n}\cdot\boldsymbol{q}$ (the last term should be compared with $\hbar/\tau$), which yields

$$\widetilde{\hat{G}^A}_{\boldsymbol{q}-\boldsymbol{k}} \approx \frac{\widetilde{P}_{-\boldsymbol{n}}}{E_A - \xi_{\boldsymbol{k}} + \hbar v\boldsymbol{n}\cdot\boldsymbol{q}}. \qquad \text{(A.8)}$$

With Eqs. (5.80) and (A.8) the $\boldsymbol{k}$ integral can be evaluated as follows:

$$\int \frac{d\boldsymbol{k}}{(2\pi)^2}\widetilde{G^A}(\boldsymbol{q}-\boldsymbol{k}, E)\sigma_i G^R(\boldsymbol{k}, E+\hbar\omega)\sigma_s = \qquad \text{(A.9)}$$

$$\int \frac{d\phi_{\boldsymbol{n}}}{2\pi} \int_{-E_F}^{\infty} N(\xi, \boldsymbol{n}) d\xi \frac{\widetilde{P}_{-\boldsymbol{n}\sigma_i} P_{\boldsymbol{n}\sigma_s}}{(E_{_A} - \xi + \hbar v \boldsymbol{n} \cdot \boldsymbol{q})(E_{_R} - \xi)} \approx \tag{A.10}$$

$$N \int \frac{d\phi_{\boldsymbol{n}}}{2\pi} \widetilde{P}_{-\boldsymbol{n}\sigma_i} P_{\boldsymbol{n}\sigma_s} \int_{-\infty}^{\infty} \frac{d\xi}{(\xi - E_{_A} - \hbar v \boldsymbol{n} \cdot \boldsymbol{q})(\xi - E_{_R})} = \tag{A.11}$$

$$N \int \frac{d\phi_{\boldsymbol{n}}}{2\pi} \widetilde{P}_{-\boldsymbol{n}\sigma_i} P_{\boldsymbol{n}\sigma_s} \frac{2\pi i}{i\hbar/\tau + \hbar\omega - \hbar v \boldsymbol{n} \cdot \boldsymbol{q}} = \tag{A.12}$$

$$\frac{2\pi N \tau}{\hbar} \int \frac{d\phi_{\boldsymbol{n}}}{2\pi} \frac{\widetilde{P}_{-\boldsymbol{n}\sigma_i} P_{\boldsymbol{n}\sigma_s}}{1 - i\tau\omega + i\tau v \boldsymbol{n} \cdot \boldsymbol{q}}. \tag{A.13}$$

In Eq. (A.11) the exact DOS $N(\xi, \boldsymbol{n})$ is replaced by its Fermi surface value N and the lower ξ-integration limit is extended from $-E_F$ to $-\infty$ under condition (5.38). Then, the ξ integral is calculated along the contour encircling the upper complex half-plane with a pole at

$$\xi = E_{_R} = E + \hbar\omega + \frac{i\hbar}{2\tau}. \tag{A.14}$$

Inserting Eq. (A.12) into Eq. (A.5) and using Eq. (5.49) for ζ, we arrive at Eq. (5.94).

A.2 Evaluation of *k* Integrals in Hikami Boxes in Eqs. (5.124) and (5.125)

To calculate the $\boldsymbol{k}$ integral in Eq. (5.124) we first expand the Green functions $\hat{G}^R_{\boldsymbol{q}-\boldsymbol{k}, E+\hbar\omega}$ and $\hat{G}^A_{\boldsymbol{q}-\boldsymbol{k}, E}$ in small Cooperon momentum $\boldsymbol{q}$ as we did in Eq. (A.8) of Appendix A.1:

$$\hat{G}^R_{\boldsymbol{q}-\boldsymbol{k}, E+\hbar\omega} \approx \frac{\hat{P}_{-\boldsymbol{n}}}{E_{_R} - \xi_{\boldsymbol{k}} + \hbar\omega + \hbar v \boldsymbol{n} \cdot \boldsymbol{q}}, \tag{A.15}$$

$$\hat{G}^A_{\boldsymbol{q}-\boldsymbol{k}, E} \approx \frac{\hat{P}_{-\boldsymbol{n}}}{E_{_A} - \xi_{\boldsymbol{k}} + \hbar v \boldsymbol{n} \cdot \boldsymbol{q}} \tag{A.16}$$

Then, the integral is converted to $\int \frac{d\phi_{\boldsymbol{n}}}{2\pi} \int N d\xi \ldots$ and the ξ-integration is done in the complex plane following the same steps as in Eqs. (A.9)–(A.12). The final result is

$$\int \frac{d\boldsymbol{k}}{(2\pi)^2} (\hat{G}^A_{\boldsymbol{k},E}\, \hat{v}^x_{\boldsymbol{k}}\, \hat{G}^R_{\boldsymbol{k},E+\hbar\omega})_{\gamma'\beta} (\hat{G}^R_{\boldsymbol{q}-\boldsymbol{k},E+\hbar\omega}\, \hat{v}^x_{\boldsymbol{q}-\boldsymbol{k}}\, \hat{G}^A_{\boldsymbol{q}-\boldsymbol{k},E})_{\gamma\beta'}$$

$$\approx \frac{4\pi N\tau^3}{\hbar^3} \overline{[\hat{P}_{\boldsymbol{n}}\hat{v}^x_{\boldsymbol{n}}\hat{P}_{\boldsymbol{n}}]_{\gamma'\beta}[\hat{P}_{-\boldsymbol{n}}\hat{v}^x_{-\boldsymbol{n}}\hat{P}_{-\boldsymbol{n}}]_{\gamma\beta'}}, \tag{A.17}$$

where $\overline{(\ldots)} = \int_0^{2\pi} \ldots d\phi_{\boldsymbol{n}}/2\pi$ is the averaging over the momentum direction $\boldsymbol{n}$. The equation above is independent of E, so that the energy integral in Eq. (5.124) is $\hbar/2\pi$, which along with Eq. (A.17) yields Eq. (5.126).

To evaluate the integrals over $\boldsymbol{k}$ and $\boldsymbol{k}'$ in Eq. (5.125) we set $\boldsymbol{q}, \omega \to 0$ [see Eq. (5.99)] in the Green functions and re-group them as follows:

$$\int \frac{d\boldsymbol{k}}{(2\pi)^2} \int \frac{d\boldsymbol{k}'}{(2\pi)^2} (\hat{G}^A_{\boldsymbol{k},E}\, \hat{v}^x_{\boldsymbol{k}}\, \hat{G}^R_{\boldsymbol{k},E+\hbar\omega} \hat{G}^R_{\boldsymbol{q}-\boldsymbol{k}',E+\hbar\omega})_{\gamma'\beta}$$

$$\times (\hat{G}^R_{\boldsymbol{q}-\boldsymbol{k},E+\hbar\omega}\, \hat{G}^R_{\boldsymbol{k}',E+\hbar\omega}\, \hat{v}^x_{\boldsymbol{k}'}\, \hat{G}^A_{\boldsymbol{k}',E})_{\gamma\beta'} \tag{A.18}$$

$$\approx \int \frac{d\boldsymbol{k}}{(2\pi)^2} [\hat{G}^A(\boldsymbol{k})\hat{v}^x(\boldsymbol{k})\hat{G}^R(\boldsymbol{k})]_{\gamma'\beta_1} \hat{G}^R_{\gamma\gamma_1}(-\boldsymbol{k})$$

$$\times \int \frac{d\boldsymbol{k}'}{(2\pi)^2} \hat{G}^R_{\beta_1\beta}(-\boldsymbol{k}')[\hat{G}^R(\boldsymbol{k}')\hat{v}^x(\boldsymbol{k}')\hat{G}^A(\boldsymbol{k}')]_{\gamma_1\beta'}. \tag{A.19}$$

Each integral can now be done in the similar way as in Eqs. (A.9)–(A.12) of Appendix A.1:

$$\int \frac{d\boldsymbol{k}}{(2\pi)^2} [\hat{G}^A(\boldsymbol{k})\hat{v}^x(\boldsymbol{k})\hat{G}^R(\boldsymbol{k})]_{\gamma'\beta_1} \hat{G}^R_{\gamma\gamma_1}(-\boldsymbol{k})$$

$$\approx -\frac{2\pi i N\tau^2}{\hbar^2} \overline{[\hat{P}_{\boldsymbol{n}}\hat{v}^x_{\boldsymbol{n}}\hat{P}_{\boldsymbol{n}}]_{\gamma'\beta_1}[\hat{P}_{-\boldsymbol{n}}]_{\gamma\gamma_1}}, \tag{A.20}$$

$$\int \frac{d\boldsymbol{k}'}{(2\pi)^2} \hat{G}^R_{\beta_1\beta}(-\boldsymbol{k}')[\hat{G}^R(\boldsymbol{k}')\hat{v}^x(\boldsymbol{k}')\hat{G}^A(\boldsymbol{k}')]_{\gamma_1\beta'}$$

$$\approx -\frac{2\pi i N\tau^2}{\hbar^2} \overline{[\hat{P}_{-\boldsymbol{n}'}]_{\beta_1\beta}[\hat{P}_{\boldsymbol{n}'}\hat{v}^x_{\boldsymbol{n}'}\hat{P}_{\boldsymbol{n}'}]_{\gamma_1\beta'}}. \tag{A.21}$$

Inserting these into Eq. (5.125) we obtain Eq. (5.127).

Appendix B

Unconventional Superconductivity in Noncentrosymmetric Proximity Superconductors

As mentioned in Section 6.2.2, proximity structures in which conventional superconductors are combined with TI or SOC materials are an example of noncentrosymmetric superconductors (NCSs). Such proximity NCSs offer diverse possibilities for engineering novel superconducting states and devices. The supplemental material below provides a theoretical formalism for understanding unconventional superconductivity in these systems.

B.1 Condensate Green's Function for a Mixed Singlet-Triplet Pairing Interaction

We begin by reviewing some relevant results for the condensate Green function of a mixed-parity superconductor in the presence of both SOC and supercurrent [134, 135]. The normal system is treated as a single-band conductor with the Hamiltonian

$$\hat{h}_{\boldsymbol{k}} = \xi_{\boldsymbol{k}} + \boldsymbol{\sigma} \cdot \boldsymbol{\gamma}_{\boldsymbol{k}}, \tag{B.1}$$

consisting of a spin-independent isotropic dispersion $\xi_{\boldsymbol{k}}$ (measured from the Fermi level) and the SOC. The superconducting state is described by the BdG Hamiltonian

$$\hat{H}_k = \begin{bmatrix} \xi_{k+q} + \boldsymbol{\sigma} \cdot \boldsymbol{\gamma}_{k+q} & \hat{\Delta}_k \\ \hat{\Delta}_k^\dagger & -(\xi_{-k+q} + \boldsymbol{\sigma} \cdot \boldsymbol{\gamma}_{-k+q})^* \end{bmatrix}. \tag{B.2}$$

Compared to the main text, here we consider a more general pair potential $\hat{\Delta}$ which comes from the pairing interaction in both even-parity, spin-singlet and odd-parity, spin-triplet channels and is given by

$$\hat{\Delta}_k = (\Delta_k + \boldsymbol{d}_k \cdot \boldsymbol{\sigma}) i\sigma_y, \qquad \Delta_{-k} = \Delta_k, \qquad \boldsymbol{d}_{-k} = -\boldsymbol{d}_k. \tag{B.3}$$

The singlet pair potential, Δ_k, is assumed to be isotropic in $\boldsymbol{k}$ space. The triplet gap function is represented by a Balian–Wertheimer $\boldsymbol{d}$ vector. As in the main text, the wave vector shift $\boldsymbol{q}$ in Eq. (B.2) accounts for the presence of a superconducting phase gradient associated with a dissipationless electric current. It is assumed to be weak enough to disregard pair-breaking effects and the resulting suppression of $\hat{\Delta}_k$.

We are interested in Green's function, $\hat{\mathcal{G}}(E, \boldsymbol{k})$, of the BdG equation:

$$(E - \hat{H})\hat{\mathcal{G}}(E, \boldsymbol{k}) = \hat{I}, \tag{B.4}$$

where $\hat{I}$ is a 4×4 unit matrix, and $\hat{\mathcal{G}}(E, \boldsymbol{k})$ is a matrix of the form

$$\hat{\mathcal{G}}(E, \boldsymbol{k}) = \begin{bmatrix} \hat{G}(E, \boldsymbol{k}) & \hat{F}(E, \boldsymbol{k}) \\ \hat{F}^\dagger(E, \boldsymbol{k}) & \hat{\bar{G}}(E, \boldsymbol{k}) \end{bmatrix}. \tag{B.5}$$

Above, each entry is a 2×2 matrix in spin space: $\hat{G}(E, \boldsymbol{k})$ and $\hat{\bar{G}}(E, \boldsymbol{k})$ are the quasiparticle Green functions, while $\hat{F}(E, \boldsymbol{k})$ and its Hermitian conjugate $\hat{F}^\dagger(E, \boldsymbol{k})$ are the anomalous (condensate) Green functions. Equation (B.4) can be written explicitly as

$$\begin{bmatrix} E_+ - \boldsymbol{\sigma} \cdot \boldsymbol{\gamma}_+ & -(\Delta + \boldsymbol{d} \cdot \boldsymbol{\sigma}) i\sigma_y \\ i\sigma_y(\Delta^* + \boldsymbol{d}^* \cdot \boldsymbol{\sigma}) & E_- + \boldsymbol{\sigma}^* \cdot \boldsymbol{\gamma}_- \end{bmatrix} \begin{bmatrix} \hat{G} & \hat{F} \\ \hat{F}^\dagger & \hat{\bar{G}} \end{bmatrix} = \hat{I}.$$

For ease of calculation, the arguments of the Green functions are omitted. We also introduce the shorthand notations

$$E_+ \equiv E - \xi_{k+q}, \qquad E_- \equiv E + \xi_{-k+q}, \tag{B.6}$$

$$\boldsymbol{\gamma}_\pm \equiv \boldsymbol{\gamma}_{\pm k+q}, \qquad \Delta \equiv \Delta_k, \qquad \boldsymbol{d} \equiv \boldsymbol{d}_k. \tag{B.7}$$

The solution for the condensate function $\hat{F}$ has the general form

$$F = (f_0 + \boldsymbol{f} \cdot \boldsymbol{\sigma}) i\sigma_y, \tag{B.8}$$

with the singlet and triplet pair amplitudes defined as

$$f_0 = D_0/\Pi, \qquad \boldsymbol{f} = \boldsymbol{D}/\Pi, \tag{B.9}$$

$$\Pi = (D_0^2 - \boldsymbol{D}^2)/(\Delta^2 - \boldsymbol{d}^2), \tag{B.10}$$

where scalar D_0 and vector $\boldsymbol{D}$ are given by

$$\begin{aligned} D_0 &= \Delta(E_+E_- + \boldsymbol{\gamma}_+ \cdot \boldsymbol{\gamma}_- - |\Delta|^2) + E_+\boldsymbol{\gamma}_- \cdot \boldsymbol{d} + E_-\boldsymbol{\gamma}_+ \cdot \boldsymbol{d} \\ &\quad + i\,(\boldsymbol{\gamma}_- \times \cdot \boldsymbol{\gamma}_+) \cdot \boldsymbol{d} + \Delta^* \boldsymbol{d}^2, \end{aligned} \tag{B.11}$$

$$\begin{aligned} \boldsymbol{D} &= (\Delta E_- + \boldsymbol{\gamma}_- \cdot \boldsymbol{d})\boldsymbol{\gamma}_+ + (\Delta E_+ + \boldsymbol{\gamma}_+ \cdot \boldsymbol{d})\boldsymbol{\gamma}_- + \Delta^2 \boldsymbol{d}^* \\ &\quad + (E_+E_- - \boldsymbol{\gamma}_+ \cdot \boldsymbol{\gamma}_- - \boldsymbol{d} \cdot \boldsymbol{d}^*)\boldsymbol{d} \\ &\quad + \boldsymbol{d} \times (\boldsymbol{d} \times \boldsymbol{d}^*) + i\Delta\,\boldsymbol{\gamma}_+ \times \boldsymbol{\gamma}_- + iE_+\boldsymbol{d} \times \boldsymbol{\gamma}_- - iE_-\boldsymbol{d} \times \boldsymbol{\gamma}_+. \end{aligned} \tag{B.12}$$

Because of the interplay of the $\boldsymbol{d}$ vector, SOC and supercurrent, the triplet pair amplitude has a rather rich structure. In particular, the terms linear in SOC $\boldsymbol{\gamma}_\pm$ account for the p-wave pairing characteristic of NCSs, whereas the vector cross-product terms are specific to intrinsic and induced non-unitary pairing.

Let us briefly review the case of an intrinsic non-unitary pairing known from the theory of triplet superfluids [133]. This case is recovered by setting $\Delta = 0$ and $\boldsymbol{\gamma}_\pm = 0$ in Eqs. (B.11) and (B.12). It is instructive to look at axial vector $i\boldsymbol{f} \times \boldsymbol{f}^*$ which is given by

$$i\boldsymbol{f} \times \boldsymbol{f}^* = \frac{(E^2 - \xi_{\boldsymbol{k}}^2)^2 - \Delta_+^2\Delta_-^2}{(E^2 - \xi_{\boldsymbol{k}}^2 - \Delta_+^2)^2(E^2 - \xi_{\boldsymbol{k}}^2 - \Delta_-^2)^2}\, i\boldsymbol{d} \times \boldsymbol{d}^*. \tag{B.13}$$

Here, $\Delta_\pm = \sqrt{\boldsymbol{d} \cdot \boldsymbol{d}^* \pm |i\boldsymbol{d} \times \boldsymbol{d}^*|}$ are the energy gaps for the pairing with the net spin projection parallel (+) and antiparallel (−) to vector $i\boldsymbol{d} \times \boldsymbol{d}^*$. In other words, axial vector $i\boldsymbol{d} \times \boldsymbol{d}^*$ characterizes net spin polarization of triplet Cooper pairs. We see that vector $i\boldsymbol{f} \times \boldsymbol{f}^*$ is simply proportional to $i\boldsymbol{d} \times \boldsymbol{d}^*$, carrying essentially the same information. An interesting question is whether polarization $i\boldsymbol{f} \times \boldsymbol{f}^*$ can occur in a superconducting condensate with a purely real $\boldsymbol{d}$-vector, for which $i\boldsymbol{d} \times \boldsymbol{d}^* = 0$. Such a possibility is rather counter-intuitive, as it goes beyond the original concept of the non-unitary pairing [133]. Nevertheless, such a possibility does exist in mixed-parity NCSs driven by an electric supercurrent. In this case, we encounter an induced triplet state whose classification bears a close similarity to the intrinsic non-unitary pairing.

B.2 Current-Induced Triplet Pairing: Charge-Spin Conversion

Hereafter, we focus specifically on proximity NCSs. This is an example of superconductors in which $\boldsymbol{d} = 0$, reflecting no intrinsic pairing interaction in the triplet channel. In this case, the triplet order parameter occurs as a combined effect of a singlet pairing and SOC. Setting $\boldsymbol{d} = 0$ in Eqs. (B.11) and (B.12), we obtain the pairing amplitudes explicitly as

$$f_0(E, \boldsymbol{k}, \boldsymbol{q}) = \Delta \frac{(E - \xi_{\boldsymbol{k}+\boldsymbol{q}})(E + \xi_{-\boldsymbol{k}+\boldsymbol{q}}) + \boldsymbol{\gamma}_{\boldsymbol{k}+\boldsymbol{q}} \cdot \boldsymbol{\gamma}_{-\boldsymbol{k}+\boldsymbol{q}} - |\Delta|^2}{\Pi(E, \boldsymbol{k}, \boldsymbol{q})}, \tag{B.14}$$

$$\begin{aligned} \boldsymbol{f}(E, \boldsymbol{k}, \boldsymbol{q}) = {} & \Delta \frac{E(\boldsymbol{\gamma}_{\boldsymbol{k}+\boldsymbol{q}} + \boldsymbol{\gamma}_{-\boldsymbol{k}+\boldsymbol{q}}) + \xi_{-\boldsymbol{k}+\boldsymbol{q}} \boldsymbol{\gamma}_{\boldsymbol{k}+\boldsymbol{q}} - \xi_{\boldsymbol{k}+\boldsymbol{q}} \boldsymbol{\gamma}_{-\boldsymbol{k}+\boldsymbol{q}}}{\Pi(E, \boldsymbol{k}, \boldsymbol{q})} \\ & + \Delta \frac{i \boldsymbol{\gamma}_{\boldsymbol{k}+\boldsymbol{q}} \times \boldsymbol{\gamma}_{-\boldsymbol{k}+\boldsymbol{q}}}{\Pi(E, \boldsymbol{k}, \boldsymbol{q})}. \end{aligned} \tag{B.15}$$

The denominator is given by

$$\Pi(E, \boldsymbol{k}, \boldsymbol{q}) = [D_0^2(E, \boldsymbol{k}, \boldsymbol{q}) - \boldsymbol{D}^2(E, \boldsymbol{k}, \boldsymbol{q})]/\Delta^2,$$

which after some algebra takes the form

$$\begin{aligned} & \Pi(E, \boldsymbol{k}, \boldsymbol{q}) = \\ & \left[(E - \xi_{\boldsymbol{k}+\boldsymbol{q}})^2 - \gamma_{\boldsymbol{k}+\boldsymbol{q}}^2 - |\Delta|^2\right] \left[(E + \xi_{-\boldsymbol{k}+\boldsymbol{q}})^2 - \gamma_{-\boldsymbol{k}+\boldsymbol{q}}^2 - |\Delta|^2\right] \\ & + |\Delta|^2 \left[(\xi_{\boldsymbol{k}+\boldsymbol{q}} + \xi_{-\boldsymbol{k}+\boldsymbol{q}})^2 - (\boldsymbol{\gamma}_{\boldsymbol{k}+\boldsymbol{q}} + \boldsymbol{\gamma}_{-\boldsymbol{k}+\boldsymbol{q}})^2\right]. \end{aligned} \tag{B.16}$$

The zeros of this function yield the spectrum of single-particle excitations. In TRS case, where $\boldsymbol{q} = 0$, $\boldsymbol{\gamma}_{-\boldsymbol{k}} = -\boldsymbol{\gamma}_{\boldsymbol{k}}$ and $\xi_{-\boldsymbol{k}} = \xi_{\boldsymbol{k}}$, the spectrum has an isotropic singlet gap $|\Delta|$. This can be easily checked by factorizing $\Pi(E, \boldsymbol{k}, 0)$ as follows:

$$\begin{aligned}\Pi(E, \boldsymbol{k}, 0) &= \left[(E-\xi_{\boldsymbol{k}})^2 - \gamma_{\boldsymbol{k}}^2 - |\Delta|^2\right]\left[(E+\xi_{\boldsymbol{k}})^2 - \gamma_{\boldsymbol{k}}^2 - |\Delta|^2\right] \\ &\quad + 4|\Delta|^2\xi_{\boldsymbol{k}}^2 \\ &= \left[E^2 - (\xi_{\boldsymbol{k}} + |\boldsymbol{\gamma}_{\boldsymbol{k}}|)^2 - |\Delta|^2\right]\left[E^2 - (\xi_{\boldsymbol{k}} - |\boldsymbol{\gamma}_{\boldsymbol{k}}|)^2 - |\Delta|^2\right].\end{aligned} \tag{B.17}$$

The latter equation yields two spin-split BCS-like spectral branches

$$E_{\boldsymbol{k}}^{(1)} = \pm\sqrt{(\xi_{\boldsymbol{k}} + |\boldsymbol{\gamma}_{\boldsymbol{k}}|)^2 + |\Delta|^2}, \qquad E_{\boldsymbol{k}}^{(2)} = \pm\sqrt{(\xi_{\boldsymbol{k}} - |\boldsymbol{\gamma}_{\boldsymbol{k}}|)^2 + |\Delta|^2}.$$

The above results hold for generic spin-splitting field $\boldsymbol{\gamma}_{\boldsymbol{k}}$ and dispersion $\xi_{\boldsymbol{k}}$. For $\xi_{\boldsymbol{k}} = -\mu$, we recover the results for proximity NCSs discussed in the main text. Also, for specific $\boldsymbol{\gamma}_{\boldsymbol{k}}$ and $\xi_{\boldsymbol{k}}$ and in the appropriate limits, we recover the results for Green's functions obtained in earlier publications, e.g. in [114–117].

By analogy with intrinsic non-unitary superconductors, axial vector $i\,\boldsymbol{f}(E, \boldsymbol{k}, \boldsymbol{q}) \times \boldsymbol{f}^*(E, \boldsymbol{k}, \boldsymbol{q})$ characterizes net Cooper-pair spin polarization. From Eq. (B.15), we find at the Fermi level ($E = 0$)

$$\begin{aligned}& i\,\boldsymbol{f}(0, \boldsymbol{k}, \boldsymbol{q}) \times \boldsymbol{f}^*(0, \boldsymbol{k}, \boldsymbol{q}) \\ &= \frac{2|\Delta|^2}{\Pi(0, \boldsymbol{k}, \boldsymbol{q})^2}(\xi_{-k+q}\,\boldsymbol{\gamma}_{k+q} - \xi_{k+q}\,\boldsymbol{\gamma}_{-k+q}) \times (\boldsymbol{\gamma}_{k+q} \times \boldsymbol{\gamma}_{-k+q}).\end{aligned} \tag{B.18}$$

We see that the pair polarization can be generated solely by a supercurrent in the absence of any normal-state spin polarization when $\boldsymbol{\gamma}_{\boldsymbol{k}}$ is a pure SOC field restricted by TRS ($\boldsymbol{\gamma}_{-\boldsymbol{k}} = -\boldsymbol{\gamma}_{\boldsymbol{k}}$). Physically, this pair polarization is the result of the charge-spin conversion, or, in other words, it is a magnetoelectric effect as opposed to an intrinsic polarization in Eq. (B.13).

Clearly, the induced polarization is odd in $\boldsymbol{q}$. We can therefore linearize Eq. (B.18) with respect to $\boldsymbol{q}$. Expanding $\boldsymbol{\gamma}_{\pm k+q} \approx \pm\boldsymbol{\gamma}_{\boldsymbol{k}} + (\boldsymbol{q}\cdot\nabla)\boldsymbol{\gamma}_{\boldsymbol{k}}$, we approximate the torque by the lowest order term:

$$\boldsymbol{\gamma}_{k+q} \times \boldsymbol{\gamma}_{-k+q} \approx 2\boldsymbol{\gamma}_{\boldsymbol{k}} \times (\boldsymbol{q}\cdot\nabla)\boldsymbol{\gamma}_{\boldsymbol{k}}.$$

In this order, the other factors in Eq. (B.18) can be taken at $\boldsymbol{q} = 0$, which yields

$$i\,\boldsymbol{f}(0,\boldsymbol{k},\boldsymbol{q})\times\boldsymbol{f}^*(0,\boldsymbol{k},\boldsymbol{q})\approx\frac{8|\Delta|^2}{\Pi(0,\boldsymbol{k},0)^2}\xi_k\,\boldsymbol{\gamma_k}\times[\boldsymbol{\gamma_k}\times(\boldsymbol{q}\cdot\nabla)\boldsymbol{\gamma_k}]. \quad \text{(B.19)}$$

Importantly, the pair spin polarization does not vanish upon averaging in momentum space. We prove this for a linear spectrum with $\boldsymbol{\gamma_k}=\alpha_{\text{so}}\boldsymbol{k}\times\boldsymbol{z}$ and $\xi_{\boldsymbol{k}}=-\mu$. In this case, $(\boldsymbol{q}\cdot\nabla)\boldsymbol{\gamma_k}=\alpha_{\text{so}}(\boldsymbol{q}\times\boldsymbol{z})$ and, therefore,

$$\begin{aligned} i\,\boldsymbol{f}(0,\boldsymbol{k},\boldsymbol{q})\times\boldsymbol{f}^*(0,\boldsymbol{k},\boldsymbol{q}) &\approx -\frac{8|\Delta|^2\alpha_{\text{so}}^3\mu}{\Pi(0,k,0)^2}(\boldsymbol{k}\times\boldsymbol{z})\times[(\boldsymbol{k}\times\boldsymbol{z})\times(\boldsymbol{q}\times\boldsymbol{z})] \\ &= -\frac{8|\Delta|^2\alpha_{\text{so}}^3\mu}{\Pi(0,k,0)^2}\boldsymbol{k}\,[\boldsymbol{k}\cdot(\boldsymbol{z}\times\boldsymbol{q})]. \end{aligned} \quad \text{(B.20)}$$

The denominator depends only on wave vector length $k=|\boldsymbol{k}|$:

$$\Pi(0,k,0)=\left(\mu^2-|\Delta|^2-\alpha_{\text{so}}^2k^2\right)^2+4|\Delta|^2\mu^2, \quad \text{(B.21)}$$

so the averaging affects only the numerator:

$$\begin{aligned} \overline{i\,\boldsymbol{f}(0,\boldsymbol{k},\boldsymbol{q})\times\boldsymbol{f}^*(0,\boldsymbol{k},\boldsymbol{q})} &= -\frac{8|\Delta|^2\alpha_{\text{so}}^3\mu}{\Pi(0,k,0)^2}\,\overline{\boldsymbol{k}\,[\boldsymbol{k}\cdot(\boldsymbol{z}\times\boldsymbol{q})]} \\ &= \frac{4|\Delta|^2\alpha_{\text{so}}^3\mu}{\Pi(0,k,0)^2}\,k^2(\boldsymbol{q}\times\boldsymbol{z}). \end{aligned} \quad \text{(B.22)}$$

Here, the bar denotes angle integral $\int_0^{2\pi}\frac{d\theta_k}{2\pi}\ldots$ over the directions of $\boldsymbol{k}$, and we used the identity $\overline{k_ak_b}=\frac{k^2}{2}\delta_{a,b}$ for the averaged product of the projections of $\boldsymbol{k}=[k_x,k_y,0]$ (where $a,b=x,y$).

To summarize, we have proved that for a Rashba-like $\boldsymbol{\gamma_k}$ the pair spin polarization is perpendicular to the direction of the supercurrent. A similar calculation can be done for other NCSs, e.g. for the representatives of the cubic crystal group. In that case, the pair spin polarization appears to be parallel to vector $\boldsymbol{q}$.

Bibliography

1. Wigner, E. P. (1931), *Gruppentheorie* (Friedrich Vieweg und Sohn, Braunschweig, Germany), pp. 251–254; (1959), *Group Theory* (Academic Press Inc., New York), pp. 233–236.
2. v. Klitzing, K., Dorda, G., and Pepper, M. (1980), New method for high-accuracy determination of the fine-structure constant based on quantized Hall resistance, *Phys. Rev. Lett.* **45**, 494–497.
3. Thouless, D. J., Kohmoto, M., Nightingale, M. P., and den Nijs, M. (1982), Quantized Hall conductance in a two-dimensional periodic potential, *Phys. Rev. Lett.* **49**, 405–408.
4. Halperin, B. I. (1982), Quantized Hall conductance, current-carrying edge states, and the existence of extended states in a two-dimensional disordered potential, *Phys. Rev. B* **25**, pp. 2185–2190.
5. MacDonald, A. H., and Streda, P. (1984), Quantized Hall effect and edge currents, *Phys. Rev. B* **29**, 1616.
6. Haldane, F. D. M.(1988), Model for a quantum Hall effect without Landau levels: Condensed-matter realization of the "parity anomaly", *Phys. Rev. Lett.* **61**, 2015–2018.
7. Kane, C. L., and Mele, E. J. (2005), Quantum spin Hall effect in graphene, *Phys. Rev. Lett.* **95**, 226801.
8. Kane, C. L., and Mele, E. J. (2005), Z_2 topological order and the quantum spin Hall effect, *Phys. Rev. Lett.* **95**, 146802.
9. Anderson, P. W. (1958), Absence of diffusion in certain random lattices, *Phys. Rev.* **109**, 14921505.
10. Novoselov, K. S., Geim, A. K., Morozov, S. V., Jiang, D., Katsnelson, M. I., Grigorieva, I. V., Dubonos, S. V., and Firsov, A. A. (2005), Two-dimensional gas of massless Dirac fermions in graphene, *Nature (London)* **438**, 197.
11. Bernevig, B. A., and Zhang, S.-C. (2006), Quantum spin Hall effect, *Phys. Rev. Lett.* **96**, 106802.

12. Qi, X.-L., Wu, Y.-S., and Zhang, S.-C. (2006), Topological quantization of the spin Hall effect in two-dimensional paramagnetic semiconductors, *Phys. Rev. B* **74**, 085308.
13. Bernevig, B. A., Hughes, T. L., and Zhang, S. C. (2006), Quantum spin Hall effect and topological phase transition in HgTe quantum wells, *Science* **314**, pp. 1757–1761.
14. König, M., Wiedmann, S., Brüne, C., Roth, A., Buhmann, H., Molenkamp, L. M., Qi, X.-L., and Zhang, S.-C. (2007), Quantum spin Hall insulator state in HgTe quantum wells, *Science* **318**, pp. 766–770.
15. Liu, C. X., Hughes, T. L., Qi, X. L., Wang, K., and Zhang, S. C. (2008), Quantum spin Hall effect in inverted type-II semiconductors, *Phys. Rev. Lett.* **100**, 236601.
16. Knez, I., Du, R. R., and Sullivan, G. (2011), Evidence for helical edge modes in inverted InAs/GaSb quantum wells, *Phys. Rev. Lett.* **107**, 136603.
17. Qian, X., Liu, J., Fu, L., and Li, J. (2014), Quantum spin Hall effect in two-dimensional transition metal dichalcogenides, *Science* **346**, 1344.
18. Tang, S. J., Zhang, C. F., Wong, D., Pedramrazi, Z., Tsai, H. Z., Jia, C. J., Moritz, B., Claassen, M., Ryu, H., Kahn, S., Jiang, J., Yan, H., Hashimoto, M., Lu, D. H., Moore, R. G., Hwang, C. C., Hwang, C., Hussain, Z., Chen, Y. L., Ugeda, M. M., Liu, Z., Xie, X. M., Devereaux, T. P., Crommie, M. F., Mo, S. K., and Shen, Z. X. (2017), Quantum spin Hall state in monolayer 1T′ - WTe_2, *Nature Phys.* **13**, 683.
19. Ugeda, M. M., Pulkin, A., Tang, S., Ryu, H., Wu, Q. S., Zhang, Y., Wong, D., Pedramrazi, Z., Martin-Recio, A., Chen, Y., Wang, F., Shen, Z. X., Mo, S. K., Yazyev, O. V., and Crommie, M. F. (2018), Observation of topologically protected states at crystalline phase boundaries in single-layer WSe_2, *Nat. Commun.* **9**, 3401.
20. Wu, S., Fatemi, V., Gibson, Q. D., Watanabe, K., Taniguchi, T., Cava, R. J., and Jarillo-Herrero, P. (2018), Observation of the quantum spin Hall effect up to 100 Kelvin in a monolayer crystal, *Science* **359**, 76.
21. Tkachov, G. (2021), Topological electronic states and thermoelectric transport at phase boundaries in single-layer WSe_2: An effective Hamiltonian theory, *J. Phys.: Condens. Matter* **33**, 325301. https://doi.org/10.1088/1361-648X/abc809.
22. Liu, C.-X., Qi, X.-L., Dai, X., Fang, Z., and Zhang, S.-C. (2008), Quantum anomalous Hall effect in $Hg_{1-y}Mn_yTe$ quantum wells, *Phys. Rev. Lett.* **101**, 146802.

23. Chang, C. Z. *et al* (2013), Experimental observation of the quantum anomalous Hall effect in a magnetic topological insulator, *Science* **340**,167.

24. Fu, L., Kane, C. L., and Mele, E. J. (2007), Topological insulators in three dimensions, *Phys. Rev. Lett.* **98**, 106803.

25. Fu, L., and Kane, C. L. (2007), Topological insulators with inversion symmetry, *Phys. Rev. B* **76**, 045302.

26. Moore, J. E., and Balents, L. (2007), Topological invariants of time-reversal-invariant band structures, *Phys. Rev. B* **75**, 121306(R).

27. Hasan, M. Z., and Kane, C. L. (2010), Colloquium: Topological insulators, *Rev. Mod. Phys.* **82**, 3045–3067.

28. Qi, X.-L., and Zhang, S.-C. (2011), Topological insulators and superconductors, *Rev. Mod. Phys.* **83**, 1057–1110.

29. Volkov, B. A., and Pankratov, O. A. (1985), Two-dimensional massless electrons in an inverted contact, *JETP Lett.* **42**, 178–181.

30. Pankratov, O. A., Pakhomov, S. V., and Volkov, B. A. (1987), Supersymmetry in heterojunctions: Band-inverting contact on the basis of $Pb_{1-x}Sn_xTe$ and $Hg_{1-x}Cd_xTe$ *Sol. State Commun.* **61**, 93 (1987).

31. Zhang, H., Liu, C.-X., Qi, X.-L., Dai, X., Fang, Z., and Zhang, S.-C., Topological insulators in Bi_2Se_3, Bi_2Te_3 and Sb_2Te_3 with a single Dirac cone on the surface, *Nat. Phys.* **5**, 438–442 (2009).

32. Yan, B., Liu, C.-X., Zhang, H.-J., Yam, C.-Y., Qi, X.-L., Frauenheim, T., and Zhang, S.-C. (2010), Theoretical prediction of topological insulators in thallium-based III-V-VI2 ternary chalcogenides, *EPL* **90**, 37002.

33. Lin, H., Markiewicz, R. S., Wray, L. A., Fu, L., Hasan, M. Z., and Bansil, A. (2010), Single-Dirac-cone topological surface states in the $TlBiSe_2$ class of topological semiconductors, *Phys. Rev. Lett.* **105**, 036404.

34. Eremeev, S. V., Koroteev, Yu. M., and Chulkov, E. V. (2010), Ternary thallium-based semimetal chalcogenides Tl-V-VI2 as a new class of three-dimensional topological insulators, *Pis'ma Zh. Eksp. Teor. Fiz.* **91**, 664 (2010) [*JETP Lett.* **91**, 594].

35. Eremeev, S. V., Koroteev, Yu. M., and Chulkov, E. V. (2010), On possible deep subsurface states in topological insulators: The $PbBi_4Te_7$ system *Pis'ma Zh. Eksp. Teor. Fiz.* **92**, 183 [*JETP Lett.* **92**, 161].

36. Jin, H., Song, J. H, Freeman, A. J., and Kanatzidis, M. G. (2011), Candidates for topological insulators: Pb-based chalcogenide series, *Phys. Rev. B* **83**, 041202.

37. Hsieh, D., Qian, D., Wray, L., Xia, Y., Hor, Y. S., Cava, R. J., and Hasan, M. Z. (2008), A topological Dirac insulator in a quantum spin Hall phase, *Nature* **452**, pp. 970–974.

38. Hsieh, D., Xia, Y., Wray, L., Qian, D., Pal, A., Dil, J. H., Meier, F., Osterwalder, J., Kane, C. L., Bihlmayer, G., Hor, Y. S., Cava, R. J., and Hasan, M. Z. (2009), Observation of unconventional quantum spin textures in topological insulators, *Science* **323**, pp. 919–922.

39. Xia, Y., Qian, D., Hsieh, D., Wray, L., Pal, A., Lin, H., Bansil, A., Grauer, D., Hor, Y. S., Cava, R. J., and Hasan, M. Z. (2009), Observation of a large-gap topological-insulator class with a single Dirac cone on the surface, *Nat. Phys.* **5**, 398–402.

40. Chen, Y. L., Analytis, J. G., Chu, J.-H., Liu, Z. K., Mo, S.-K., Qi, X. L., Zhang, H. J., Lu, D. H., Dai, X., Fang, Z., Zhang, S. C., Fisher, I. R., Hussain, Z., and Shen, Z.-X. (2009), Experimental realization of a three-dimensional topological insulator, Bi_2Te_3, *Science* **325**, 178–181.

41. Sato, T., Segawa, K., Guo, H., Sugawara, K., Souma, S., Takahashi, T., and Ando, Y. (2010), Direct evidence for the Dirac-cone topological surface states in the ternary chalcogenide $TlBiSe_2$, *Phys. Rev. Lett.* **105**, 136802.

42. Kuroda, K., Ye, M., Kimura, A., Eremeev, S. V., Krasovskii, E. E., Chulkov, E. V., Ueda, Y., Miyamoto, K., Okuda, T., Shimada, K., Namatame, H., and Taniguchi, M. (2010), Experimental realization of a three-dimensional topological insulator phase in ternary chalcogenide $TlBiSe_2$, *Phys. Rev. Lett.* **105**, 146801.

43. Chen, Y. L., Liu, Z. K., Analytis, J. G., Chu, J.-H., Zhang, H. J., Yan, B. H., Mo, S.-K., Moore, R. G., Lu, D. H., Fisher, I. R., Zhang, S. C., Hussain, Z., and Shen, Z.-X. (2010), Single Dirac cone topological surface state and unusual thermoelectric property of compounds from a new topological insulator family, *Phys. Rev. Lett.* **105**, 266401.

44. Souma, S., Eto, K., Nomura, M., Nakayama, K., Sato, T., Takahashi, T., Segawa, K., and Ando, Y. (2012), Topological surface states in lead-based ternary telluride $Pb(Bi_{1-x}Sb_x)2Te_4$, *Phys. Rev. Lett.* **108**, 116801.

45. Kuroda, K., Miyahara, H., Ye, M., Eremeev, S. V., Koroteev, Yu. M., Krasovskii, E. E., Chulkov, E. V., Hiramoto, S., Moriyoshi, C., Kuroiwa, Y., Miyamoto, K., Okuda, T., Arita, M., Shimada, K., Namatame, H., Taniguchi, M., Ueda, Y., and Kimura, A. (2012), Experimental verification of $PbBi_2Te_4$ as a 3D topological insulator, *Phys. Rev. Lett.* **108**, 206803.

46. Fu, L. (2009), Hexagonal warping effects in the surface states of the topological insulator Bi_2Te_3, *Phys. Rev. Lett.* **103**, 266801.

47. Liu, C.-X., Qi, X.-L., Zhang, H. J., Dai, X., Fang, Z., and Zhang, S.-C. (2010), Model Hamiltonian for topological insulators, *Phys. Rev. B* **82**, 045122.
48. Alpichshev, Z., Analytis, J. G., Chu, J.-H., Fisher, I. R., Chen, Y. L., Shen, Z. X., Fang, A., and A. Kapitulnik, (2010), STM imaging of electronic waves on the surface of Bi_2Te_3: Topologically protected surface states and hexagonal warping effects, *Phys. Rev. Lett.* **104**, 016401.
49. Kuroda, K., Arita, M., Miyamoto, K., Ye, M., Jiang, J., Kimura, A., Krasovskii, E. E., Chulkov, E. V., Iwasawa, H., Okuda, T., Shimada, K., Ueda, Y., Namatame, H., and Taniguchi, M. (2010), Hexagonally deformed Fermi surface of the 3D topological insulator Bi_2Se_3, *Phys. Rev. Lett.* **105**, 076802.
50. Li, Y.-Y., Wang, G., Zhu, X.-G., Liu, M.-H., Ye, C., Chen, X., Wang, Y.-Y., He, K., Wang, L.-L., Ma, X.-C., Zhang, H.-J., Dai, X., Fang, Z., Xie, X.-C., Liu, Y., Qi, X.-L., Jia, J.-F., Zhang, S.-C., and Xue, Q.-K. (2010), Intrinsic topological insulator Bi_2Te_3 thin films on Si and their thickness limit, *Adv. Mater.* **22**, 40024007.
51. Kim, D., Syers, P., Butch, N. P., Paglione, J., and Fuhrer, M. S. (2013), Coherent topological transport on the surface of Bi_2Se_3, *Nat. Commun.* **4**, 2040.
52. Murakami, S. (2007), Phase transition between the quantum spin Hall and insulator phases in 3D: Emergence of a topological gapless phase, *New J. Phys.* **9**, 356.
53. Weng, H., Fang, C., Fang, Z., Bernevig, B. A., and Dai, X. (2015), Weyl semimetal phase in noncentrosymmetric transition-metal monophosphides, *Phys. Rev. X* **5**, 011029.
54. Huang, S.-M. *et al* (2015), A Weyl fermion semimetal with surface Fermi arcs in the transition metal monopnictide TaAs class, *Nat. Commun.* **6**, 7373.
55. Lv, B. Q. *et al* (2015), Experimental discovery of Weyl semimetal TaAs, *Phys. Rev. X* **5**, 031013.
56. Su-Yang Xu, S.-Y. *et al* (2015), Discovery of a Weyl fermion semimetal and topological Fermi arcs, *Science* **349**, 613–617.
57. Armitage, N., Mele, E., and Vishwanath, A. (2018), Weyl and Dirac semimetals in three-dimensional solids, *Rev. Mod. Phys.* **90**, 015001.
58. Bardeen, J., Cooper, L. N., and Schrieffer, J. R. (1957), Theory of superconductivity, *Phys. Rev.* **108**, pp. 1175–1204.
59. Sato, M., and Ando, Y. (2017), Topological superconductors: A review, *Rep. Prog. Phys.* **80**, 076501.

60. Culcer, D., Cem Keser, A., Li, Y.-Q., and Tkachov, G (2020), Transport in two-dimensional topological materials: Recent developments in experiment and theory, *2D Mater.* **7**, 022007.

61. Majorana, E. (1937), Teoria simmetrica dell'elettrone e del positrone. *Nuovo Cim.* **14**, 171.

62. Sarma, S. D., Freedman, M., and Nayak, C. (2015), Majorana zero modes and topological quantum computation, *NPJ Quantum Inf.* **1**, 15001.

63. Read, N., and Green, D. (2000), Paired states of fermions in two dimensions with breaking of parity and time-reversal symmetries and the fractional quantum Hall effect, *Phys. Rev. B* **61**, 10267.

64. Mackenzie, A. P., and Maeno, Y. (2003), The superconductivity of Sr_2RuO_4 and the physics of spin-triplet pairing, *Rev. Mod. Phys.* **75**, 657.

65. Luke, G. M. *et al* (1998), Time-reversal symmetry-breaking superconductivity in Sr_2RuO_4, *Nature* **394**, 558.

66. Kashiwaya, S., Kashiwaya, H., Kambara, H., Furuta, T., Yaguchi, H., Tanaka, Y., and Maeno, Y. (2011), Edge states of Sr_2RuO_4 detected by in-plane tunneling spectroscopy, *Phys. Rev. Lett.* **107**, 077003.

67. Wang, H., Ma, L., and Wang, J. (2018), Tip-induced or enhanced superconductivity: A way to detect topological superconductivity, *Sci. Bull.* **63**, 1141.

68. Andreev, A. F. (1964), The thermal conductivity of the intermediate state in superconductors, *Sov. Phys. JETP.* **19**, 1228.

69. Fu, L., and Kane, C. L. (2008), Superconducting proximity effect and Majorana fermions at the surface of a topological insulator, *Phys. Rev. Lett.* **100**, 096407.

70. Qi, X. L., Hughes, T. L., and Zhang, S. C. (2010), Chiral topological superconductor from the quantum Hall state, *Phys. Rev. B* **82**, 184516.

71. Chung, S. B., Qi, X. L., Maciejko, J., and Zhang, S. C. (2011), Conductance and noise signatures of Majorana backscattering, *Phys. Rev. B* **83**, 100512.

72. Wang, J., Zhou, Q., Lian, B., and Zhang, S. C. (2015), Chiral topological superconductor and half-integer conductance plateau from quantum anomalous Hall plateau transition, *Phys. Rev. B* **92**, 064520.

73. He, Q. L. *et al* (2017), Chiral Majorana fermion modes in a quantum anomalous Hall insulator-superconductor structure, *Science* **357**, 294.

74. Shen, J. *et al* (2020), Spectroscopic fingerprint of chiral Majorana modes at the edge of a quantum anomalous Hall insulator/superconductor heterostructure, *PNAS* **117**, 238.

75. Jackiw, R., and Rebbi, C. (1976), Solitons with fermion number 1/2, *Phys. Rev. D* **13**, 3398–3409.

76. Berry, M. V. (1984), Quantal phase factors accompanying adiabatic changes, *Proc. Roy. Soc. London* **392**, 45–57.

77. Mahan, G. D. (2000), *Many-Particle Physics* (Kluwer Academic/Plenum Publishers, New York).

78. Rammer, J. (2004), *Quantum Transport Theory* (Westview Press, Boulder, CO).

79. Kohmoto, M. (1985), Topological invariant and the quantization of the Hall conductance, *Ann. Phys. (Berlin)* **160**, 343–354.

80. Nielsen, H. B., and Ninomiya, M. (1981), Absence of neutrinos on a lattice: (i). Proof by homotopy theory. *Nucl. Phys. B* **185**, 20–40.

81. Lu, H. Z., Shan, W. Y., Yao, W., Niu, Q., and Shen, S. Q. (2010), Massive Dirac fermions and spin physics in an ultrathin film of topological insulator, *Phys. Rev. B* **81**, 115407.

82. Shan, W. Y., Lu, H. Z., and Shen, S. Q. (2010), Effective continuous model for surface states and thin films of three-dimensional topological insulators, *New J. Phys.* **12**, 043048.

83. Bernevig, B. A. (2013), *Topological Insulators and Topological Superconductors* (Princeton University Press).

84. Roth, A., Brüne, C. Buhmann, H., Molenkamp, L. W., Maciejko, J., Qi, X.-L., and Zhang, S.-C. (2009), Nonlocal transport in the quantum spin Hall state, *Science* **325**, 294–297.

85. Gusev, G. M., Olshanetsky, E. B., Kvon, Z. D., Levin, A. D., Mikhailov, N. N., and Dvoretsky, S. A. (2012), Nonlocal transport near charge neutrality point in a two-dimensional electron-hole system, *Phys. Rev. Lett.* **108**, 226804.

86. Chu, J., and Sher, A. (2008), *Physics and Properties of Narrow Gap Semiconductors* (Springer, New York).

87. Shi, L.-K., and Song, J. C. W. (2019), Symmetry, spin-texture, and tunable quantum geometry in a WTe_2 monolayer, *Phys. Rev. B* **99**, 035403.

88. Abramowitz, M., and Stegun, I. (1964), *Handbook of Mathematical Functions with Formulas, Graphs, and Mathematical Tables* (National Bureau of Standards, Washington, D.C.).

89. Fisher, D. S., and Lee, P. A. (1981), Relation between conductivity and transmission matrix, *Phys. Rev. B* **23**, 6851–6854.

90. Winkler, R. (2003), *Spin-Orbit Coupling Effects in Two-Dimensional Electron and Hole Systems* (Springer, Berlin).

91. Chu, R.-L., Shan, W.-Y., Lu, J., and Shen, S.-Q. (2011), Surface and edge states in topological semimetals, *Phys. Rev. B* **83**, 075110.
92. Kourtis, S., Li, J., Wang, Z., Yazdani, A., and Bernevig, B. A. (2016), Universal signatures of Fermi arcs in quasiparticle interference on the surface of Weyl semimetals, *Phys. Rev. B* **93**, 041109(R).
93. Hikami, S., Larkin, A. I., and Nagaoka, Y. (1980), Spin-orbit interaction and magnetoresistance in the two dimensional random system, *Prog. Theor. Phys.* **63**, 707–710.
94. Suzuura, H., and Ando, T. (2002), Crossover from symplectic to orthogonal class in a two-dimensional honeycomb lattice, *Phys. Rev. Lett.* **89**, 266603.
95. McCann, E., Kechedzhi, K., Fal'ko, V. I., Suzuura, H., Ando, T., and Altshuler, B. L. (2006), Weak-localization magnetoresistance and valley symmetry in graphene, *Phys. Rev. Lett.* **97**, 146805.
96. Ortmann, F., Cresti, A., Montambaux, G., and Roche, S. (2011), Magnetoresistance in disordered graphene: The role of pseudospin and dimensionality effects unraveled, *Europhys. Lett.* **94**, 47006.
97. Lu, H.-Z., Shi, J., and Shen, S.-Q. (2011), Competition between weak localization and antilocalization in topological surface states, *Phys. Rev. Lett.* **107**, 076801.
98. Tkachov, G., and Hankiewicz, E. M. (2011), Weak antilocalization in HgTe quantum wells and topological surface states: Massive versus massless Dirac fermions, *Phys. Rev. B* **84**, 035444.
99. Ostrovsky, P. M., Gornyi, I. V., and Mirlin, A. D. (2012), Symmetries and weak localization and antilocalization of Dirac fermions in HgTe quantum wells, *Phys. Rev. B* **86**, 125323.
100. Tian, C. (2012), Anomalous quantum diffusion and the topological metal, *Phys. Rev. B* **86**, 121304(R).
101. Gusynin, V. P., and Sharapov, S. G. (2005), Unconventional integer quantum Hall effect in graphene, *Phys. Rev. Lett.* **95**, 146801.
102. Tkachov, G., and Hankiewicz, E. M. (2013), Spin-helical transport in normal and superconducting topological insulators, *Phys. Status Solidi B* **250**, 215.
103. Wilczek, F. (1987), Two applications of axion electrodynamics, *Phys. Rev. Lett.* **58**, 1799.
104. Qi, X.-L., Hughes, T. L., and Zhang, S.-C. (2008), Topological field theory of time-reversal invariant insulators, *Phys. Rev. B* **78**, 195424.

105. Wu, L., Salehi, M., Koirala, N., Moon, J., Oh, S., and Armitage, N. P. (2016), Quantized Faraday and Kerr rotation and axion electrodynamics of a 3D topological insulator, *Science* **354**, 1124–1127.

106. Dziom, V., Shuvaev, A., Pimenov, A., Astakhov, G. V., Ames, C., Bendias, K., Boettcher, J., Tkachov, G., Hankiewicz, E. M., Bruene, C., Buhmann, H., and Molenkamp, L. W. (2017), Observation of the universal magnetoelectric effect in a 3D topological insulator, *Nat. Commun.* **8**, 15197.

107. Gorkov, L. P. (1958), On the energy spectrum of superconductors, *J. Exp. Theor. Phys.* **34**, pp. 735–739 [*Sov. Phys. JETP* **7**, pp. 505–508].

108. de Gennes, P. G. (1964), Boundary effects in superconductors, *Rev. Mod. Phys.* **36**, pp. 225–237.

109. Tkachov, G. (2005), Nonmetallic thermal transport in low-dimensional proximity structures with partially preserved time-reversal symmetry in a magnetic field, *Phys. C* **417**, 127.

110. Fagas, G., Tkachov, G., Pfund, A., and Richter, K. (2005), Geometrical enhancement of the proximity effect in quantum wires with extended superconducting tunnel contacts, *Phys. Rev. B* **71**, 224510.

111. Rohlfing, F., Tkachov, G., Otto, F., Richter, K., Weiss, D., Borghs, G., and Strunk, C. (2009), Doppler shift in Andreev reflection from a moving superconducting condensate in Nb/InAs Josephson junctions, *Phys. Rev. B* **80**, 220507.

112. Kopnin, N. B., and Melnikov, A. S. (2011), Proximity-induced superconductivity in two-dimensional electronic systems, *Phys. Rev. B* **84**, 064524.

113. Kopnin, N. B., Khaymovich, I. M., and Mel'nikov, A. S. (2013), Predicted multiple cores of a magnetic vortex threading a two-dimensional metal proximity coupled to a superconductor, *Phys. Rev. Lett.* **110**, 027003.

114. Stanescu, T. D., Sau, J. D., Lutchyn, R. M., and Das Sarma, S. (2010), Proximity effect at the superconductor–topological insulator interface, *Phys. Rev. B* **81**, 241310(R).

115. Yokoyama, T., (2012), Josephson and proximity effects on the surface of a topological insulator, *Phys. Rev. B* **86**, 075410.

116. Tkachov, G. (2013), Suppression of surface *p*-wave superconductivity in disordered topological insulators, *Phys. Rev. B* **87**, 245422.

117. Black-Schaffer, A. M., and Balatsky, A. V. (2013), Proximity-induced unconventional superconductivity in topological insulators, *Phys. Rev. B* **87**, 220506(R).

118. Vasenko, A. S., Golubov, A. A., Silkin, V. M., and Chulkov, E. V. (2017), Odd-frequency superconductivity induced in topological insulators with and without hexagonal warping, *J. Phys.: Condens. Matter* **29**, 295502.

119. Hugdal, H. G., Linder, J., and Jacobsen, S. H. (2017), Quasiclassical theory for the superconducting proximity effect in Dirac materials, *Phys. Rev. B* **95**, 235403.

120. Zhang, D., Wang, J., DaSilva, A. M., Lee, J. S., Gutierrez, H. R., Chan, M. H. W., Jain, J., and Samarth, N. (2011), Superconducting proximity effect and possible evidence for Pearl vortices in a candidate topological insulator, *Phys. Rev. B* **84**, 165120.

121. Koren, G., Kirzhner, T., Lahoud, E., Chashka, K. B., and Kanigel, A. (2011), Proximity-induced superconductivity in topological Bi_2Te_2Se and Bi_2Se_3 films: Robust zero-energy bound state possibly due to Majorana fermions, *Phys. Rev. B* **84**, 224521.

122. Sacepe, B., Oostinga, J. B., Li, J., Ubaldini, A., Couto, N. J. G., Giannini, E., and Morpurgo, A. F. (2011), Gate-tuned normal and superconducting transport at the surface of a topological insulator, *Nat. Commun.* **2** 575.

123. Veldhorst, M., Snelder, M., Hoek, M., Gang, T., Guduru, V. K., Wang, X. L., Zeitler, U., Wiel, W. G. V. D., Golubov, A. A., and Hilgenkamp, H. (2012), Josephson supercurrent through a topological insulator surface state, *Nature Mat.* **11**, 417–421.

124. Wang, M.-X., Liu, C., Xu, J.-P., Yang, F., Miao, L., Yao, M.-Y., Gao, C. L., Shen, C., Ma, X., Chen, X., Xu, Z.-A., Liu, Y., Zhang, S.-C., Qian, D., Jia, J.-F., and Xue, Q.-K. (2012), The coexistence of superconductivity and topological order in the Bi_2Se_3 thin films, *Science* **336**, 52–55.

125. Williams, J. R., Bestwick, A. J., Gallagher, P., Hong, S. S., Cui, Y., Bleich, A. S., Analytis, J. G., Fisher, I. R., and Goldhaber-Gordon, D. (2012), Unconventional Josephson effect in hybrid superconductor-topological insulator devices, *Phys. Rev. Lett.* **109**, 056803.

126. Cho, S., Dellabetta, B., Yang, A., Schneeloch, J., Xu, Z., Valla, T., Gu, G., Gilbert, M. J., and Mason, N. (2013), Symmetry protected Josephson supercurrents in three-dimensional topological insulators, *Nat. Commun.* **4**, 1689.

127. Oostinga, J. B., Maier, L., Schüffelgen, P., Knott, D., Ames, C., Brüne, C., Tkachov, G., Buhmann, H., and Molenkamp, L. W. (2013), Josephson supercurrent through the topological surface states of strained bulk HgTe, *Phys. Rev. X* **3**, 021007.

128. Snelder, M., Veldhorst, M., Golubov, A. A., and Brinkman, A., (2013), Andreev bound states and current-phase relations in three-dimensional topological insulators, *Phys. Rev. B* **87**, 104507.

129. Tkachov, G., and Hankiewicz, E. M. (2013), Helical Andreev bound states and superconducting Klein tunneling in topological insulator Josephson junctions, *Phys. Rev. B* **88**, 075401.

130. Linder, J., and Balatsky, A. (2019), Odd-frequency superconductivity, *Rev. Mod. Phys.* **91**, 045005.

131. Cayao, J., and Black-Schaffer, A. M. (2017), Odd-frequency superconducting pairing and subgap density of states at the edge of a two-dimensional topological insulator without magnetism, *Phys. Rev. B* **96**, 155426.

132. Tanaka, Y., Sato, M., and Nagaosa, N. (2012), Symmetry and topology in superconductors: Odd-frequency pairing and edge states, *J. Phys. Soc. Jpn.* **81**, 011013.

133. Leggett, A. J. (1975), A theoretical description of the new phases of liquid ^{3}He, *Rev. Mod. Phys.* **47**, 331.

134. Tkachov, G. (2017), Magnetoelectric Andreev effect due to proximity-induced nonunitary triplet superconductivity in helical metals, *Phys. Rev. Lett.* **118**, 016802.

135. Tkachov, G. (2018), Probing the magnetoelectric effect in noncentrosymmetric superconductors by equal-spin Andreev tunneling, *J. Phys.: Condens. Matter* **31**, 055301.

136. Edelstein, V. M. (1995), Magnetoelectric Effect in Polar Superconductors, *Phys. Rev. Lett.* **75**, 2004.

137. Ivanov, D. A. (2001), Non-Abelian statistics of half-quantum vortices in p-wave superconductors, *Phys. Rev. Lett.* **86**, 268.

138. Kitaev, A. Y. (2001), Unpaired Majorana fermions in quantum wires, *Phys. Usp.* **44**, 131.

139. Kwon, H. J., Sengupta, K., and Yakovenko, V. M. (2004), Fractional ac Josephson effect in p- and d-wave superconductors, *Eur. Phys. J. B* **37**, 349.

140. Sau, J. D., Lutchyn, R. M., Tewari, S., and Das Sarma, S. (2010), Generic new platform for topological quantum computation using semiconductor heterostructures, *Phys. Rev. Lett.* **104**, 040502.

141. Oreg, Y., Refael, G., and von Oppen, F. (2010), Helical liquids and Majorana bound states in quantum wires, *Phys. Rev. Lett.* **105**, 177002.

Index